Praise for
Essential Cob Construction

To save our precious Earth, we need to build with our precious earth. This book describes and details the practice of place-based cob construction, an ancient method that is more timely now than ever before. It is an affordable and ecological way to create better housing for more people throughout the world.

—Paul Hawken, environmentalist, entrepreneur, New York Times best-selling author of *The Ecology of Commerce, Drawdown,* and *Regeneration*

This comprehensive work explores cob's advantages as a building material and its applications for ecological and sustainable architecture. It describes all steps from mix testing, building design, and calculations through the realization of a building. It is the best book I have come across in this field.

—Professor Dr. Gernot Minke, founding director, Building Research Institute, University of Kassel, Germany, author, *Building with Earth*

At last!—the book we've been waiting for, by just the right authors! Anthony, Massey, and Michael are the perfect team who know the joy of getting your hands dirty and building with friends, study and appreciate the rich and venerable traditions of earthen architecture, and can bring it all into the 21st century with the engineering that modern construction requires. Well done!

—Bruce King, PE, director, Ecological Building Network, co-author, *Build Beyond Zero* and *The New Carbon Architecture*

Absolutely fabulous! This simple description applies whether you are reading this book for inspiration, for details, for scientific data, or for ideas. It is so rare that a book captivates me, draws me in, and satiates my thirst for information. I have already recommended it (before it is released) to both builders and building officials as it offers so much more than I could ever provide from my own experience. Thank you Michael, Anthony, and Massey for writing what really is the ESSENTIAL book on cob construction.

—Gord Baird, co-owner/builder, Eco-Sense, co-author, *Essential Composting Toilets*

A comprehensive guide and reference every builder should have. Answering questions about design, function, strength, and safety, *Essential Cob Construction* will do a lot to make cob an accessible part of the building lexicon.

—John Curry and Kindra Welch, cob builders and designers, ClaySandStraw

In our work to advance alternatives to conventional, wasteful building materials, building science-based texts like these are essential. Filled with practical tips and inspiring insights, *Essential Cob Construction* shows us the way and puts the power of restorative shelter back into our hands.

—Miya Kitahara, Built Environment program manager, StopWaste, specializing in local policy for embodied carbon

Essential Cob Construction is a comprehensive and timely addition to the literature on cob building. In addition to the wealth of experience and practical information for builders, much-needed technical and structural guidance is provided to enable the design, permitting, and construction of safe and durable cob buildings as cob gains wider and more diverse use.

— David Eisenberg, director, Development Center for Appropriate Technology

Essential Cob Construction is the cob manual the building industry needs. The depth of technical information here will help designers, builders, and code officials in all climates to make informed choices about cob construction. The step-by-step guides to engineering, design, and building are clear and well laid-out. Filled with excellent photos, this manual brings together the mind and heart of a very human building material, and offers hope that we are moving forward with ancient and modern approaches to our built environment.

— Sukita Reay Crimmel, founder, Claylin Earthen Floors, co-author,
Earthen Floors: A Modern Approach to an Ancient Practice

Essential Cob Construction is a well-crafted book that weaves practical techniques and hands-on know-how with up-to-date code information, insightful charts, and essential building principles. Although focused primarily on cob construction, it is densely packed with easy-to-grasp content that is relevant to all forms of natural building.

— Athena Steen, director, The Canelo Project, co-author, *The Straw Bale House*

I bet the authors had no idea what they were getting themselves into when they started to write this book. I am glad they persevered, because now we all have the benefit of a wide-ranging compendium that brings the art, craft, and science of one of the most ancient forms of earth building fair and square into the modern era. Although the focus is North America, this will have wide application elsewhere. Thank you, from the earth building community.

— Graeme North MNZM FNZIA, architect, inaugural chair, Earth Building Association of
New Zealand, chair, New Zealand Technical Committee for Earth Building Standards

Anthony, Michael, and Massey have given us an excellent resource for dreaming, designing, and building with cob. Not only that, they've managed to work in a healthy amount of history, engineering methodology, and helpful advice that ties this blend of clay, sand, and straw to many other earthen building techniques used around the world. Bravo!

--Tim Krahn, P. Eng., author, *Essential Rammed Earth Construction*

Cob construction has been tested and proven in the real world for centuries—millennia even. However, in this day and age of increasingly onerous regulation and certification, it can be an uphill battle to work with this sustainable, expressive material. I suspect that *Essential Cob Construction* will become the go-to bible for amateurs and professionals alike. I look forward to adding it to my bookshelf, where I am sure it will become a dog-eared keystone text for my practice.

— Féile Butler, MRIAI, conservation architect, Natural Building Consultant

What an excellent resource for both the natural building and conventional building communities. This is a groundbreaking book that will help to inspire and inform anyone who reads it with an interest in cob and/or natural building.

—Adam Weismann and Katy Bryce, co-founders, Clayworks, authors, *Building with Cob* and *Clay and Lime Renders Plasters and Paints*

This is an impressive and well-organized collection of information that aligns well with what I've been learning through my 15 years of cob work, and addresses the gaps in my own knowledge. Because the content is so clear and applicable, this will become my top recommendation to students who attend our intensive cob workshops. As both a builder and a teacher, it's truly the book I've been hoping for.

—Greg Allen, owner/educator, Mud Dauber School of Natural Building

The authors of *Essential Cob Construction* have crafted a masterpiece of knowledge. This book contains an encyclopedic amount of information, covering all essential aspects of cob construction, from its historical and cultural context to its technical performance, practical applications, and architectural potential. As we face the social and ecological challenges of the decades ahead, efforts such as *Essential Cob Construction* may well be what save us.

—Jacob Deva Racusin, director, Building Science and Sustainability, New Frameworks, author, *Essential Building Science*

This is the book we have been waiting for! It not only provides the most thorough information yet on cob building, but also does a beautiful job of making the technical sides of cob building accessible. Drawing from their own decades of experience as well as centuries of building practice and more recent engineering and testing, the authors have presented information spanning from building science to testing mixes to budgeting projects in ways that are easy to understand. This book is a huge step towards making cob building more accessible in the modern building world.

—Sasha Rabin, founder, Earthen Shelter, Director of Natural Building and Advocacy, Quail Springs Permaculture

Essential Cob Construction brings modern knowledge to an age-old material, and empowers readers to build climate-appropriate cob walls. Finally, a book on cob that includes construction timing, work site planning, tool lists, and recipes! Kudos!

—Sigi Koko, founder, Down to Earth Design, buildnaturally.com

This comprehensive book goes beyond *essential*. It covers virtually every aspect of cob construction and is written by a team who have decades of experience with muddy, hands-on building and engineering design. They have written what is literally the current state of the art, from the building science to the prescriptive codes that will allow cob to more easily be employed legally in many jurisdictions. All of this is lavishly illustrated with photos and diagrams showing both the beauty and the nitty gritty how-to of cob construction.

—Kelly Hart, greenhomebuilding.com, www.naturalbuildingblog.com, author, *Essential Earthbag Construction*

sustainable
building
essentials

essential
COB
CONSTRUCTION
a guide to design, engineering, and building

Anthony Dente, PE, Michael G. Smith, and Massey Burke

new society
PUBLISHERS

New Society
Sustainable Building Essentials Series

Series editors

Chris Magwood and Jen Feigin

Title list

Essential Cob Construction, Anthony Dente, PE, Michael G. Smith, and Massey Burke

Essential Green Roof Construction, Leslie Doyle

Essential Rammed Earth Construction, Tim Krahn

Essential Rainwater Harvesting, Rob Avis and Michelle Avis

Essential Composting Toilets, Gord Baird and Ann Baird

Essential Natural Plasters, Michael Henry and Tina Therrien

Essential Earthbag Construction, Kelly Hart

Essential Cordwood Building, Rob Roy

Essential Light Straw Clay Construction, Lydia Doleman

Essential Sustainable Home Design, Chris Magwood

Essential Building Science, Jacob Deva Racusin

Essential Prefab Straw Bale Construction, Chris Magwood

Essential Hempcrete Construction, Chris Magwood

See www.newsociety.com for a complete list of new and forthcoming series titles.

THE SUSTAINABLE BUILDING ESSENTIALS SERIES covers the full range of natural and green building techniques with a focus on sustainable materials and methods and code compliance. Firmly rooted in sound building science and drawing on decades of experience, these large-format, highly illustrated manuals deliver comprehensive, practical guidance from leading experts using a well-organized step-by-step approach. Whether your interest is foundations, walls, insulation, mechanical systems, or final finishes, these unique books present the essential information on each topic including:

- Material specifications, testing, and building code references
- Plan drawings for all common applications
- Tool lists and complete installation instructions
- Finishing, maintenance, and renovation techniques
- Budgeting and labor estimates
- Additional resources

Written by the world's leading sustainable builders, designers, and engineers, these succinct, user-friendly handbooks are indispensable tools for any project where accurate and reliable information is key to success. GET THE ESSENTIALS!

Cover design by Diane McIntosh.
Cover photo credits clockwise from left: Ray Main, Seabrook Munko, Gina Casey, Kindra Welch.
Illustrations by Dale Brownson.
Printed in Canada. First printing November 2023.

Inquiries regarding requests to reprint all or part of *Essential Cob Construction* should be addressed to New Society Publishers at the address below. To order directly from the publishers, please call (250) 247-9737, or order online at www.newsociety.com

Any other inquiries can be directed by mail to:
New Society Publishers
P.O. Box 189, Gabriola Island, BC V0R 1X0, Canada
(250) 247-9737

LIBRARY AND ARCHIVES CANADA CATALOGUING IN PUBLICATION

Title: Essential cob construction : a guide to design, engineering and building / Anthony Dente, Michael Smith, and Massey Burke.

Other titles: Cob construction : a guide to design, engineering and building

Names: Dente, Anthony, author. | Smith, Michael, 1968- author. | Burke, Massey, author.

Series: Sustainable building essentials.

Description: Series statement: Sustainable building essentials | Includes bibliographical references and index.

Identifiers: Canadiana (print) 20230486290 | Canadiana (ebook) 20230486312 | ISBN 9780865719682 (softcover) | ISBN 9781771423588 (EPUB) | ISBN 9781550927627 (PDF)

Subjects: LCSH: Earth houses—Design and construction. | LCSH: Cob (Building material)

Classification: LCC TH4818.A3 D46 2024 | DDC 693/.22—dc23

Funded by the Government of Canada   Financé par le gouvernement du Canada | Canadä

New Society Publishers' mission is to publish books that contribute in fundamental ways to building an ecologically sustainable and just society, and to do so with the least possible impact on the environment, in a manner that models this vision.

WE DEDICATE THIS BOOK TO THE unnamed and uncounted builders throughout the ages whose labor made cob and other natural building traditions come to life ... and to the pioneers of the natural building revival who put their ears to the ground to remember something that was almost forgotten, opening a door between the past and the future (especially Ianto Evans, Linda Smiley, Athena and Bill Steen, and Carole Crews) ... and to the engineers, researchers, architects, and code developers whose work has secured a lively future for these timeless materials and techniques (among them Bruce King, David Eisenberg, John Fordice, Martin Hammer, Fred Webster, Art Ludwig, Mark Aschhiem, and Kevin Donahue).

Gratitude

T HIS BOOK REPRESENTS the work of hundreds of individuals. Several people contributed enormously by drafting sections of text: Martin Hammer (who wrote the first draft of Chapter 7), Art Ludwig, David Wright, and the CobBauge team: Matthew Fox, Jim Carfrae, and Steve Goodhew.

Our gratitude goes to the many colleagues who reviewed portions of the manuscript and/or shared their personal experiences and knowledge: Greg Allen, Gord Baird, Katy Bryce, Féile Butler, Elke Cole, Carole Crews, John Curry, Sandy Curth, Myles Danforth, Kevin Donahue, David Eisenberg, Amanda Fischer, Yelda Gin, Mohammed Gomaa, Martin Hammer, Kyle Holzheuter, Stephen Hren, Leslie Jackson, Dan Johnson, Rebecca Kennedy, Tim Krahn, Ben Loescher, Bernhard Masterson, Mark Mazziotti, Hana Mori, Molly Murphy, Graeme North, Rob Pollacek, Sasha Rabin, Jim Reiland, Emily Reynolds, Conrad Rogue, Jess Shockley, Athena Steen, everyone at Terran Robotics, Bob Theis, Anni Tilt, Adam Weismann, and Kindra Welch.

Thanks also to those who offered photographs for our use: Danielle Ackley, Greg Allen, Katy Bryce, Féile Butler, Kevin Cain, CobBauge, Elke Cole, Sukita Crimmel, John Curry, Sandy Curth, Kiko Denzer, Hoyt Dingwall, John Fordice, Kyle Holzheuter, Scott Howard, Stephen Hren, John Hutton, Ya-yin Lin, Art Ludwig, Hana Mori, Seabrook Munko, Ananth Nagarajan, Graeme North, John Orcutt, Rob Pollacek, Sasha Rabin, Conrad Rogue, Laura Sandage, Tom Shaver, Jess Shockley, Athena Steen, Carlos Ventura, Catherine Wanek, WASP, Adam Weismann, Kindra Welch, and Rob West.

Many other people have been instrumental to the research efforts described in this book. They include everyone at the Cob Research Institute, everyone at Verdant Structural Engineers (especially Kevin Donahue, Kelsey Holmes Foster, Rachel Tove-White, Francisco Ordenes, Elli Terwiel, Colt Bender, and Wilzen Bassig), Quail Springs Permaculture, Art Ludwig, Jess Tong, Linda Safarik-Tong and Roger Tong, and all cob researchers and students including Daniel C. Jansen, Tonya Nielson, Mark Aschhiem, Hana Mori, Dezire Q'anna Perez-Barbante, and Julia F. Sargent.

To everyone at New Society Publishers, especially Rob West, Chris Magwood, and Linda Glass, whose dedication not only to this book but to the entire *Essentials* series has gone a long way toward ensuring that reliable information about natural building is available to everyone.

And to our partners and families, of course: Kelly, Louie, Oscar, Cathy, Tomo, Amani, David, Mimi and Milton, and all of the Burke-Fordyces. Thanks for your patience and support. We love you!

A Note on References

This book ended up much longer than originally planned, and we hope the value of the information it contains is worth the number of trees harvested for its printing. Many excellent books on cob and related subjects already exist; we have made an effort to minimize duplication by referring the reader to these other sources wherever possible. To save space and a few trees, we often mention these books only by title in the main text; the authors' names and publication information are to be found in the Resources section at the end of the book.

Contents

Chapter 1

The History of Cob

Earth is one of the oldest and most common building materials. It has been used in one form or another wherever the local geology provides clay soils suitable for construction—which is most places on most continents. The known history of earthen building goes back at least 10,000 years: adobe houses dating from 8000 BC have been discovered in Turkmenistan, rammed earth foundations from 5000 BC exist in Assyria, and the 4000-year-old Great Wall of China was constructed primarily of rammed earth. Many surviving earthen buildings around the globe have been in use continuously for centuries and provide living laboratories for what techniques and designs work best in each region. And the earthen building tradition is far from dead. The UN estimates that 30% of the world's population today live in homes made of unfired earth.[1] People in many parts of the world find building with earth to be a practical way to meet their present-day needs, and, as we will see, these techniques are still evolving and adapting to a modern context. This book focuses on cob, and especially on the ways information recently acquired through scientific testing of this ancient building system can be used to improve the durability, performance, and acceptance of cob buildings.

Villagers in the Indian state of Punjab building a cob house in 2020. This is one of many parts of the world where ancient earth-building traditions have been passed down generation to generation to the present day.
CREDIT: SANSKAR HARDAHA/ALAMY STOCK PHOTO

To contextualize cob within the field of traditional earthen building, it is useful to look at the three major approaches that have developed over the millennia for constructing load-bearing walls primarily from earth. Perhaps the simplest system, variously called *cob, monolithic adobe,* or *coursed adobe* in English, is to mix clay subsoil with water, straw, or other fiber, and sometimes sand or gravel until it is firm but still workable, and then shape the resulting compound into a wall while still wet. Historically, each layer, or *lift,* of this fiber-stabilized mud was allowed to dry before the next one was added. The second technique is known in English as *adobe* (a word which derives from the Middle Egyptian via Arabic and then Spanish), or *mud brick.* A very similar combination of clay soil, water, and fiber is mixed to a plastic consistency, then formed into blocks using simple molds. These mud bricks are left to dry in the sun and then stacked to make a wall, with a thin layer of earthen mortar in between. The third system is *rammed earth.* Slightly dampened earth, usually with a high proportion of sand and/or gravel, is compacted inside a sturdy form. When the form is full to the top it can be removed immediately and reassembled upward to construct the next section of wall.

Cob, adobe, and rammed earth are all still viable today, and each of them is currently enjoying a resurgence in various parts of the globe. Of these three systems, cob is probably the least well known in North America and has been the last to receive serious attention in the areas of engineering research, code development, and modernization of the building process to make it more accessible to contractors and comprehensible to building officials. Recent efforts have begun to address that lack. But before delving into the present and future of cob construction, we'll take a quick glance at the history of this remarkable building system.

The First "Age of Cob"

The term *cob* (probably from the Old English for a *loaf* or *lump*) originally referred to monolithic earthen building methods native to Devon and the West Country of England. The clay subsoil there is particularly well-suited to earthen building and provided the raw material for many of the iconic whitewashed, thatch-roofed houses that are popularly associated

This picturesque cottage in Lincolnshire, with its decorative reed thatch, hip roof, and whitewashed lime plaster, is typical of cob houses in England and southeastern Ireland.

CREDIT: FÉILE BUTLER

with the UK countryside. It is estimated that there are still at least 20,000 cob houses in Devon, and an equal number of barns, outbuildings, and boundary walls.² These structures were built by mixing clay soil with water and straw in a pit using people or cattle as mixers, shoveling it up onto a stone foundation and compacting it in place by walking on the fresh cob and beating the sides of the wall with a paddle, then trimming and finishing the walls with lime plaster. The buildings constructed in this way were very thick-walled (2' to 3' thick, [60–100cm] or even thicker) and were usually one or two stories tall.

There are many other living (and extinct) traditions of earthen construction in which moist straw-reinforced clay soil is used to build a wall without additional structure or formwork. These systems have many names in different parts of the world: *mud wall* in Ireland, *bauge* in France, *tourton* in Belgium, *lehmweller* in Germany, *daga* in Mali, *swish* in Ghana, *tamboho* in Madagascar, *zabur* in Yemen, *pahsa* in Turkey, *tawf* in Iraq, *chineh* in Iran, *dorodango* in Japan, and *coursed adobe* in the American Southwest. Building systems similar to English cob evolved in countless regions, including much of Northern Europe from Ireland to Ukraine, France and Germany (in both countries, at least 50,000 cob buildings remain in use today ³),Spain, Portugal, Italy, many parts of Africa, including the Maghreb and the Sahel, the Middle East and the Arabian Peninsula, parts of East Asia, and Mexico. Because the terminology is so diverse and poorly understood and the earthen building techniques in many areas have been little studied by academics, the architecture of these regions is often erroneously said to be of mud brick—and in some cases, the techniques of cob, adobe, and rammed earth overlap in the same region or even in the same building, making the situation even more confusing. The West African nations of Morocco, Mauritania, Mali, Niger, and Burkina Faso are especially renowned for their mud architecture, which ranges from simple one-room homes and granaries to majestic mosques and public buildings such as the Great Mosque of Djenné in Mali, which is believed to be the largest earthen building in the world. The 16th-century walled city of Shibam in South Yemen, known as the "Manhattan of the Desert," contains hundreds of mud "skyscrapers" (towering up to 11 stories tall) built out of both cob and adobe brick. In Afghanistan, Pakistan, and India, people continue to build cob homes and public buildings following traditions that are many centuries old.

Indigenous peoples in what is now the Southwestern US built with coursed adobe from at latest 1200 AD until the 17th century, when the Spanish introduced the adobe block.⁴

In a Dogon village in Mali, these granaries are constructed with cob walls, which help maintain even temperatures and humidity levels for the storage of millet and other grains. CREDIT: SCOTT HOWARD

The Taourirt Kasbah in Ouarzazate, Morocco. This fortified complex of almost 300 rooms was built with a combination of cob, rammed earth, and mud brick from the 17th through 19th centuries. It served as residence and administrative center for a powerful ruling family.
CREDIT: CATHERINE WANEK

By the 1300s the method was sufficiently developed to allow the construction of pueblos with walls 1' (30 cm) thick and up to five stories high. The earliest parts of Taos Pueblo were constructed this way approximately 900 years ago, making this multi-story apartment building the oldest continuously inhabited building in North America.[5] Many archaeological sites and surviving structures in New Mexico, Arizona, and Utah show clear evidence of coursed adobe,[6,7] including Casa Grande in Arizona and Horsecollar Ruin in Utah.

English colonists carried the technique with them to New Zealand, Australia, and the Northeastern US, where a number of cob homes were constructed in the 18th and 19th centuries.[8] Cob buildings have proven practical in many climates, ranging from windswept northern coastlines to hot arid deserts.

Cob building remained a vigorous tradition throughout much of the UK until the late 19th century. By then, the construction of new cob homes had begun to wane due to rising labor costs, the increasing affordability of industrial materials such as brick and concrete block,

and the growth of building codes focused on standardized, rather than vernacular, building practices. One of the last major pre-revival cob buildings built in Devon was a house designed by the Arts and Crafts architect Ernest Gimson and completed in 1912.[9] After WWI, the war economy turned to reconstruction, and, on both sides of the Atlantic, newly industrialized construction methods quickly supplanted what remained of traditional ways of building. Decades passed with no new cob construction in either the UK or North America. After flourishing since time immemorial, the cob tradition in the UK appeared to have come to an end.

A Cob Renaissance

The dearth of new cob construction in the UK led to a loss of associated knowledge and skills. Cob houses are durable, and many beloved English cottages remained inhabited after maintenance ceased, but nearly a lifetime intervened between the end of World War I and the resurgence of interest in traditional building techniques in the 1970s, '80s, and '90s. As a result, much local knowledge of how to build and maintain cob was lost and had to be relearned. In the UK, where the cob revival was motivated first by the needs for historical preservation and deferred maintenance, this re-learning was largely a resurrection of the old practices combined with more mechanized mixing and other adaptations.

In 1978, restorationist Alfred Howard built the first new English cob building in several decades—a bus shelter in the Devon village of Down St. Mary. Howard and many other builders first honed their craft by restoring old cob structures before venturing into new construction. The first cob house of the revival (actually a part-cob reconstruction of a crumbling stone barn) was built by Kevin McCabe in 1995. Known as "the King of Cob," McCabe has since tackled increasingly ambitious projects, culminating in the 2010s with Dingle Dell, a 13,500 ft² (1,250 m²) complex for which he mixed over 2,000 tons of cob and built a

quarter of a mile of cob walls 3' (1 m) thick and up to 29' (9 m) high. The building was designed to PassivHaus thermal standards and achieved Code for Sustainable Homes Level 6, the highest level of certification in the UK.

In parallel with the revival in the UK, a new cob tradition was born on the West Coast of the US. Welsh-born landscape architect Ianto Evans, along with Linda Smiley and Michael G. Smith, started the Cob Cottage Company in Oregon in 1993 with the goal of researching and teaching earthen building techniques adapted to the Pacific Northwest. Since cob knowledge was traditionally passed down orally and little had been written down, and because information about contemporary global earthen building traditions was hard to access in a pre-internet world, US cob techniques were essentially reinvented based on limited historical information from the UK and considerable creative input from early adopters in the US.

"The Laughing House" at Cob Cottage Company headquarters in the Oregon rainforest is a typical example of Oregon Cob: small, highly sculptural, and constructed largely of site-harvested and salvaged materials. CREDIT: MICHAEL G. SMITH

Oregon Cob therefore diverged substantially from traditions in the UK and elsewhere, emphasizing manual small-batch mixing for maximum quality control and artistic expression through sculptural designs. Another innovation that distinguishes Oregon Cob from most other cob traditions is that each lift is not allowed to dry completely before the next is added. The top of the wall is kept moist during construction, which allows manual integration of each lift with its neighbors. This produces something approaching a truly monolithic wall, which is better able to resist lateral forces, such as earthquakes.

Interest in the US was initially sparked by environmental concerns, the high cost of conventional housing, and the desire for healthier lifestyles, so the first generation of Oregon Cob buildings tended to be small, owner-built, non-permitted, and constructed largely of found and salvaged materials. In the final years of the 20th century, builders trained by the Cob Cottage Company introduced Oregon Cob in many parts of the world, notably in Canada, Mexico, Argentina, and Thailand, where the technique took root and prospered.

Over the last three decades, builders in the US, UK, and around the world have been experimenting with new techniques, borrowing knowledge from other earthen building traditions as these become increasingly connected, and sharing their successes and failures through workshops, conferences, publications, and on-line forums. More recently, researchers at nonprofits and universities have begun systematically testing the structural, thermal, and other properties of cob in an effort to understand best practices for cob construction and to write building codes and standards. Hundreds of new cob homes have been built—in many countries and climate zones—both by owner/builders and, increasingly, by professional builders as well.

The CRI and the IRC

The Cob Research Institute (CRI) is a nonprofit organization started in 2008 with the mission to make cob legally accessible to all who wish to build with it. It was founded by California architect John Fordice, who fell in love with cob after attending a hands-on Cob Cottage Company workshop in 1996. Frustrated by the difficulty of obtaining legal permission for cob buildings, Fordice passed the hat at a Natural Building Colloquium and raised enough money to file for official nonprofit status. He assembled a volunteer Board of Directors and began combing through the international literature on the engineering and regulation of earthen buildings, while researching the necessary testing and other steps toward approval of a cob code.

In 2013, CRI Board members Massey Burke and Anthony Dente collaborated with engineering faculty and students at the University of San Francisco to study physical properties such as compressive strength and modulus of rupture of cob mixes with varying amounts and lengths of straw. This was the start of a series of research collaborations with dozens of individuals, universities, and testing facilities. An express intention in all of CRI's research is to find safe ways to build with cob that meet the strict evidence requirements of building codes while maintaining cob's character as a user-friendly, low-environmental-impact building system. Modern cob builders avoid the use of stabilizers such as Portland cement and asphalt emulsion, both of which are commonly added to rammed earth and adobe walls. A major goal is to codify building techniques that require minimal external inputs and little or no mechanization so that they are accessible to people in a very wide range of socio-economic conditions, and can be legally implemented by people and communities with limited resources.

Our initial foray into cob structural testing stimulated our interest in taking a cob building through a full Alternative Materials and Methods (AMMR) permit application in a Bay Area jurisdiction. Our motivations were: 1) to set a precedent for permitted cob in a relatively stringent permitting context, so that it would be likely to be useful in other less-demanding jurisdictions; and 2) to practice translating vernacular cob knowledge into the formal language and numbers required by a building department. The opportunity for such a project was provided by the Tong family in Berkeley. After nearly three years of collaboration between Massey, Anthony, and David Lopez of the Berkeley Building Department, we secured a permit for a small backyard studio in 2016. We suspect that this building permit holds the world record for the most intellectual capital invested per square foot. The results of the process were compiled into a white paper that can be found on the CRI website.[10] This process helped clarify what kind of information and further testing would be needed to create a complete cob code, as well as helping us understand how to communicate about cob with building officials.

In the last few years, CRI has collaborated on the construction of eight full-scale cob wall panels for testing. These employed a range of reinforcing strategies—from straw-only to steel mesh to a rebar grid similar to those used to reinforce concrete walls. Each panel was attached to a testing frame that applied force to the tops of the walls in back-and-forth cycles to simulate the effects of earthquakes. Some were also tested for resistance to out-of-plane forces. Two additional full-scale cob walls were built in a laboratory in Texas and subjected to rigorous fire testing. Many smaller samples have been tested for density, compressive strength, flexural strength, and insulation value. Outstanding

contributors to these efforts include Art Ludwig of Oasis Design, Sasha Rabin of Quail Springs Permaculture, and students and faculty at Santa Clara University and the California Polytechnic State University, San Luis Obispo.

This ongoing program of laboratory testing and the collection of existing earthen building standards from around the world gave CRI the data necessary to write and defend the first prescriptive building code for cob anywhere in the world. In 2019, under the direction of lead code writer Martin Hammer, CRI submitted their code as a proposed Appendix to the International Residential Code (IRC). The IRC is a model code published and updated on a triennial cycle by the International Code Council (ICC). It is the basis for building codes for one-and two-family dwellings, townhouses, and accessory structures throughout the United States (except Wisconsin). CRI's proposal was approved by a two-thirds majority vote of International Code Council voting members, most of whom are building and fire officials.

Ancient materials meet modern technology when cob is tested in a laboratory. In this photo, the cob wall in the background is being tested for fire-resistance in a facility in Texas. See more on this test in Chapter 3.
CREDIT:
ANTHONY DENTE

The result was *Appendix AU: Cob Construction (Monolithic Adobe)* in the 2021 IRC (referred to in this book simply as "Appendix AU"). This model building code is reprinted in its entirety in the appendix of this book, along with official commentary intended to make it more comprehensible.

Unlike the main body of the code, adoption of Appendices to the IRC is optional; each Appendix must be specifically adopted by a jurisdiction such as a state, county, or city in order to become a part of its building regulations. The public can influence this process by expressing a need for such a code to their local building department, elected officials, or overseeing state agency. Other natural building systems, including strawbale and light straw-clay, have undergone the same process, first becoming Appendices to the IRC, and then being adopted into state and local building codes. For example, *IRC Appendix AS: Strawbale Construction* was approved as part of the 2015

IRC and has since been adopted by at least six states and nine city or county jurisdictions.

As board members of CRI, all three authors of this book were intimately involved with the last decade of cob testing and with the writing of Appendix AU. Anthony was a major contributor to much of the testing design and was the primary drafter of the structural sections of the code. Massey was involved with early testing of cob's physical properties. Michael was an integral part of the code writing and editing process; as a founding director he played a critical role in the formative early years of the Cob Research Institute. We all continue to be involved in ongoing testing in an effort to refine the code to make it more useful to designers, builders, homeowners, and building officials in more diverse geological and climatic areas.

The Next Age of Cob

We hope that this book will help pave the way for a new golden age of cob. The time seems right. Climate change, devastating wildfires, and the desire for affordable, healthy housing are among the strongest forces driving renewed interest in this ancient technique. New scientific information is available which in some cases validates and at other times allows us to improve on traditional earthen building methods. The model building code, which will continue to evolve through amendments supported by further testing, will reduce legal and bureaucratic barriers to cob construction in the US (and hopefully elsewhere, by example). We hope you will choose to design and construct beautiful buildings out of this amazing material, and in so doing join the growing global community of cob advocates, specialists, and enthusiasts.

Groups such as this one at Spirit Pine Sanctuary in Southern California have come together at hands-on workshops all over the world to learn cob construction. Cob Cottage Company founders Linda Smiley and Ianto Evans are reclining in the foreground.
Credit: Art Ludwig

Chapter 2

Rationale and Appropriate Use

Frequently Asked Questions

What is cob? Cob, known also as *monolithic* or *coursed adobe,* is a mix of clay soil, sand, and straw. It is mixed onsite to a wet, doughy consistency and shaped in situ in vertical layers called *lifts*. Sourcing the materials for cob is usually hyper-local: clay soil often comes from the building site and/or from nearby excavations or quarries, straw comes from local or regional grain farms, and sand comes from regional quarries and local aggregate yards and/or is inherent in the clay soil.

Why build with cob? The UK cob revival and the birth of a US cob-building tradition were driven in great part by the growing awareness of several interwoven ecological problems—notably, climate change, global deforestation, and wildlife habitat loss. Other factors include the rising cost of housing and the proliferation of illnesses associated with toxic building materials and unhealthy interior environments. Cob can offer many personal benefits as well: opportunities for artistic expression, biophilic materials, healthy living spaces, and the satisfaction of making your own shelter are just a few. It is very easy to sculpt the material into almost any desired shape—curved walls and arched openings and niches are simple to shape with cob, as are more elaborate three-dimensional forms. The ease with which even novice builders can express themselves artistically is one of the primary attractions of cob—especially in its North American variation, which emphasizes careful handwork for greater control of the material.

How much does it cost? Cob has a reputation for being radically inexpensive. The hard costs of making cob walls can be kept very low because cob is relatively easy to learn, takes advantage of low-cost raw materials, and doesn't require a lot of expensive tools. However, this is only true if you are using your own or volunteer labor, and not assigning it a monetary cost. If you are planning to pay for labor at normal construction rates, costs for building with cob will be comparable to conventional building. Additionally, the walls of a building only account for a fraction of the overall budget; other components such as site prep, design and permits, foundation, roof, and mechanical systems typically require skilled labor and are responsible for the lion's share of a project's cost.

Does this mean that we are discouraging a community approach to cob building?

Cob lends itself to curvilinear forms and whimsical designs. Kindra Welch designed this mountain retreat for family members in northern New Mexico. The cob walls were built primarily in work parties by previously untrained volunteers. CREDIT: CLAY SAND STRAW

Absolutely not! Cob is a fabulous material for community engagement. Just be aware that there are considerations and tradeoffs with this approach, as with any building system.

Cob is an especially viable option for those with access to land and plenty of available time and/or community support, but limited building skills and financial resources. Because cob is labor intensive, a contractor-built cob house may not be less expensive than a conventional custom home, but a much higher proportion of the construction funds will go to local artisans rather than supporting industries with dubious labor practices and environmental impacts. Locally harvested materials are immune to supply chain disruptions stemming from global politics and interruptions in international manufacturing and transportation. So, building with earth is a choice to support one's local community and ecology while reducing impacts in both nearby and distant areas of the globe.

How long does it take? One of cob's main disadvantages compared to either rammed earth or adobe—let alone conventional building systems—is that the speed of wall construction is constrained by drying conditions. The amount of wall height that can be added in a day is generally limited to between 1 and 2 feet, depending on the specifics of the mix, wall thickness, and weather. Cob construction is easiest, fastest, and most pleasant in warm, dry weather. However, it is not limited to those conditions: even in cold, wet weather, a lift of cob that sits overnight protected from rain gains a surprising degree of stiffness and resistance to slumping.

Although it is possible to erect the walls of a single-story cob building in a week, most experienced builders prefer to leave more time between lifts in order to reduce problems such as slumpage and cracking. Building more slowly can reduce the amount of trimming and

remedial work needed, which brings down the total labor needed for the walls. Depending on climate, design, size of building crew, and other factors, professional builders typically take between three and eight weeks to complete the cob walls of a building.

Environmental Benefits

All construction has a significant environmental impact, but cob building is one of the gentlest methods from the perspectives of both climate effects and resource conservation. The materials used in cob construction—clay subsoil, sand, and straw—require very little fossil fuel or electrical energy to extract, harvest, transport, and refine compared with most other building materials. This means that cob has very low embodied carbon (also known as *upfront carbon emissions*.) Cob walls replace what would otherwise typically be wood, concrete, or brick—all of which embody or emit considerably more carbon and create far more damage to forest lands and other habitats. The clay soil that forms the bulk of a cob wall can often be harvested close to or on the building site (sometimes simply from excavation for foundations and site work), reducing the energy spent on transportation. At end-of-life, cob components can be composted and/or returned to earth. In short, a cob building's ecological footprint is typically quite small compared with almost any conventional building material.

Table 2.1 indicates that cob is slightly carbon-negative (i.e., carbon-storing) due to the carbon-storage capacity of straw. However, these figures do not include any carbon calculations for cob mixing, so mechanically-mixed cob is likely carbon neutral or slightly carbon-emitting. In addition, these figures only apply to the cob portions of the building, so close attention to minimizing the embodied carbon in other parts of the building is still

Table 2.1: ICE (Inventory of Carbon and Energy)[1] Values for Embodied Energy and Carbon in Materials Relevant to Cob Construction

Material	Embodied energy in MJ/kg	Carbon storage in kg: CO_2/kg	Embodied CO_2e (CO_2 equivalent) in kg: CO_2e/kg
Clay soil	0.083	0	0.0052
Sand	0.081	0	0.0051
Straw	0.24	1.62[2]	-1.61 (stored)
Cob[3]	0.085 to 0.094	0.016 to 0.113	-0.011 to -0.108 (stored)
Concrete block	0.72	0	0.08

required. This is especially true for concrete foundations, which have a very large carbon footprint and are often larger in cob buildings than in conventional buildings due to wall width and mass.

Performance of Cob

Cob and moisture: In general, cob is moisture-resistant when it is kept off the ground, has a good roof, and is finished with vapor-permeable materials that do not trap water inside the walls. Exposed cob erodes slowly in wind-driven rain and other weather conditions, but it does erode, so weather-resistant finishes may be needed depending on the exposure of the building site and the owner's appetite for maintenance.

Structural qualities: When each lift of cob is carefully integrated with the previous layer, cob

The living area of Bliss Haven, a cob home designed by Kindra Welch and built by her company Straw Clay Wood. An exterior view is shown on the cover of this book. For its energy and water efficiency, low-impact materials, and indoor environmental quality, this building received a 5-star rating from Austin (Texas) Energy's Green Build Program. CREDIT: KINDRA WELCH

walls become *semi-monolithic,* with an almost continuous matrix of straw linking the whole wall together into a single structural mass. The weakest points of the system are the boundaries between lifts, though these junctions are not as weak as in masonry buildings built of the same materials, such as adobe. In high seismic zones, cob generally requires added reinforcing, usually in the form of rebar, mesh, or other tensile materials inside the walls.

Thermal performance: Cob can be mixed to a range of densities, which yield different thermal behaviors, but in general cob is a *thermally massive* material. This means that it acts like a thermal battery that needs to be charged. A good thermal design for cob is climate-specific: in some climates, cob works well as a standalone material, but in many contexts, an energy-efficient cob building requires additional insulation. Cob also complements and benefits from passive solar design strategies. We dive deeply into cob's thermal performance in Chapter 3.

Fire resistance: Cob is extremely fire-resistant; it has a 2-hour fire rating under ASTM E119. This makes it appropriate for use in situations where fire safety is of utmost importance, such as boundary fences and the shared walls between dwelling units in a multiplex.

Appropriate Uses

When compared to other earthen building systems, cob has both advantages and disadvantages. Unlike adobe and rammed earth building, cob doesn't require forms or molds, which makes for a very simple, low-tech construction process. On the other hand, cob is labor-intensive, which makes it either expensive or slow to construct. It is also physically massive, which limits design options in seismic areas. Because it is thermally massive, cob is not

suited to all climate zones without additional insulation. Considerations for when to use cob include access to suitable clay subsoil, climate, building size and purpose, and seismic zone, as well as the culture of the local building department (if permitting is required).

Soil availability: Cob is one of the most clay-intensive forms of wall construction, and even a very small building can use a full dump truck's worth of clay soil. Fortunately, clay soils that can be used for cob construction are common in most regions of the world. By some estimates, clay soils suitable for earthen building make up 74% of the Earth's crust.[4] Exceptions include regions that are very young geologically, such as most of the Hawaiian Islands. In these areas, there hasn't been enough time for bedrock to weather into clay minerals. Other places where clay can be scarce or unavailable include glaciated areas, where the earth's surface has been churned up in unpredictable ways, and in river valleys, where clay-bearing strata in alluvial soils may be buried under many feet of clay-poor deposits. The mere presence of clay in the soil is not the only consideration: some clay soils make stronger cob than others, and some require extensive processing (sifting to remove rocks or soaking to soften and hydrate clay), so both the qualities of the resulting cob and the efficiency of mixing vary depending on the clay source.

Climate: Cob is a dense, heavy material with relatively poor insulation qualities. In areas where temperatures remain cold for long periods of time, or remain hot even at night, cob will need to be insulated to perform efficiently. Passive solar designs that take advantage of winter sun to heat the interior of the building make a big difference in a cob building's efficiency. Current building code energy requirements

for mass walls in most climate zones in the US require added insulation for typically-sized cob walls in residential buildings.

Seismic zones: Cob reinforced with straw fiber alone may not perform well in strong seismic events. Since building mass magnifies seismic forces on a building, appropriate attention to structural design is required. According to the current cob code in the US, cob buildings in high seismic areas require an engineered design and added reinforcing, such as steel rebar or mesh.

The local building department: Before there was a cob building code, supportive officials within building departments could often be found who were willing to help shepherd projects through the approval process. Unfortunately, these individuals can't be found in all building departments. Since cob is

radically different from conventional building systems, cob projects are sometimes met with resistance by building officials. The existence of the cob model code and its future adoption by more jurisdictions should make the permitting process much more routine. However, at the current juncture, it is still wise to take the culture of your building department into consideration when deciding whether to seek a permit for a cob building.

Building size: Because current cob construction methods are relatively slow, modern cob techniques have been used primarily for smaller structures. This is partially cultural: many contemporary cob practitioners are drawn to the simplicity of hand mixing and placement and the unique opportunity for community participation that a less-mechanized, labor-intensive method of building offers. However, in this cultural moment, cob's minimal carbon impact,

One of the entrances to the central Medina in Marrakesh, Morocco. Thick cob and adobe walls like these surround the entire historical center of the city. In hot desert climates, massive earthen structures help to cool and shade adjacent streets and buildings. CREDIT: CATHERINE WANEK

inherent supply chain resilience, and demonstrated resistance to wildfire have the potential to transform the demand for cob construction more rapidly than we might anticipate, and the technique has unexplored potential at larger scale. We expect that further advances in technical research combined with completely new innovations like 3D printing will radically transform the cob landscape in the next decades. However, at least until we have more data on reliable reinforcement strategies, the size of cob buildings should be limited to one or, at most, two stories. Structural testing of cob wall systems is ongoing and recommendations will be adjusted over time.

Chapter 3

Building Science

THE PRINCIPLES of building science help us understand how to manage the flows of air, heat, and moisture (both liquid and vapor) through a building. To perform well, a building needs to be designed and constructed with attention to all of these flows, as well as to the contexts of site and climate.

Heat flow is of particular interest for a cob building that includes conditioned space. Building materials typically have one of two thermal performance characteristics: *insulation* or *thermal mass.* Many natural building systems combine both characteristics, but most are either predominantly insulative or predominantly massive. Because cob is primarily a massive material with poor insulating properties, one of the first questions to consider is whether or not insulation will be needed in addition to the mass of the cob. A brief overview of thermodynamics is helpful for understanding what these terms mean.

Heat can be transferred in three ways: by *conduction, convection,* and *radiation.*

Conduction refers to the physical transfer of energy from one molecule to another through direct contact—the molecules literally run into each other and "hand off" their energy, always moving from warmer to cooler. Highly conductive materials such as metals have molecules that are close together, making energy transfer faster. Metal feels colder than wood not because it is a lower temperature, but because it conducts heat away from your hand more quickly. High-mass walls such as concrete, brick, and cob typically conduct energy more rapidly than less-dense walls.

Convection refers to the ability of a fluid (liquid or gas, including water and air) to carry

energy with it from one place to another. This is why airflow and air leakage are important to consider when designing a good thermal envelope: as a fluid, air is very good at escaping a home—taking with it your hard-earned heat! Conversely, fluids like air do *not* conduct heat well when still. The smaller the pockets of air, the better they insulate, because the air volumes in small pockets are too small to sustain air movement (*convection currents*) that would transfer heat. The best insulating materials tend to be those containing many tiny pockets of trapped air (think goose down in a sleeping bag or cellulose insulation in an attic).

Radiation refers to the transfer of energy via electromagnetic waves across space from a warmer object to a cooler one—such as the sun to the earth. Radiation explains how a sun-warmed rock or cob wall heats your skin even if you are not touching it, or, conversely, how a cold window or wall seems to "suck" heat out of your body. Unlike conduction and convection, radiation does not depend on physical contact between the warmer and cooler objects. Radiation passes most easily through a vacuum, less well through a gas, and even less well through solid objects.

Thermal Properties of Cob

We can now return to the thermal properties of cob and the question of insulation. Materials that insulate do not conduct heat well because they are relatively light and low density—their physical structures offer molecules relatively little opportunity to transfer heat through direct contact. Many insulative materials also block convective heat transfer because they contain *dead air spaces*—that is, little pockets of still air that are isolated from one another and are too small to transfer heat internally through convection.

Highly insulative natural materials include straw, wool, hemp, and cotton, as well as more processed materials such as loose cellulose, wood fiberboard, and mineral wool. In the US, a material's insulative capacity is indicated by its *R-value* (resistance to heat flow). In much of the rest of the world, it is measured by its *U-value* (a measure of thermal conductivity). The density of a material and its R-value tend to be inversely correlated. So, straw, which is not dense, has a high R-value, making it a good insulator. Cob, on the other hand, is very dense, so it has a low R-value.

R-values can refer to either *total* or *effective* R-value, which is the measure of how insulative an entire wall assembly is, or *R-value per inch* of thickness of a material. The higher the R-value (or the lower the U-value), the more insulative the material: good insulators such as straw, wool, fiberglass, cellulose, and rigid foam have values from about R-2 per inch to R-5 per inch. In contrast, high-density cob, at 110 pcf, has an R-value of around 0.22 per inch. For context, Table 3.1 lists densities of common insulating and massive building materials.

Energy efficiency codes equate thermal performance with high R-value, but R-value is not the whole story. The thermal mass of a wall can also have a significant effect on its performance. Thermal mass materials are denser than insulative materials, containing fewer or no air pockets. As a result, massive materials are more thermally *conductive* than insulative materials. But there is nuance to how heat flows through these materials. Thermal mass materials have high *heat capacity*, which is the measure of how much heat a material can hold. Mathematically, the heat capacity of a material is the product of its *specific heat* (the amount of heat a material can hold per unit of mass) times the density of the material times its volume.

A high heat capacity means that a lot of energy input is required to raise the temperature of the material. As a result, a mass wall material like cob will transfer heat from the warmer side of the wall to the cooler side, but not very quickly. How fast this happens depends both on the conductivity of the material and the temperature differential between the interior and exterior wall surfaces. Informal observations suggest that a good rule of thumb is about 1 inch per hour of heat transfer through cob: for example, heat radiated by the setting sun at 5 pm onto the outside surface of a 12"-thick (30 cm) cob wall will not reach the inside surface until 5 am the next morning.

This mental model is a good starting point for understanding the basic behavior of thermal mass. In practice, however, mass walls are rarely in a steady state thermal condition; many

Table 3.1: Density of Various Building Materials

Material	Density (pcf)
Fiberglass insulation	0.5–1.0
Cellulose insulation	1.5–2.0
Cork	7
Straw bales	7–9
Light straw-clay	10–50
Pine (seasoned)	28
Gypsum board	48
Clay (dry)	63
Cob and adobe (dry)	70–115
Sand (dry)	90–120
Fired brick	100–120
Concrete	110–144
Sandstone	147
Granite	175

variables contribute to their actual thermal performance. For example, as the temperature outside the wall drops from 90 degrees at 5 pm to 70 degrees at 10 pm, the passage of heat into the wall slows down. If the nighttime low temperature outside drops below the temperature of the wall's exterior face, the heat flow will reverse and never reach the inside of the building at all. This buffering action of thermal mass has the effect of confining daytime/nighttime temperature swings to the exterior portion of the wall, keeping the interior surface temperature stable. This behavior is often referred to as the *thermal flywheel effect.*

Gord and Ann Baird of Eco-Sense set out to measure exactly how these heat flows behave in real life. When building their cob home on Vancouver Island in British Columbia, Canada, they implanted multiple temperature sensors in the walls, as well as inside and outside the building. Measurements taken over a full year showed that most heat does not make it through the wall: the exterior portions of the walls generally absorb and release heat only to the exterior air, and the interior parts absorb and release heat to the interior air. Only a small amount of heat was transferred through the wall from exterior to interior or vice versa. This behavior was likely in part a function of the moderate climate (the coldest temperature recorded during the year was 19°F [-7°C] and the warmest 99°F [37°C]). Another factor was their modified cob mix that included pumice as an insulating additive.[1]

The conductive behavior of thermal mass materials drives how the wall affects indoor air temperatures, but it's radiation that drives how that *feels* to building occupants. Once heat reaches the interior surface of a mass wall by conduction, some of it is radiated from the surface to other masses, including humans. Unless we are in direct physical contact with a thermal mass, we feel its warmth (or coldness) primarily through radiation. The size and shape of a thermal mass are important because its surface area-to-volume ratio determines how quickly or slowly the mass will radiate heat: a large volume of mass with less surface area, like a cob bench, will change temperature more slowly than the same volume of mass spread over a bigger surface area, like an earthen plaster. More or less responsive masses can be useful in different contexts, so this palette of options is important to bear in mind when designing with massive materials.

As we mentioned at the start of this chapter, one of the first questions to consider is whether insulation will be needed in addition to the mass of the cob. A wall system that has insulation on the exterior *and* mass on the interior has a good combination of thermal characteristics for many climate zones. With this type of assembly, the insulation restricts heat transfer through the wall (outside to inside in hot weather, and inside to outside in cold weather), and the mass releases and absorbs heat from the interior of the building, keeping the interior temperature stable.

Prescriptive building and energy codes for mass walls require added insulation in many climate zones. However, both mass and insulation are not always necessary. Depending on the climate, adding insulation to cob walls may just add unnecessary time and expense to a project. Where the outside temperature is always fairly comfortable (in a place like San Diego, for example), a massive cob wall will always maintain a comfortable temperature—even without insulation. Climate zones with large day-to-night temperature swings around a comfortable average are also well suited to mass walls with no added insulation, because of the buffering flywheel effect described above. Even buildings in cold but dependably sunny areas, like the high-elevation areas of Colorado, Utah, and New Mexico do not necessarily need as much insulation as prescriptive codes require when mass walls are combined with good passive solar design. In such situations, whole-building energy performance modeling using computer models may allow you to make a case that extra insulation is not needed. See more on this option below.

On the other hand, climates that remain cool, cold, or cloudy for extended periods benefit from a well-insulated building envelope. Uninsulated mass walls in these climate zones tend to stabilize at a cold temperature and require extra heat input to keep the interior spaces comfortable. Conversely, uninsulated mass buildings in climates that seasonally remain hot day and night (>75°F/24°C) are often uncomfortable because the mass walls absorb large quantities of heat and radiate it to the

building's occupants. Thermal design is covered in more detail below, in "Optimizing Cob's Thermal Performance in Your Climate Zone."

Mass-Enhanced R-Value

Because the actual thermal *performance* of thermal mass walls is often much better than their designated R-values would suggest, they can also be assigned an *effective R-value* or *mass-enhanced R-value*. The most detailed research to date on the effective R-value/U-value of adobe mass walls was conducted by researchers at the University of New Mexico Energy Institute.

New Mexico was a very early adopter of ASHRAE Standard (90-75) as well as energy code Chapter 53 of what was then the Uniform Building Code. Both of these Standards were intended to address the energy crisis of the 1970s by setting new rules for building energy efficiency, and both included prescriptive

A Trombe wall helps to heat a cob building at Quail Springs Permaculture. The lower part of the wall is colored black and covered with a glass skin. This arrangement serves as a solar collector and transmits heat to the inside of the wall. See "Cob-Specific Design" in Chapter 7 for more on passive solar heating and cooling strategies.
CREDIT: SASHA RABIN

R/U values for mass walls. While implementing these Standards, the New Mexico design and building community found many problems with the thermal models used in the prescriptive codes. In response, state-funded researchers developed a computer model to better understand the thermal behavior of adobe mass walls in response to a range of variables, including diurnal temperature swings, *insolation* (absorbed solar radiation), wind speed, radiation to the night sky, and wall color. The computer model combined the physical characteristics of the wall (density, conductivity) with dynamic conditions at the interior and exterior surfaces of the wall. Six adobe wall types were modeled: two wall thickness variations and three exterior insulation variations. The adobe walls were either 10" (25 cm) or 14" (36 cm) thick, and each wall thickness was modeled with no insulation, 1" (2.5 cm) of foam insulation, and 2" (5 cm) of foam insulation. All walls had ¾" (2 cm) of stucco on the exterior and ½" (1.3 cm) of gypsum plaster on the interior. (More details can be found on The Earthbuilders' Guild [TEG] website.[2])

The result of the research was a list of effective U-values organized by 11 New Mexico climate regions. These effective U-values were intended to substitute directly for the broader "steady state" U-value prescribed by the energy codes. For each wall type, all the effective U-values in different climate regions were usually as good or better than the steady-state U-value assigned by ASHRAE. For the full data set please refer to the documents on the TEG website, but some of the most salient conclusions include:

1. Effective U-values are significantly different from steady-state U-values. Thus, consideration of thermal storage and solar design is important.

2. The effective U-value for east, west, and south walls is strongly impacted by color—in some instances it may be more effective to paint a wall than to insulate it.

3. The economically optimum amount of insulation varies depending on the orientation of the wall and is different from what would be predicted using steady-state U-values. Generally, less insulation is required when effective U-values are used.

4. Effective U-values can easily be substituted in a building code. The result is a performance code that allows a broader range of construction methods. In particular, uninsulated mass walls are shown to be much more efficient when all factors are considered.

This project was probably the most extensive earthen mass wall thermal modeling effort to date, but since the model was confined to New Mexico climate regions, its direct application is limited to that state. However, because New Mexico climates represent a wide range of heating and cooling conditions, it does provide guidance for thermal modeling in other states, countries, and climate zones.

The New Mexico data is of particular interest to cob builders because, in contrast to nearly all other mass wall modeling to date, the information is specific to earthen walls. This matters because not all mass materials are created equal. Concrete and stone are both denser than cob; they make a building feel cold in winter and hot in the summer. Compared to their concrete counterparts, cob and adobe buildings are generally experienced as warm in winter and cool in summer. Their lower density explains this in part, and earth's relationship to moisture likely plays a role as well; moisture behaves very differently in cob vs stone or cementitious materials. Because thermal mass data from concrete is not an accurate indicator of cob's thermal behavior, more specific modeling of cob buildings is still needed. Other effective U-value research exists, including some preliminary dynamic U-value modeling from Oak Ridge National Laboratories in Tennessee[3], but none is as detailed as the New Mexico research.

An interesting footnote is that the New Mexico adobe building community is still flourishing today, whereas the formally small but growing community of adobe builders in Southern California has largely disappeared because California's energy codes do not address these issues. Compliance with California's requirements for wall thickness and/or added insulation were considered too expensive or otherwise impractical by adobe builders there—just as they were in New Mexico before that state's codes were revised.

Optimizing Cob's Thermal Performance in Your Climate Zone

Cob is a mix of insulating (straw) and massive (clay and sand) ingredients. Its thermal performance is highly dependent on the density of the mix. As defined by the IRC, thermal mass materials have densities greater than 20 pcf

(320 kg/m[3]) and insulative values under R-1 per inch. For cob building, the practical density range is generally 70–110 pcf (1121–1762 kg/m[3]), though we do know people testing mixes as low as 50 pcf (800 kg/m[3]). Below 50 pcf, the material should be considered light straw-clay, a non-structural wall infill system governed by IRC Appendix AR, rather than cob (see Lydia Doleman's book, *Essential Light Straw Clay Construction* for information about this very versatile building system.) When designing with cob, one of the biggest questions is how to employ the thermal characteristics of cob to best advantage in your climate conditions.

Testing conducted by the Cob Research Institute (CRI) documented that cob insulation values and densities range from R-0.22 per inch at 110 pcf (the density of "typical" cob) to around R-0.60 per inch at 70 pcf (near cob's lightest, highest-fiber extreme). To put these numbers in the context of other natural wall systems, Table 3.2 shows the density/R-value continuum of some common natural wall systems, beginning with straw only (strawbale) and ending with clay and aggregate only (rammed earth.) As you can see, cob is somewhere in the middle; "heavy straw-clay," "low-density cob," and "cob and adobe" are all closely related variants. These terms are not strictly technical, but they point out useful distinctions—both with regard to mixing and building practices and thermal and structural design considerations—in the range of what is commonly called cob. The "heavy straw-clay" category in particular captures a gray area between light straw-clay and cob, where the mixes are dense enough and contain enough clay to be self-supporting and shapeable but are not thought to be able to carry any additional load (though this assumption needs more testing).

The distinctions between *high-density* and *low-density cob* and *heavy straw-clay* are not

Table 3.2: Density and Insulation of Natural Wall Systems

Material	Density	R-value per inch
Strawbale[1]	7–9 pcf	1.55 flat, 1.85 on edge
Light straw-clay[2]	10–50 pcf	1.80–0.84
Heavy straw-clay[3]	50–70 pcf	0.84–0.60
Low-density cob[3]	70–90 pcf	0.60–0.40
Cob[3] (high-density) and adobe	90–110 pcf	0.40–0.22
Rammed earth[4]	110–120 pcf	0.22–0.1

[1] Strawbale densities and R-Values are from IRC Appendix AS – Strawbale Construction.

[2] Light straw-clay densities and R-Values are from IRC Appendix AR – Light Straw-Clay Construction.

[3] Cob mix densities and R-values are from research conducted by the Cob Research Institute.

[4] Rammed earth density values are from Burroughs, S. "The Relationship Between the Density and Strength of Rammed Earth." *Proceedings of the Institution of Civil Engineers,* 162:3, 2009.

recognized in Appendix AU, which sets no upper or lower density limits on cob (although the commentary to Section 109.2 mentions that cob's "practical density range" is between 70–110 pcf). Table 3.3 and Figure 3.1 show the code-defined relationship between mass wall densities and their R-values, beginning at 20 pcf, which is the lowest density defined as a mass wall by the IRC. In the US, it is standard for thermal resistance to be calculated in R-value per inch of material thickness, and the current prescriptive insulation requirements are all written in this format. Due to the effects of mass-enhanced R-value (explained above), it is probably more accurate to assign total R-values to full-width cob wall systems, rather than to extrapolate from a test conducted on, for example, a 3.5"-thick (9 cm) sample to a per-inch R-value. More research on this subject is needed, as we suspect that this effect—and other unknown factors—may contribute to discrepancies in available data. For example, there are differences between the R-values found in Table 3.3 and Figure 3.1 and those shown in Table 3.7 from the CobBauge project.

When permitting a cob building, there are two pathways to demonstrate thermal compliance. The first is to follow prescriptive guidelines for the minimum R-value of mass walls. Recognizing the value of thermal mass, the IRC and other prescriptive codes require that mass wall systems have a total R-value that is less than non-mass walls in the same climate (see Table 3.4). However, the discussion above about the research conducted in New Mexico on effective U-values illustrates the limitations of assigning a static R-value to mass walls. Nowadays, it is common for designers to use *energy performance models* to document thermal code compliance. In this approach, the designer hires a consultant to make a thermal computer model of the entire building. This allows for

some horse-trading among building components and can result in thinner or less-insulative walls balanced by a more insulative roof, for example. Be aware that the computer models commonly used for this purpose have been designed and tested with more conventional materials such as concrete in mind. It is possible to model earthen walls with the software, but the modeler needs to be experienced or at least innovative to do a good job, and the computer algorithms need to be calibrated with the results of physical experiments.[4]

Table 3.3: R-Values of Different Earthen Wall Densities

Wall Material	Density (PCF)	R-Value/ inch wall thickness (ft'·°F·h/BTU/in)
Light Straw Clay	20	1.48[1]
Light Straw Clay	30	1.22[1]
Light Straw Clay	40	1.01[1]
Light Straw Clay	50	0.84[1]
Cob	75	0.54[2]
Cob	110	0.22[2]

[1] R-Value data is from Table AR103.2.3 of Appendix AR in the 2024 IRC.
[2] R-Value data is from Section AU109.2 of Appendix AU in the 2024 IRC.

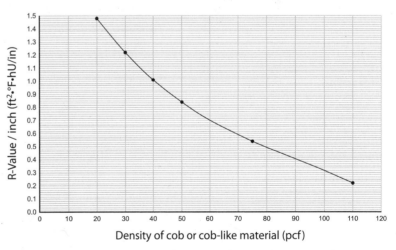

Note: The curve shown is a cubic fit curve based on limited testing. Dots shown are tested data points included in IRC appendices AR and AU.

Fig. 3.1: *Earthen wall density vs r-value.* Credit: Wilzen Bassig, Verdant Structural Engineers

Table 3.4: Minimum R-Values for Framed and Mass Walls per Climate Zone[1]

Climate Zone	Wood Frame Wall R-Value[2]	Mass Wall R-Value[3]
0 and 1	13 or 0&10ci	3/4
2	13 or 0&10ci	4/6
3	20 or 13&10ci or 0&15ci	8/13
4 except Marine	30 or 20&5ci or 13&10ci or 0&20ci	8/13
5 and Marine 4	30 or 20&5ci or 13&10ci or 0&20ci	13/17
6	30 or 20&5ci or 13&10ci or 0&20ci	15/20
7 and 8	30 or 20&5ci or 13&10ci or 0&20ci	19/21

ci = continuous insulation

[1] These requirements come from Table N 1102.1.3 (R402.1.3) of the IRC.

[2] The first value is cavity insulation; the second value is continuous insulation. Therefore, as an example, "13&5" means R-13 cavity insulation plus R-5 continuous insulation.

[3] The second R-value applies where more than half of the insulation is on the interior of the mass wall. Per Section AR104.1 of the 2024 IRC, walls with light straw clay infill of densities of greater than or equal to 20 pounds per cubic foot (480.6 kg/m 3) shall be classified as mass walls.

As Table 3.4 shows, the IRC and other prescriptive codes require some insulation for mass walls in all climate zones. Luckily, all cob walls do have *some* inherent insulation, and optimized densities can be tailored to specific climate zones. For example, if you are in Climate Zone 4 non-marine or below (nearly half of the contiguous US), there is a pathway to prescriptive code compliance that doesn't dictate an excessively thick cob wall. What it does require (in Zones 3 and 4) is a low-density mix—around 70–75 pcf. Table 3.5 shows that in the coldest Climate Zones (Zones 6–8), cob

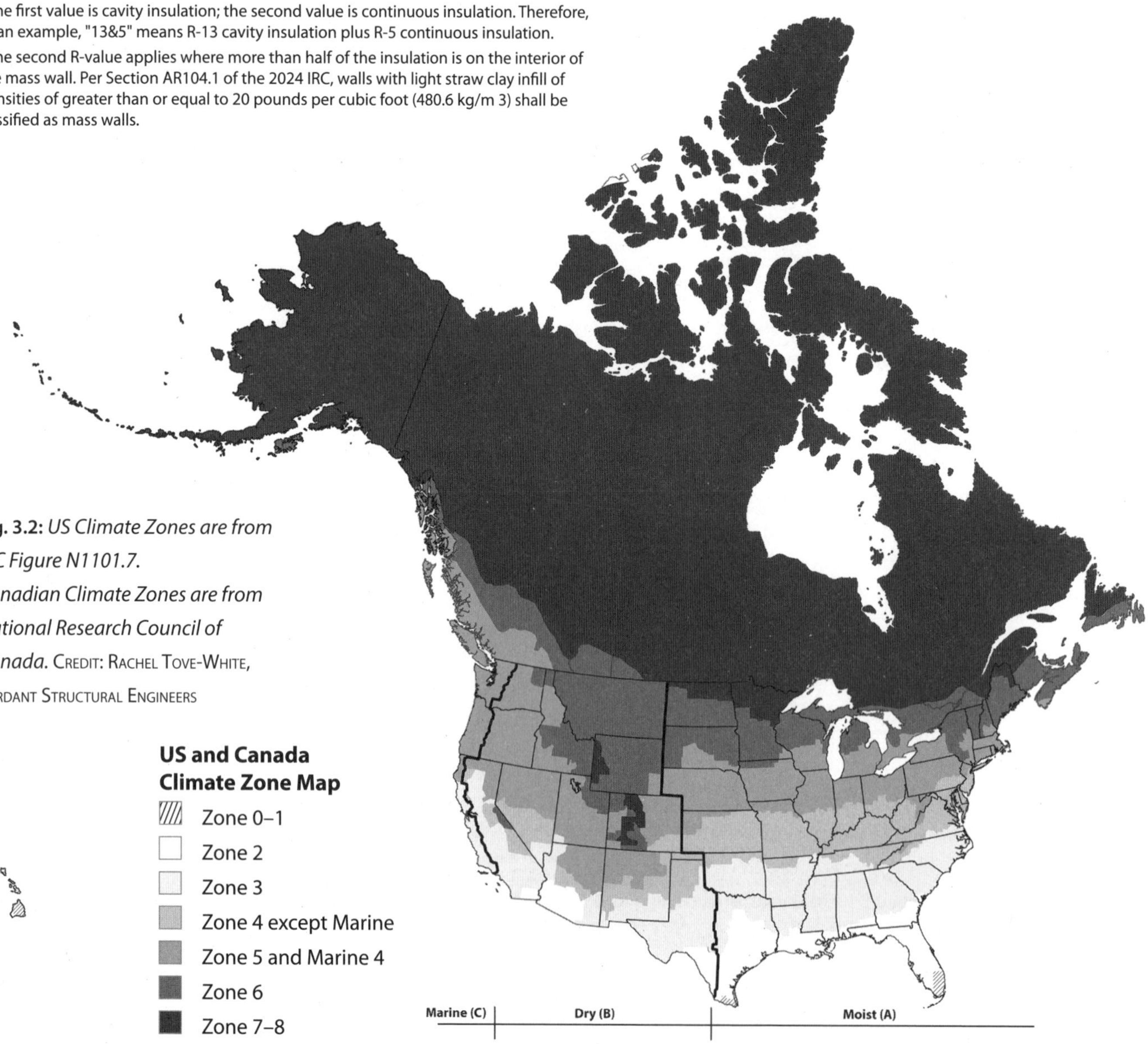

Fig. 3.2: *US Climate Zones are from IRC Figure N1101.7. Canadian Climate Zones are from National Research Council of Canada.* CREDIT: RACHEL TOVE-WHITE, VERDANT STRUCTURAL ENGINEERS

US and Canada Climate Zone Map
- Zone 0–1
- Zone 2
- Zone 3
- Zone 4 except Marine
- Zone 5 and Marine 4
- Zone 6
- Zone 7–8

Marine (C)   Dry (B)   Moist (A)

walls need to be more than 2′ (60 cm) thick—even when low-density cob is used. At some point, the extra work and materials required to build such thick walls (and the large seismic forces experienced by such massive walls) make it more practical to keep the walls thinner and insulate them with another material. This is why in places like the UK and France builders are pursuing approaches like the one described in the sidebar below, "CobBauge: A Structural and Insulating Cob Wall."

Not all areas of the world have climate zone maps available with detailed correlations to mass-wall insulation requirements. However, the information in Table 3.5 can be applied anywhere in the world if you know a few

things about the local climate. The Climate Zones used in the IRC are the same as the International Energy Conservation Code (IECC) Climate Zones, which are defined objectively using *heating* and *cooling degree days* (HDD and CDD) as the criteria. HDD and CDD are quantifications of the amount of energy needed to heat and cool a building in a particular place. Specifically, HDD is the sum of the positive differences between 65°F (18°C) and mean daily temperature over an entire year. CDD is the sum of the positive differences between mean daily temperature and 65°F (18°C) over an entire year. This results in a set of climate regions numbered 0–8, from tropical to arctic, which are further subdivided based on

Table 3.5: Required Insulation Thickness per Climate Zone[1]

Climate Zones	Base Wall Density (PCF)	Required Wall Thickness w/No Add'l Insulation[3] (in)	Additional Exterior Insulation Required (in)		
			Light Straw Clay[4], 20 pcf R= 1.48/in	Straw[5], R= 1.85/in	Hempcrete[6], 12.5 pcf R=2.1/in
Climate Zones 0 & 1 Required Mass Wall R-Value: 3[2]	110 75	13.6 5.6	Not needed Not needed	Not needed Not needed	Not needed Not needed
Climate Zone 2 Required Mass Wall R-Value: 4[2]	110 75	16.2 7.4	Not needed Not needed	Not needed Not needed	Not needed Not needed
Climate Zones 3 & 4 (non-Marine) Required Mass Wall R-Value: 8[2]	110 75	34.4 14.8	2.7 Not needed	2.2 Not needed	1.9 Not needed
Climate Zones 5 & 4 Marine Required Mass Wall R-Value: 13[2]	110 75	57.1 23.3	6.1 2.6	4.9 2.1	4.3 1.9
Climate Zone 6 Required Mass Wall R-Value: 15[2]	110 75	66.2 27.0	7.5 4.0	6.0 3.2	5.3 2.8
Climate Zones 7 & 8 Required Mass Wall R-Value: 19[2]	110 75	84.4 34.4	10.2 6.7	8.1 5.4	7.2 4.7

Base Wall Density (PCF)	Base Wall R-Value/in (ft²•°F•h/BTU/in)	Base Wall Thickness (in)	Total Earth Plaster Thickness[7] (in)	R-Value of Base System (ft²•°F•h/BTU)
110	0.22	16	2	3.96
75	0.54	16	2	9.08

[1] This table assumes that the additional insulation is placed exterior to the cob. If the insulation is added to the inside of the wall, more thickness will be required to meet IRC thermal standards.

[2] This requirement can be found in Table N11.02.1.3 (R402.1.3) of the 2024 IRC.

[3] This wall thickness satisfies the insulation requirements of the IRC. Other sections may have additional structural requirements that govern the wall thickness.

[4] R-Value taken from Table AR103.2.3 in Appendix AR of the 2024 IRC.

[5] On-edge R-Values are used for the table. The R-Value for straw laid flat is R = 1.55/in, taken from Section AS108.1 in Appendix AS of the 2024 IRC.

[6] R-Value taken from Table BA106.2 in Appendix BA of the 2024 IRC.

[7] R-Value of earth plaster is taken as 0.22 ft²•°F•h/BTU/in.

rainfall and humidity into *moist, dry,* and *marine* (which means dry summer, wet winter): A, B, C.

If someone outside the US wants to know their IECC Climate Zone (also known as the "ASHRAE 169 Climate Zone" in reference to the ASHRAE Standard that contains the zone definitions), they can simply apply the criteria listed in Table 3.6 to determine it for themselves. The most current form of the Standard, which at the time of writing is ASHRAE 169-2021, is usually behind a paywall, but earlier versions can be accessed for free—such as Addendum A from ASHRAE 169-2020, a revision to the 2020 Standard with updated data.[5] This document lists cities all around the world and their corresponding IECC/ASHRAE climate zones. For example, Beijing, China is listed as Zone 4A, and Mexico City as Zone 3A.

Other than the groundbreaking CobBauge project described below, we know of few physical precedents for combining cob with insulation. However, there is a range of strategies that we believe should be considered and implemented more widely. These options fall into three basic categories: 1) attaching other natural wall systems that wrap around the outside or inside of cob walls, 2) applying appropriate conventional or natural insulation products to the exterior or interior of cob walls, and 3) creating cavities within the walls that are filled with an insulating material. For options 1 and 2, thermal modeling suggests that it is usually more effective to apply the insulation to the exterior rather than the interior of the wall.

Various insulating natural wall systems can be combined with cob. Light straw-clay and other light earth mixtures can be added to a cob wall by attaching framing directly to the cob and/or to the foundation and roof framing. Massey carried out one such experiment, wrapping a cob wall with lightweight framing and light straw-clay infill, which proved easy to implement. (It would be useful to repeat the experiment while monitoring the moisture content of the light straw-clay adjacent to the cob wall to make sure it is able to dry before decomposition begins.) Hempcrete can either be formed next to a cob wall in the same manner (as Stephen Hren demonstrated in his hempcrete-insulated cob home in North Carolina[6]), or else sprayed on. As Table 3.5 shows, only fairly thin layers of these more insulative materials are needed in most climate zones. Even in the coldest climates (Zone 8), a 16"-thick (41 cm) high-density cob wall surrounded by 10¼" (26 cm) of light straw-clay or 7¼" (18 cm) of hempcrete would satisfy the code. It is also possible to combine cob walls with full-sized straw bales, but we feel that this is a less practical option, both because the connection is awkward and the resulting wall thickness is extreme.

Even lighter insulation materials can be used as long as they are contained effectively. For example, a cavity can be framed outside the cob wall and surrounded with wire mesh, reed

Table 3.6: IECC Climate Zone Definitions

Zone Number	Thermal Criteria	
	IP Units	SI Units
0	10,800 < CDD50°F	6000 < CDD10°C
1	9,000 < CDD50°F < 10,800	5000 < CDD10°C < 6000
2	6,300 < CDD50°F ≤ 9,000	3500 < CDD10°C ≤ 5000
3	CDD50°F ≤ 6,300 and HDD65°F ≤ 3,600	CDD10°C ≤ 3500 and HDD18°C ≤ 2000
4	CDD50°F ≤ 6,300 and 3,600 < HDD65°F ≤ 5,400	CDD10°C ≤ 3500 and 2000 < HDD18°C ≤ 3000
5	CDD50°F < 6,300 and 5,400 < HDD65°F ≤ 7,200	CDD10°C < 3500 and 3000 < HDD18°C ≤ 4000
6	7,200 < HDD65°F ≤ 9,000	4000 < HDD18°C ≤ 5000
7	9,000 < HDD65°F ≤ 12,600	5000 < HDD18°C ≤ 7000
8	12,600 < HDD65°F	7000 < HDD18°C

Source: Table C301.3 of the 2021 International Energy Conservation Code

mats, lath, or wood siding. This cavity can then be filled with the vapor-permeable insulation of your choice, either a loose material (cellulose, wool, rice hulls, or a mixture of rice hulls with lime or clay slip) or batts of wool, cotton, or fiberglass. Careful planning is needed to protect these insulation materials from condensation and moisture damage, which can negatively affect the cob walls.

Insulative plasters are another option worth considering. For example, the Italian company Diasen produces a very lightweight mixture of hydraulic lime, cork, and clay. This product is quite expensive, but it has a remarkably high insulation value of nearly R-4 per inch of thickness. Just 2.5" (6.4 cm) of this material is enough to bring a 16"-thick (41 cm) low-density cob wall into compliance with energy codes in even the coldest climate zones. Other products are available, or you can make your own insulating plaster by combining lightweight materials such as perlite, pumice fines, and cork with lime and/or clay.

The third option for insulating cob involves creating cavities in the middle of a cob wall to be filled with an insulating material. This technique is already being employed by builders working with 3D earthen printing (see "3D Printing and Automated Construction" in Chapter 13). It could also be done in a more standard cob wall by, for example, building around hollow PVC pipes (they would be drawn upward as the wall gains height and eventually removed). While this seems like the simplest approach, since the insulating material only needs to fill a space rather than to hold itself together, it has several limitations. Because the insulation is discontinuous, it will have a less beneficial effect than an uninterrupted layer. Also, cavities—especially large ones—will weaken the wall structurally. This strategy is therefore probably practical only in low-seismic zones and non-structural walls.

This 3D-printed earthen wall demonstrates one strategy for adding insulation to a cob wall. The wall is constructed with vertical air spaces that can be infilled with natural insulation materials, including rice hulls, as shown here. This wall was part of the Gaia project, which was printed using WASP printers in Italy in 2018. See the sidebar in Chapter 13 for more on 3D printing. CREDIT: WASP

CobBauge: A Structural and Insulating Cob Wall

By Dr. Matthew Fox, Dr. Jim Carfrae, and Professor Steve Goodhew

One of the principal issues concerning the effective utilization of earth-building techniques is the need to satisfy building codes. Although there is plenty of anecdotal evidence that cob buildings often feel warm in winter and cool in summer, building codes require more demonstrable data. When thermal properties are measured in a laboratory, traditional cob does not offer sufficient insulation to meet the thermal conductivity or resistance requirements of many building codes. As a result, new cob buildings designed for cooler climate zones have been required to add insulation materials such as wood fiber, typically installed on the outside of the wall after the cob is complete.

A relatively new method of improving the thermal performance of cob is the *CobBauge construction system* ("*bauge*" is a French word for cob), resulting from an EU-funded partnership between researchers in the UK and France that was based at the University of Plymouth.

The CobBauge system is a dual-layer composite wall: the inner layer of traditional, dense cob delivers structural load-bearing performance, and the outer layer of low-density light earth provides insulation. This wall system satisfies both the structural and thermal requirements of building codes in the UK and France. As has been discussed elsewhere in this chapter, different climates demand different approaches to the balance of mass and insulation in an earthen wall. The CobBauge system will work well in any cool and some cold temperate climates. With modifications, it can work in very cold parts of North America as well. The R-value of 19.5 achieved in our prototype building is sufficient to pass energy code requirements for all climate zones in the US.

Initially, the project's researchers attempted to produce a monolithic cob wall system that could satisfy both structural and thermal demands with a single mixture of subsoil and fiber. Six different fiber types were ☞

Prototype CobBauge wall in the labs at the University of Plymouth, showing the two-part composite wall consisting of a layer of structural cob (on right) and a layer of insulating light-earth (on left).
CREDIT: LLOYD RUSSELL

combined with 12 different clay-rich subsoils in various proportions in an attempt to produce an optimized mixture. The fibers were wheat straw, flax straw, hemp straw, flax shiv, hemp shiv, and reed. Unfortunately, it quickly became apparent that this approach was not feasible. The key difficulty was the inability of any mixture with an overall density low enough (650 kg/m^2, or 40 pcf) to be acceptable for the thermal building code to also perform well structurally. No single material was able to fulfill both functions.

Our second approach was much more successful. This is a dual wall system consisting of a layer of cob able to withstand a building's structural loads alongside a conjoined layer of lower-density light earth to satisfy the thermal code. By varying the densities and thicknesses of both layers, it is possible to alter the overall thermal resistance of the wall and its bearing capacity.

There is a close correlation between the density and thermal conductivity of a range of fibers and cob mixes. At the lower end of the curve, the pure fibers such as straw and hemp give thermal conductivity values of between 0.045 W/mK and 0.06 W/mK. The best-insulating mixture of clay slip and fiber contained 50% hemp shiv by dry weight of soil (or 1 bucket of clay slip to 3 buckets of shiv) and had a thermal conductivity of 0.09 W/mK and a density of 230 kg/m^3 (14.4 pcf). The low-density mixes used for the thermal layer of a CobBauge wall most often have a density of approximately 350 kg/m^3 (21.8 pcf) and a thermal conductivity of around 0.11 W/mK.[7] Denser structural cob mixes unsurprisingly have much higher thermal conductivity values. The CobBauge structural layer has a density of 1,400 to 1,600 kg/m^3 (87–100 pcf) and a conductivity of around 0.64 W/mK. These figures are used in Table 3.7 to calculate the total thermal conductivity of the dual-wall system.

The calculated U-value for the entire wall, including interior earthen plaster and exterior lime render, is 0.29 W/m^{2}K, which equates to a US imperial R-value of 19.5 ft^2·°F·h/BTU. This is the first load-bearing wall formed entirely from earth and natural fibers to meet the UK building standard's thermal requirement of 0.3 W/m^{2}K or less. Each time the CobBauge system is used, these figures will vary slightly depending on the properties of local soils and fibers. It is useful to measure the conductivity of each mix during the design process. In addition, the project engineer will likely require structural load testing of the structural mix.

Whilst a 600 mm (24") total wall thickness was chosen for our pilot projects, the freeform nature of the material offers opportunities to vary the thickness of each layer depending on what properties are required. By increasing or reducing the thickness of either component, you can ☞

Table 3.7: Typical U- and R-values for CobBauge Wall System, Including Finishes

Composite cob wall elements	Density lbs/ft^3 (kg/m^3)	Thickness in. (m)	U-value (thermal conductivity) per thickness BTU·in/(h·ft^2·°F) (W/m·K)	R-value (thermal resistance) per thickness ft^2·°F·h/BTU·in (m·K/W)	R-value per wall element ft^2·°F·h/BTU (m^2·K/W)
Internal surface		0	n/a	n/a	0.68 (0.12)
Internal earthen plaster		0.8 (0.02)	3.05 (0.44)	0.35 (2.5)	0.28 (0.05)
High-density cob	88.8 (1423)	11.8 (0.3)	4.44 (0.64)	0.23 (1.57)	2.67 (0.47)
Light earth	21.2 (340)	11.8 (0.3)	0.76 (0.11)	1.31 (9.1)	15.50 (2.73)
Exterior lime render		0.8 (0.02)	4.16 (0.60)	0.21 (1.5)	0.17 (0.03)
Exterior surface		0	n/a	n/a	0.34 (0.06)
Total		25.2 (0.64)			19.64 (3.45)

tune the thermal resistance to obtain the value required by local building codes.

Two different arrangements of the wall components are possible. Either the thermal layer is placed around the outside of the structural layer (which is the normal solution) or the sequence is reversed, with the thermal layer on the internal face of the building. External insulation is akin to wrapping a blanket around the building, maintaining a steady inside temperature controlled by the higher heat capacity of the structural layer. Placing the thermal insulating layer on the inside speeds up the thermal response time of the building, enabling a faster warm up—but also a faster cool down.

One critical challenge for the success of the dual-wall system was to ensure good bonding between the layers. This was not easy given the dissimilarity of the two materials—one with a fiber ratio of approximately 2.5% of the dry weight of soil, the other with a fiber ratio of around 50% of the dry weight of soil. The solution was to install the two layers next to each other inside prefabricated formwork, using an angled placement tool that allows the structural layer to be placed first in each 200–250 mm (8–10") lift. (See "Formwork" in Chapter 14, for more details of the CobBauge forming system.) When the placement tool is withdrawn, the resulting cavity is filled with the lower density thermal layer. In this way the two layers of the wall sit next to each other in a wet state. In theory no additional bonding material is needed because the clays in each mixture join to form an unbreakable whole, but we advise laying strands of hemp straw across the structure as a sort of natural wall tie (see photo). Note that, as this system has been developed in the UK, no testing on its use in seismic conditions has been performed to date.

Two prototype buildings have so far been completed. One, in Normandy, France, demonstrates how a two-story building can be constructed. The second is a classroom on the city center campus of the University of Plymouth that is designed for meetings and small group tutorials. A third building is currently under construction in the east of the UK in Norfolk by Hudson Architects. This will be the first CobBauge building constructed in a competitive market.

All three buildings are being monitored and the findings will be published. The initial stage of monitoring focuses on the speed of drying and its ramifications for construction timing, such as when best to apply the internal and external finishes. After construction is complete, heat flux sensors are used to measure *in-situ* thermal transmission values over a number of weeks during the heating season. Temperature sensors will eventually reveal a complete view of how the walls behave both in summer and winter. It will be interesting to compare ☞

This photo shows the placement tool in use within the formwork. Structural cob is being added to the interior side of the wall, on the right. After the tool is removed, the resulting cavity will be filled with light earth. In the foreground are strands of hemp fiber to be laid across the structure as a natural wall tie. CREDIT: MATTHEW FOX

each building's actual thermal performance with the calculations made during the design stage based on laboratory measurements.

To give designers and engineers information about how this natural wall system performs in relation to increasing expectations of "airtight" construction, all the CobBauge buildings have been subject to air pressure testing. The UK prototype was measured to have an air permeability value of 4 $m^3/(h \times m^2)$, which satisfied the UK code requirement for a non-domestic building of ≤ 5 $m^3/(h \times m^2)$. Other sensors measure relative humidity, air temperature, CO_2 levels, airborne particulates, and volatile organic compounds (VOCs) to determine the air quality in relationship to indoor environmental standards. Data such as this serves to build confidence in our system as a viable construction material for new buildings.

Moisture Management and Control

Moisture management is an important consideration for all buildings. Cob is not necessarily more vulnerable to moisture penetration than other wall systems—and in some ways, it is *less* vulnerable—but cob walls do require attention to moisture detailing.

How does a cob wall (or any other kind) get wet? Moisture gets into buildings in several ways: carried in with materials themselves during the building process (such as wet cob or plaster, green lumber, etc.), precipitation wetting walls or leaking through the roof, groundwater rising through capillary action, plumbing leaks, and water vapor diffusion from moist air either generated inside the building or leaking through the building envelope from outside. In both the natural and conventional building communities, there is increasing consensus that trying to make a building that moisture will never get into is essentially impossible; so it is important to design buildings that allow moisture *out*.

It's helpful to understand the difference in behavior between liquid water and water vapor. Water in all its states is a molecule with two positively charged hydrogen atoms and one negatively charged oxygen atom (H_2O). Each water molecule is only about 0.3 nanometers in diameter: one billion laid end to end would be about one foot long. When water is in liquid form, its positive and negative charges cause the molecules to clump together and behave as much larger entities. When moisture is in vapor form, the molecules are too spread out for their charges to cause this clumping, so they act as lone particles, which are much smaller than the molecule aggregates of liquid water.[8] This difference means that managing the liquid and vapor forms of water requires somewhat different strategies.[9] For a deeper dive into managing moisture in buildings, see *Essential Building Science* by Jacob Deva Racusin. *The Natural Building Companion's* moisture chapter is also a great resource, as are John Straube's many articles on the subject of buildings and moisture (see Resources).

Moisture management strategies for cob buildings are relatively simple, both for keeping moisture out and for dealing with moisture once it gets in. For keeping water out, the *good hat + a good pair of boots* approach is still the most important: a cob building needs a good foundation to prevent rising damp (which includes raising the bottom of the wall above the splash zone) and a good roof with a sufficient roof overhang for your climate. Additional strategies for keeping moisture out include good flashing details at doors and windows, installing plumbing in such a way that any leaks are obvious and easy to address, and, in areas with extreme wind-driven rain, adding some form of exterior rain screen.

We assume that no matter what we do, moisture will eventually get inside a cob wall, so it is important to build in such a way that it can get out again. Moisture enters a wall in both liquid and vapor forms, but it generally leaves in vapor form. Accordingly, most strategies for

Fig. 3.3:

Recommended flashing detail for windows in cob walls.

Credit: Rachel Tove-White, Verdant Structural Engineers

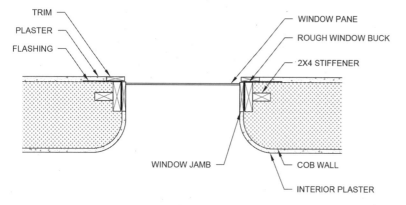

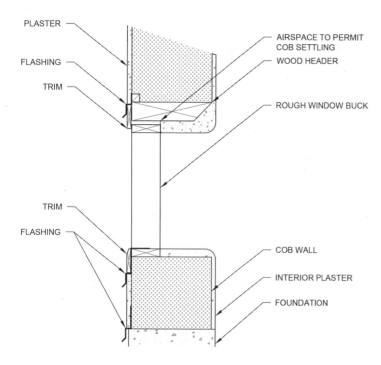

letting moisture out involve maintaining the *vapor permeability* of the walls. Finishes with low permeability (i.e., waterproof finishes) are not recommended. (See Chapter 15.)

While cob is not waterproof, it is not as vulnerable to moisture as people generally expect. There are several reasons for this. The first is that most of the component materials in a cob wall *will* not be harmed by the presence of water: neither clay nor sand is going to mold or rot when wet. The straw in the mix will rot if left in high-moisture conditions for a long time; but it is protected to a degree by the presence of clay and the absence of oxygen. Clay minerals are *hydrophilic*, which means that they like to bond with water molecules. So, as liquid water enters the wall, the clay "grabs" the water molecules, pulling them away from the straw in the mix. This same helpful action assists in the protection of other elements in cob walls, including wood and probably steel.

The hydrophilic nature of clay also helps keep the *core* of the wall dry. Because clay particles and water molecules are both polar—meaning that each has a positively and a negatively charged end—they like to stick to each other and can form something approaching a "membrane" of linked clay and water molecules. (This is the same concept as sealing a pond by lining it with a layer of clay.) This membrane-like assembly slows the incursion of more water into the wall, which explains why a clay exterior plaster is sufficient protection in many settings as long as the wall is not exposed to sustained windblown rain. Eventually (typically, after many years) the earthen plaster will erode or degrade to the point where it needs to be replaced, but this does not imply that damage has been sustained by the cob itself.

It is important to be aware that the protective capacity of clay minerals is only effective up to a certain moisture content. Beyond this point, the wall is still structurally stable and may appear undamaged, but the matrix is wet enough to allow straw decomposition. We have seen this occur in a cob landscape wall with a minimal roof cap after a few years in a moderately wet climate; after cutting into this wall, we found that the straw had begun to decompose. This example is mainly to illustrate that clay is not supernatural in its protective capacities, as people sometimes assume. On the other hand, when cob walls are well-protected from the top and bottom, the straw inside has frequently been observed in excellent condition decades and even centuries after construction.

If moisture problems are left unaddressed for too long, the straw will eventually rot away and/or the entire base of the wall can become saturated. Fully saturated cob will lose much of its compressive strength, and an entire wall can collapse suddenly. This scenario is unlikely unless the wall is in continuous contact with a source of moisture (like wet earth piled up against the base of the wall) or is coated with an inappropriate material that traps moisture.[10] Structural collapse due to saturation has mostly been observed in old cob and adobe structures that were later covered with cement stucco or another "waterproof" coating that trapped water inside the wall and caused moisture to build up over time. Previously dried cob re-absorbs water slowly, as anyone who has tried to rehydrate old cob has discovered, and basic good design is enough in most cases and in most climates to prevent severe moisture issues. Good maintenance is also extremely important. Leaky roofs, leaky gutters, and leaky plumbing are among the most common sources of water intrusion into old cob buildings. And cracked plasters (especially those with low vapor permeability, like Portland cement stucco) can allow water into the wall and then keep it trapped inside.

Above: *This adobe block was removed from a 1940s building in Walnut Creek, CA, that Anthony and Massey worked on. The straw in the block was in pristine condition after more than 70 years. The adobe is surrounded by cement mortar which, though not recommended practice, was common in adobe buildings of that era.* Credit: Anthony Dente

When the plaster was removed from this cob wall near Dublin, Ireland, the straw showed little or no sign of decomposition after more than 150 years. The clay in cob seems to preserve straw indefinitely, even in wet climates. Credit: Féile Butler

In hurricane-prone areas with extremely intense rainfall combined with high winds, normal design precautions may not be enough to protect cob walls. We have seen adobe walls in southern Mexico that sustained severe damage from the intensity of wind-blown rain during a hurricane. We have also noticed increased erosion rates of exposed earthen walls in California, as the occurrence of atmospheric river storms (a combination of high-intensity rainfall with strong winds) becomes more common in the context of climate change. These types of storms should be taken into account when choosing finish materials for cob walls that are particularly exposed. See Figure 15.1 for a recommended *rainscreen siding* detail that will protect cob walls from windblown rain.

When considering strategies for weather resistance, keep in mind that not all clays are created equal. Some types of clay have a much higher degree of weather resistance than others; the reasons appear to relate both to clay mineralogy and to the proportions of variously sized aggregates in a clay soil. These differences are more obvious in clay plasters than in cob because plasters are more likely to be exposed to weather. A good rule of thumb is that if a clay soil absorbs water slowly when you are processing it into a mix (that is, if it requires extended soaking to achieve workability) then it will likely behave the same way in a wall.

Water in vapor form is not a big concern for cob walls as long as the finish materials are permeable enough to release the vapor. Once cob walls have completed their initial drying process, they will absorb water vapor when the air touching them has a higher relative humidity than the cob itself, and release vapor when the air near the wall is drier than the cob. Cob walls can "store" quite a lot of water vapor without any noticeable change in the wall or the interior conditions of the building. One of the beneficial results is that cob tends to regulate indoor air humidity. Buildings containing earthen materials are actually much less likely to suffer from moisture-related interior air quality problems—including mold growth—than those made of less-permeable conventional materials.

The Bairds' cob house in British Columbia illustrates cob's resilience to water vapor exceptionally well. Over a full year, their Eco-Sense project conducted the most detailed moisture and temperature monitoring that we are aware of in a cob building: sensors were installed in and near both north and south earthen mass walls to record the environmental variables of relative humidity (RH), temperature, and dew point—inside the building, outside the building, and in the interior of the walls. These observations underscore cob's unique ability to

An example of 10-year-old exterior clay plaster. This wall is frequently exposed to both rain and wind, but not often to sustained intense wind-driven rain. Notice the areas of exposed straw and coarser surface texture, which are typical of several years of weathering in moderate exposure.

moderate indoor air humidity and to maintain a moisture balance inside the wall that preserves the straw in the cob matrix. Because cob can keep water vapor in a dispersed state, condensation (and resulting moisture damage) within a cob wall are not likely to be a problem.

Eco-Sense Cob Home Performance Report (excerpt)[11]

Authors: Christina Goodvin, Gord Baird, and Ann Baird

"The results demonstrate the exceptional performance of the walls in moderating humidity through all situations and seasons, responding within minutes to the changing respective indoor and outdoor environments. The cob wall acted like an impassible barrier with sponges on each side that could absorb and release moisture without condensation. The walls maintain the suggested moisture equilibrium of 0.4% to 6.0% and the yearly average indoor relative humidity was 54.9% (range of 47–62%), and indoor temperature average was 20.5°C (69°F). The walls have an active and observable effect of taking in and releasing moisture and heat to maintain a balance. The cob walls of the Baird home behaved as a selectively permeable membrane responding quickly and effectively to environmental changes to assist in maintaining a desirable and healthy indoor environment … During the period of data collection for this report the annual regional average humidity was 79.23%. The Eco-Sense measured yearly average for the outside humidity is 78.7%; the corresponding inside humidity of the home had an average of 54.9%. The range of RH within the cob home for the period of study was … well within the range that is considered comfortable as stipulated by ASHRAE Standard 55 (relative humidity range of 20–65%)."

Other points of particular interest:

- The wide daily difference between outside RH minimum and maximum and the contrasting very narrow range seen with the indoor RH.
- Data from a large 4-hour indoor gathering that demonstrated in real time the instantaneous ability of the earthen walls to moderate the RH.
- Data clearly illustrating that on the home's inside wall surface (the wall surface with the highest vapor pressures) the moisture level never exceeded 3%, even with greater moisture loading associated with indoor heating and activity.
- Demonstrated ability of the walls to control air humidity yet maintain water content levels that are virtually static, remaining well below the levels required for insect life (>14%), and fungal growth (>20%). Even with fluctuations in RH next to both interior and exterior wall surfaces, the wall itself maintained a wall water content of below 6.0%. These moisture levels remained static over the year that measurements were taken.
- Even when wall temperatures dropped below the dewpoint, no condensation was observed inside the walls. Cob's porous nature evidently allows for large volumes of adsorbed water which is easily attached and released. The walls are so effective at adsorption that absorption does not take effect at these temperatures and hence water does not condense on or within the walls.

Also from the report: "*Adsorption* is the tendency of a hydrophilic surface to capture and hold polarized water vapor molecules in the air. Most building materials have internal pores that adsorb water molecules continuously depending on the relative humidity. Once pore surfaces have adsorbed as much vapor moisture as they can, the pores themselves will begin to collect and store water from the air within their spaces via capillary suction, also known as *absorption*."

Steel Corrosion in Cob

The use of steel to reinforce cob walls is a relatively new practice; it has aroused some concerns about the durability of this material combination over time. As discussed above, the clay in cob is a hydrophilic material that attracts moisture to itself and pulls it away from other elements, such as straw and wood. The evidence suggests that the same action serves to protect embedded steel.

To understand the potential for steel corrosion in cob, it is helpful to explore the science of steel corrosion in Portland cement (PC) concrete, which is better understood. PC concrete has the ability to wick moisture via capillary action. At the same time, PC concrete's high alkalinity forms a protective coating around the steel that prevents it from reacting with oxygen and rusting for a very long time. Clay soils, though not as alkaline as PC concrete (pH 11), still typically tend toward alkalinity (commonly pH 7.5–10), which may contribute a degree of protection to embedded metal. Most laboratory procedures for testing steel corrosion in concrete materials, such as ASTM C0876-22B: Standard Test Method for Corrosion Potentials of Uncoated Reinforcing Steel in Concrete (Half-Cell Potential), are very specific to PC concrete. Since there is no standard for testing steel corrosion in clay, simulating the many-year corrosion process is very difficult. For the same reasons that many attributes of cob are under-studied compared to other materials—a lack of funding and industry focus—no known lab tests have been conducted on steel corrosion in cob or other earthen wall systems.

In the process of writing this book, we surveyed 14 experienced cob builders from around the globe—representing many different climate zones—about their observations of steel corrosion. Each of these builders had opened cob walls or other earthen building elements containing steel between 5 and 20 years after construction. The only reports of significant deterioration from corrosion were in cases where light-gauge metal such as chicken wire was embedded close to the surface of a clay plaster. In these cases, the clay coverage was usually much less than an inch—thin enough that the metal was likely wet for sustained periods and had more access to air for oxidation. Heavier gauge metal components such as welded wire mesh, rebar, and nails do sometimes show surface rust, but we know of no examples of structural deterioration of heavier gauge metal elements in a cob wall. In most cases where there were no other moisture issue in the wall (leaky roof, etc.), metal elements showed little to no corrosion.

To reduce the likelihood that enough moisture and oxygen will reach the steel to cause corrosion, we recommend surrounding all embedded steel elements with a minimum of 1.5" (4 cm) of cob. To be on the safe side until the science is better understood, Appendix AU requires the use of corrosion-resistant galvanized steel for all metal reinforcing less than 6-gauge ($^{13}/_{64}$" [5 mm] in diameter.) New Zealand's Earth Building Standards contain specifications for polypropylene geogrid,[12] which can be an alternative to light-gauge steel in some circumstances such as horizontal reinforcing. Another reinforcement option with similar strength to steel is *fiber-reinforced polymer* (FRP). These bars are made of fiber strands (typically of glass, basalt, or carbon) embedded in a plastic polymer matrix. We recommend consulting an engineer before using FRP bars in higher seismic regions because, although they are lighter weight and have a higher tensile strength than steel, they have less ductility, an important feature in most earthquake-resistant designs.

Air Control

The topic of airflow is very important for good building performance and can be confusing. It is frequently said that cob and other types of natural wall systems need to "breathe," but what does this mean? "Breathability" is sometimes interpreted to mean that air leakage is desirable, but this is incorrect. More accurately stated, cob walls need to remain *vapor-open*, meaning that there should be no impermeable barriers to the passage of water vapor through and out of the wall. Air leakage through gaps in the building envelope is in fact responsible for huge amounts of heat loss and can actually drive moisture into the walls and roof cavity.

All kinds of buildings are prone to unwanted heat gain and loss through air leakage; cob buildings are no exception. Although cob walls themselves are airtight, unwanted airflow can occur through gaps between cob and other elements such as doors, windows, and roof framing. The tendency of cob and clay plasters to shrink as they dry is a particular concern. For example, a ⅛" (3 mm) shrinkage crack around the top of a 20' × 20' (6 × 6 m) room adds up to 120 square inches (774 cm²)—nearly the equivalent of a one-square-foot hole! The transition between walls and roof assembly is one of the most vulnerable spots for air leakage and is also one of the hardest spots to air-proof, so particular attention is needed in this area.

There are several strategies for preventing air leakage. Problematic seams can be sealed by installing *air fins* that act as bridges across areas of possible air leakage. Air fins can be made from various vapor-permeable materials, such as 30-lb roofing felt covered by metal lath and plaster, and are usually stapled to the wood framing member and lapped over the potential air seam onto the adjacent cob wall.[13] Other strategies include trim designed to prevent air leakage, lath and plaster over the juncture between dissimilar materials, and various approaches to caulking cracks, including clay mixtures. Good air control does not necessarily require toxic materials, but it does require thoughtful attention to detail. For more options, design details, and a comprehensive discussion of air control, we recommend the following books: *Essential Building Science, The Natural Building Companion, Essential Rammed Earth,* and *Essential Light Straw Clay Construction.*

Fire Safety

Simply stated, cob doesn't burn. The only flammable ingredient in cob is straw, and straw can't burn without sufficient oxygen—which is not present in a cob wall because the straw is encased in clay soil. The authors have witnessed many scenes during building workshops where skeptical students have tossed earthen samples of various kinds into the evening campfire, only to become complete converts when the samples emerged unaffected. Cob walls have held up well in many of the recent massively destructive wildfires in California and elsewhere.

Following a recent laboratory test for fire resistance, we now have data to support a 2-hour fire-resistance rating for cob walls. In a 2021 project led by Art Ludwig and Sasha Rabin, Anthony managed the structural design of two ASTM E119 tests on the requisite 10' × 10' (3 × 3 m) test specimens. A 2-hour test was chosen because this is the maximum fire rating required for residences in the US. Two different cob wall assemblies were tested: the first had a density of around 110 pcf (1762 kg/m³), and the second ranged from 50 pcf (800 kg/m³) at the top through 70 pcf (121 kg/m³) in the middle to 90 pcf (1441 kg/m³) at the base. The intention was to determine the fire-resistance of a wide variety of mixes—including very high-straw mixes, the ones most likely to elicit concern about fire.

The ASTM E119 test requires that one face of the wall be heated to 2,000°F (1093°C) for 2 hours. The test is designed to assess both flammability and heat transfer, so heat is applied to one surface of the wall and measured on the other. During the 2-hour test period, the unheated face of the wall, which was approximately 1' (30 cm) thick, rose by only about 5°F (3°C)—which was less than the temperature change for the concrete slab in the room during the regular sun cycle of the day! After 2 hours of heating, each wall was compressively loaded to failure, and then sprayed for 2 minutes with a fire hose. This prolonged blast of high-pressure water was the part of the test we were most concerned about, but both walls passed with flying colors. A 2-hour cob fire rating was approved by the ICC at its public comment hearing in 2022 and will be included in the 2024 addition of Appendix AU.

Building codes require fire-resistance ratings for walls (and other building elements) only in particular circumstances. In the US, small-scale residential structures (one- and two-family dwellings, townhouses, and their accessory structures) require 1-hour fire-resistance rated walls when the wall: a) is less than 5 feet from a property line (3 feet when sprinklered); or b) separates units in a duplex or townhouse. Two-hour walls are needed for townhouses when not sprinklered. Walls separating a dwelling from a garage must have ½" (12 mm) gypsum board on the garage side of the wall. A cob wall can now be proposed in any of these situations.

Sasha Rabin building cob test walls for an ASTM E119 fire rating test at a fire testing facility in Texas in 2021. The project was a collaboration between Quail Springs Permaculture, Oasis Design, Verdant Structural Engineers, CRI, and Earthen Shelter.
CREDIT: JOHN ORCUTT

Two test walls were constructed with varying densities and straw content. During the test, each wall was placed against a furnace and subjected to temperatures of 2,000°F for 2 hours.
CREDIT: ANTHONY DENTE

Above: *After two hours in the furnace, the wall was rotated and the hot side tested for firehose resistance. Here a lab employee blasts the hot wall with a firehose for 2 minutes.*
CREDIT: ANTHONY DENTE

Cob wall following successful ASTM E119 fire rating test. Both test walls emerged somewhat damaged by the firehose but still structurally sound. This test provided justification for cob's 2-hour fire rating in the 2024 edition of the IRC.
CREDIT: ANTHONY DENTE

Remember that just because your *walls* don't burn, it doesn't mean that your *roof* won't. Roofing materials, roof overhangs, roof/attic ventilation, and windows and doors are all vulnerable to ignition and/or the entry of fire. Exposed wooden roof framing and some roofing materials are vulnerable to ignition. Wildfire can also enter a building through roof/attic vents, as well as through windows and doors. Non-combustible windows, certain types of glass, and manual or automatic fire shutters are partial solutions to this problem. Interior fire-sprinklers are required by many codes and can extinguish internal fires. But they can also cause significant damage to a home's contents and finishes and are sometimes ineffective in firestorms, especially when neighborhood water pressure is low due to firefighting demands or water system damage. See Chapter 12 for a discussion of fire-safe roofing options, as well as Chapter 7 for Wildland/Urban Interface (WUI) codes and guidelines.

Chapter 4

Materials

THE THREE INGREDIENTS OF COB are typically characterized as clay, sand, and straw. While this is accurate from a mixing perspective, it is an oversimplification. It is useful to examine in more detail the complexities of clay and the important characteristics of sand and straw.

Understanding Clay and Clay Soils

Clay, or, more accurately, *clay soil*, is the binder in a cob mix—the element that holds everything else together. Clays, and the clay soils that they are contained in, are formed as the result of rainwater weathering bedrock over a long period of time. Technically speaking, clay is a family of minerals known as *hydrous aluminum phyllosilicates* with an extremely small particle size (less than 2 microns, or 0.002 mm).[1] These minerals are complex and still not fully understood, and a whole book could be written about them for earthen builders. CRAterre, the Getty Conservation Institute, and other groups have begun the process of linking clay mineralogy with earthen building performance, but there is much work still to be done.

The most common source of clay used in cob mixes is clay soil, a mineral subsoil below the topsoil with enough native clay content to be suitable as the binder in an earthen wall mix, but builders do sometimes purchase "pure" bagged clays for cob construction. In practice, clay soil types vary widely in several parameters, including the percentage of clay in the soil, the type or types of clay minerals present, and the types and amounts of other materials naturally occurring in the soil. The clay mineral component of the soil is the driver of much of the soil's behavior and workability.

According to the Getty Conservation Institute, when "clay" was first defined scientifically in the 19th century, the size distinction was set in part because particles of this size were smaller than 19th-century microscopes could discern. So at that time, the term *clay minerals* was a rather non-technical grouping referring to all submicroscopic crystalline materials. As it turns out, most of these tiny particles of silicate materials share additional characteristics specific to clay minerals, but not all particles smaller than 0.002 mm are clay: about 10–20% of this group of particles are inert quartz, metal oxides, and carbonates. The advent of reliable X-ray diffractometers in the early 20th century allowed researchers to make distinctions between clay minerals and other particles of similar size.

Clay minerals are composed of two basic molecule shapes: tetrahedral molecules and octahedral molecules. The different families of clay minerals are distinguished primarily by how these two molecular shapes are arranged in the crystalline structure of the clay. Both tetrahedral and octahedral minerals produce strong bonds between adjacent molecules in a flat plane, making a "sheet" of bonded molecules, with weaker bonds between one sheet and another. This sheet formation gives clay its characteristic slippery, plastic behavior, and it also gives the mineral group one of its common names: *phyllosilicates*, which means *leaf* (sheet) *silicate*. Clays are also collectively called *hydrous aluminosilicates* because their two most common elements are silicon and aluminum in a matrix of oxygen ions and because they like to bond to water and tend to hold some water molecules interspersed between layers. Even raw (unfired) clay that appears very dry still contains some of this interspersed water, which is sometimes referred to as *structural water*. A major distinction between raw clay such as that in cob walls and fired ceramics is that the structural water is driven off during the firing process, which fixes clay in an inert crystalline structure that will not soften when rehydrated.

The layered crystalline arrangement of tetrahedral and octahedral molecules varies among different clay types, which divides clay minerals into sub-classifications relevant to their behavior as building materials. There are several of these sub-classifications (*smectite clays, illite clays, kaolinite clays*, etc.), but what is important for earthen builders to know relates to their degree of shrink/swell capacity, also known as the *relative expansiveness* of the clay. Clays with a 1:1 tetrahedral/octahedral layering structure (including kaolinites) are relatively

non-expansive, meaning that they do not swell very much when wet nor shrink very much when they dry out. Clays with a 2:1 tetrahedral/octahedral layering structure, on the other hand, have the ionic capacity to absorb much more water between their layers, and therefore swell and shrink more when absorbing and releasing water. These include illites and smectites, particularly *bentonite*, an expansive clay commonly used for sealing ponds. How these different clay types relate to the structural performance and weather resistance of earthen buildings is not fully known, but we suspect that there is a close correlation. For small and medium-scale residential projects, it is not necessary to know which type of clay mineral family is present in your clay soil because field testing and sample mixes will indicate when clays are too expansive to use or are otherwise unsuitable. But understanding that different clay minerals lend their particular characteristics to a cob mix can help explain the results of testing.

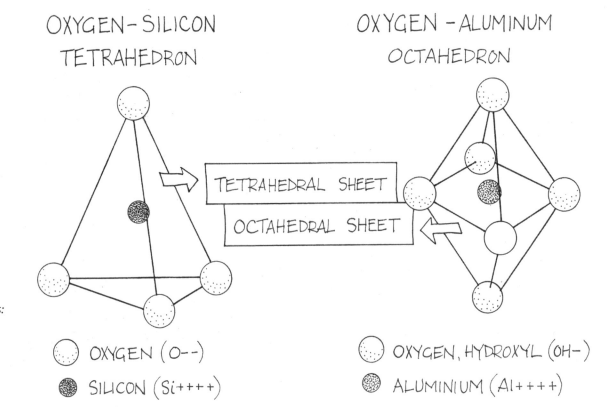

Fig. 4.1:
The two shapes of clay molecules: tetrahedral *and* octahedral.
Credit:
Dale Brownson

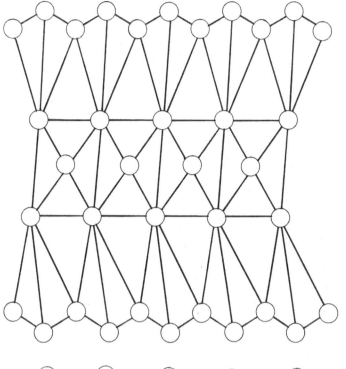

TETRAHEDRAL
SHEET

OCTAHEDRAL
SHEET

TETRAHEDRAL
SHEET

EXPANSIVE CLAYS

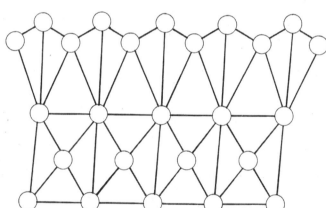

TETRAHEDRAL
SHEET

OCTAHEDRAL
SHEET

**NON-EXPANSIVE
CLAYS**

Fig. 4.2: *The layered structures of expansive vs non-expansive clays.*
Credit: Dale Brownson

In addition to clay minerals, clay soils contain other ingredients such as sand and gravel of various sizes, silt, and sometimes organic material. While a wide range of clay soils can be used in cob mixes, it is useful to have a sense of their non-clay ingredients because that will determine what else needs to be added to make a good cob mix. Table 4.1 gives soil particle sizes

Table 4.1: Soil Particle Sizes as Defined by ASTM D2487

Gravel	0.2" (4.75mm)–3.0" (75mm)
Sand	0.003" (0.075mm)–0.2" (4.75mm)
Silt	Smaller than 0.003" (0.075mm), non-plastic, no strength when dry
Clay	Smaller than 0.003"[1] (0.075mm), plastic, strong when dry

[1] Most other sources specify 0.002mm as the maximum size of clay particles.

as defined by ASTM D2487: Standard Practice for Classification of Soils for Engineering Purposes. The size boundary between sand and gravel as defined by the ASTM is ⅕ of an inch. However, since ¼" is a much more common metric for sifting and ordering sand and gravel in the US (see "Sand" below), we tend to use ¼" in practice as the defining boundary between sand and gravel.

It is rare to find pure clay in the field, but pure clay is not necessarily the most desirable option for earthen building. To make a good cob mix, many clay soils need additional sand to add compressive strength and reduce their shrink/swell behavior. Certain types of naturally occurring subsoils, such as decomposed granite, already contain a good balance of clay minerals and sand and do not need any added ingredients except for straw. Other clay soils are too sandy to use alone but can make good base material if more clay can be added.

Silt can either be a help or a hindrance: silt particles are too small to contribute compressive strength to a cob mix like sand does, and they lack the binding qualities of clay minerals. So, excessive silt tends to weaken cob, making the mix crumbly and dusty. On the other hand, very pure clay soils are difficult to work with. A little bit of silt makes it easier for water to penetrate the clay matrix and acts as a lubricant in the mixing process.

Over the centuries, earth builders, gardeners, and potters have devised many simple ways to assess the proportions of materials in a clay soil; there's *the shake test, the worm test*, and many others (see references at the end of this section). When beginning to evaluate a soil for cob building, it is frequently sufficient to simply get the soil wet and play with it. Your hands will give you quite a lot of information about how much clay is in a soil by feeling how sticky it becomes when hydrated and kneaded. When you rub some hydrated soil between your fingers, you will be able to feel how much coarse aggregate is present.

The shake test is the most frequently described basic field test for clay soils, and it can be useful for assessing aggregate content, but in general, we do not find it to be very helpful for assessing clay content, mostly because the clay is often visually indistinguishable from the silt in the same parent soil. We have occasionally witnessed shake tests that indicated a total lack of clay in soils that were clearly sticky and made perfectly good mix samples.

Clay soil tests fall into three general categories: 1. Field tests, which help you assess a clay soil initially to decide whether you want to continue with making samples; 2. Technical tests (in a lab or more controlled environment) which assess specific qualities of a clay, such as shrinkage; and 3. Mix tests, or making mix samples with a clay soil plus the other typical ingredients of a cob mix. Because there are so many different field tests, we have chosen to highlight one useful and simple test here (*the palm*

Fig. 4.3: *The shake test.* CREDIT: DALE BROWNSON

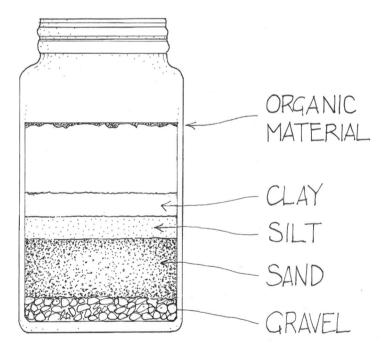

ORGANIC MATERIAL

CLAY

SILT

SAND

GRAVEL

The Palm Test

The *palm* test is an easy-to-use and versatile example of a field testing method for clay. It's Michael's favorite method for rapidly determining the clay content of a soil. Add a little water to the soil you want to test and work it with your hands for a few minutes until the lumps are broken and it feels uniformly hydrated (the moisture level should be similar to modeling clay). Remove any rocks larger than a pea from a sample the size of a golf ball. Roll the sample into a ball and press it hard onto the palm of your flat hand. Then turn your hand over and count how many times you can flex your hand before the ball falls off. It will take some practice before you can do this test easily and consistently, and your results may vary from someone else's due to the size of your hand and the amount of water you add. But in general, if the ball won't stick to your hand while you flex five times, it probably has too little clay to make strong cob. In the range between 5 and 10, the soil may or may not be suitable for cob depending on its silt content. (A very silty clay soil that measures 10 on the palm test will make weaker cob than a sandy soil that only achieves a 5.) Above 10, the soil is almost certainly usable for cob, and will probably need additional sand. The palm test assesses the stickiness of a clay soil, which lets you know how much volume of other ingredients (sand and straw) you can combine with that soil and still end up with a strong end product. It does not reveal several other important characteristics of the clay soil, including expansiveness, hardness, and weather resistance. These properties can best be revealed by making test mixes, forming them into blocks, and letting them dry (see Chapter 6 for more details on making test samples).

test, see sidebar) and then point you toward other sources for additional information. For information on field testing and mix testing, good resources include *The Hand-Sculpted House, Building with Cob, Building with Earth, Light Earth Building,* and *Essential Rammed Earth Construction.* For technical clay tests, see *Building with Earth* and the literature from the Getty Institute on restoring earthen buildings.[2] For even more deeply technical information on clay mineralogy and microstructure, please refer to the Clay Minerals Society.[3]

For some cob projects, a combination of tests from categories 1 and 3 will suffice, but for projects that require building permits, technical tests will be necessary. See Chapter 6 for more on categories 2 and 3. In general, the fewer structural demands you have of your cob, the wider the range of soils and mixes that will serve your purposes. As loads increase, the suitable range of soils narrows.

Sourcing Clay Soil

Sourcing clay soil for cob is frequently as simple as excavating your foundation: because the parent materials of clay soil are very common, the mineral soil below the organic top layer in most parts of the world contains clay of some kind. But there are several exceptions to this generalization. In very geologically young areas, like Iceland and much of Hawaii, the bedrock has not yet had time to weather fully into clay soils. In recently glaciated areas, like nearly all of Canada, parts of the Pacific Northwest and much of the Midwest and Northeastern US, the earth's surface was churned up by the movement of the ice, and glacial deposits were left behind. This makes the upper layers of mineral soil very unpredictable, and clay can be difficult to find as a result. Glaciers and other erosive forces can also strip the soil from an extensive area, leaving mainly bedrock and boulders exposed. And in alluvial soils in river flood plains, soil components have often been separated by the action of the river and then laid down in fairly uniform deposits which can be dispersed either horizontally or vertically. For example, the farm where Michael lives has about 30 feet of gravelly, sandy loam with too little clay to make good cob; under this is a layer of almost pure clay hundreds of feet deep. Volcanic-based material can also be very poor in clays, and even so-called volcanic clays have an unusual composition and may not be suitable for earth building.

The specifics of local soils and their suitability for cob vary enormously from place to place. For example, in the Coast Ranges of California, where the authors live and do much of their work, the geological and tectonic history of the hills results in clay soils that vary widely within very small areas; by contrast, the famous red dirt of Oklahoma, Arkansas, and Missouri is fairly consistent for hundreds of square miles and is the result of deposits in a huge shallow inland sea. Keep in mind that clay content very often increases as you go deeper in the soil;

it's good practice to dig several deep holes on your building site to see the different layers of soil and what they contain. Similarly, if you are sourcing clay offsite, don't take for granted that two piles of clay delivered from the same location will be the same (though they often are).

Cob buildings use quite a lot of soil. A small 120-ft^2 (11 m^2) outbuilding can easily require 10 yd^3 (8 m^3) of clay soil, or one large dump truck full. A modest full-sized home (around 1,100 ft^2 [100 m^2]) can use up to 50 yds^3 (38 m^3). There are many circumstances in which site excavation does not produce sufficient clay soil for a cob project. It could be that the soil is not suitable, or that the volume produced onsite from leveling and trenching isn't enough. The site soil may have suitable clay but also a high percentage of rocks and gravel that make it hard to work with. (It is possible to separate the clay from the rocks by sifting, or by converting the soil to slip and allowing the sand and silt to settle to the bottom [a process called *levigation*]. Sifting/screening is practical at scale if set up thoughtfully, but levigation is only practical for the small quantities needed for a finish plaster.) In these situations, where can one acquire enough suitable clay soil?

There may be other good reasons to dig a hole close by on the same property. A *borrow pit* from which soil is extracted for cob building could later be converted into a small pond. Grading for other buildings, roads, and parking areas sometimes generates plenty of soil. And even if digging and grading are not called for on the same property as the cob building, they may very well be happening close by. A common way to source soil for cob is to call local excavation contractors; because contractors often have to pay to dump excess soil, you can sometimes get soil delivered to your site for free or for the cost of transportation. But don't take the contractor's word that what she has

Students at a natural building workshop in Japan harvest clay soil from near the building site. The fact that the soil holds together in large clumps is a good indicator that it contains clay.
CREDIT:
MICHAEL G. SMITH

available is in fact suitable clay soil; she may not know what you are looking for, or she may hope that you will accept whatever she brings. Make sure you see the material yourself and test it before you have 20 tons of it dumped on your lawn!

Contacting excavation contractors is a good place to start if you are building in a relatively populous area where grading and construction are likely to be happening close by. If you live in a less densely populated area, you may still get lucky by finding a neighbor with extra clay soil; or, you may have to harvest your own. Road cuts can be a great place to find clay since access is easy and the subsurface layers of soil are exposed. A project in Northern California that was close to public land was able to get a free permit from the Forest Service to pick up soil that was eroding down onto roads. Another good offsite source of clay soil is the local rock quarry; *quarry fines* (the leftover material from processing sand and gravel) are frequently a good clay/sand mix and can often be obtained for the cost of transportation.

Other good places to look for clay include anywhere that water comes to the surface: springs, ponds, rivers, and wetlands. This is because clay's expansive properties frequently keep water from soaking into the ground. But think carefully before excavating in these locations; they are often ecologically sensitive habitats for rare animals and plants. Sometimes there is a clay layer below the surface, which traps and holds water where you can't see it. The best clue for this is vegetation type. Especially in arid and Mediterranean climates, plant species with higher water needs often indicate the location of hidden clay deposits. Get to know those indicator plants in your bioregion.

A few other visual clues may help you when prospecting for clay. If you see cracks in the soil, that is nearly always the result of expansive clay.

Colorful soil may indicate the presence of clay because clays often bond with other minerals that impart their bright colors. Natural clay deposits may be almost any color from reds, pinks, and oranges to yellow, blue, white, and black. When making clay plasters and paints, a little bit of clay goes a long way. In your travels, you might want to start collecting buckets of specially colored clays for future projects. But be careful; the process can be addictive!

Sand

Sand, whether native to the parent soil or added to the mix, is what provides cob with *compressive strength,* or the ability to be load-bearing. Sand is essentially very small rocks of varying hardness, depending on the parent material. Not all sands are created equal: they range from very coarse, such as masonry or concrete sand, to very fine, such as 90-mesh sandblasting sand; from very round, such as river or beach sand, to very sharp, such as mined and crushed plaster sand; and from very consistent in size, such as sandblasting sand, to containing a wide range of particle sizes, such as most plaster or masonry sands. The three characteristics that matter most when selecting sand for a cob mix are size, shape, and degree of particle gradation. While it is possible to make cob from many sand types, the "ideal" cob sand is coarse, angular, and *well-graded,* which means: the largest particle size is on the large end of what is considered sand (between ⅛" and ¼" diameter); the shape of individual particles is angular rather than round, and there is a wide range of particle sizes present.

Sand suitable for cob is generally coarser than the sand used for finish floors and plasters. For both cob and earthen finishes, well-graded sand is the most desirable, but for cob, a larger maximum particle size is preferable than for finish materials. Sand is often characterized by its largest particle size: ¼" *minus* means that the largest particles measure up to ¼", though most of the particles are smaller. Good cob and base coat earthen floors can be made with either ¼" or ⅛" minus sand. Plaster sand is generally ⅛" minus.

Sand may or may not be washed depending on its source; some masonry sands from aggregate yards can be a little dusty or silty. Cob can generally handle some dust in the sand, but dusty sands should be avoided for earthen finishes. As with clay soils, the smaller the load demand on your cob project, the more forgiving your choice of sand can be. As demand loads increase, it becomes more important to work with sand that is closer to the ideal.

However, from an ecological perspective, it is preferable to choose the lowest-impact rather than the structurally "ideal" sand, if these choices are in conflict. Sand is the highest-impact ingredient in cob because this heavy material is usually mined elsewhere and trucked to the site. There is no simple solution to this. It is unusual to be able to source sand from your building site—unless you have decomposed granite or glacial till subsoil or have access to river sand. River sand can be a valuable asset, but care is needed when harvesting near rivers to avoid damaging important habitat for fish eggs and other water fauna. Sand sold by aggregate yards is frequently mined from small local quarries, so it is often possible to visit your local rock quarry/sand mine and assess both the qualities of the sand and the impact of the mine. Remember that some clay soils have a naturally high native sand content and do not require much—or any—additional sand. If you are importing clay from off-site, take the extra time to find clay soil with a good sand component, if at all possible.

Cob builders most frequently purchase sand from a local landscaping or hardware store, either in bags or in bulk, or from a concrete aggregate provider. Most aggregate yards stock *concrete sand*, which is coarse and well-graded, not too expensive, and deliverable by truckload or supersack. If you'd like to haul your own sand, most aggregate yards will load up your truck and sell you the sand by weight, though some have a minimum weight for self-loading. But for anything other than a very small or low-sand cob project, you will need so much sand that it is usually worth the cost of having many yards delivered all at once. See "Calculating Volumes of Materials" in Chapter 10 for advice on how to determine the amount of sand you will need.

Straw

Straw refers to the stems of cereal crops that are left over after the edible grains have been harvested. Straw is suitable for building because it is composed mostly of carbon and silicon with very little nitrogen, and therefore is not prone to decomposition (if kept dry). In contrast, *hay* is harvested green for animal feed and contains quite a lot of nitrogen. Hay provides good nutrition not only for cattle and horses but also for microorganisms, and is therefore much more likely to decompose, especially when it gets damp (as it would if used in a cob mix).

As with the other two primary cob ingredients, straw is highly variable. In the modern industrial context, building straw is typically a "byproduct" of growing one of the primary grain crops: wheat, barley, rice, and oats. All four of these plants are annual grasses that have been selected through thousands of years of human cultivation into tall, stout, and productive versions of their wild ancestors. These days, grain crops are mostly grown in large monocultures, which produce huge amounts of straw as well as grain. All this straw is difficult to fully reincorporate into the soil before the next planting cycle, so farmers usually have a lot left over. As a result, straw is relatively inexpensive in parts of the world that practice large-scale grain growing. This is slowly changing because straw is increasingly in demand for other purposes, such as erosion control and biofuel feedstock, and as climate change impacts agricultural production. However, in contrast with strawbale or light straw-clay, cob is not a straw-heavy building system, so the reduced availability of straw is not likely to become a limiting factor in cob construction.

Straw serves many functions in a cob mix: it adds tensile strength, which creates resistance to pulling, bending, and twisting forces; it adds insulation; it keeps the wall stable and helps minimize slumping while the cob is drying; and it helps transport water out of the wall during the drying process.

Because straw is a small component of cob by volume, a few bales go a long way. For most projects, straw can be purchased from the nearest feed store. To ensure an adequate supply and better pricing for very large projects, you may want to order through the feed store ahead of time or buy directly from a farm.

Bales come in various shapes and sizes depending on the machine used to make them. In the US, the most common are rectangular two- and three-string bales. For cob building, the only difference is that two-string bales are smaller and easier to move around. All common types of straw are suitable for cob, though they do have different characteristics (see more on this below). Availability of different kinds of straw will depend on bioregion and harvest cycles.

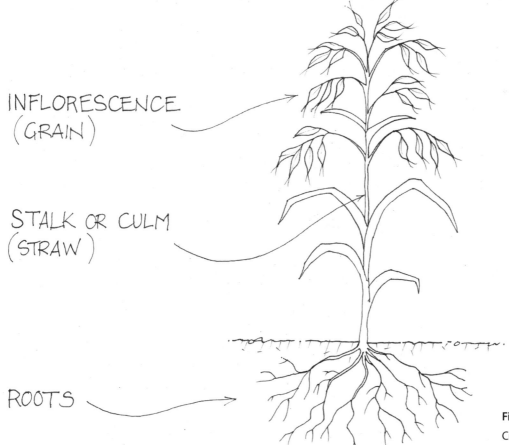

INFLORESCENCE (GRAIN)

STALK OR CULM (STRAW)

ROOTS

Fig. 4.4: *The anatomy of straw.*
CREDIT: DALE BROWNSON

Straw is not the only fiber that can function as the tensile element in cob: builders in many parts of the world have used other fibers native to their areas, including hemp fibers, pine needles, wild grasses, and the fibrous byproducts of mescal (from agave) and sorghum syrup production (see the beginning of Chapter 6 for a longer list). Where straw is available, it does tend to be the preferred fiber because it is strong, flexible, easy to use, and needs zero processing before incorporation into a cob mix.

The four dominant types of straw have slightly different characteristics. Rice straw is usually very long and flexible, but we have encountered short, stubby rice straw from time to time, possibly related to drought conditions. Wheat straw is usually relatively short and stiff, but we've seen wheat straw that is nearly 2' long. Oat and barley straw tend to be somewhere in between the characteristics of rice and wheat straw—both moderately long and moderately flexible. All straw types work well for cob, but depending on the specifics of your design, you may want to choose one type over another if several varieties are available. The length and flexibility of rice straw make it great for sculptural applications, but it clumps and tangles more easily than stiffer, shorter straw, which makes it harder to mix mechanically. A sticky clay soil with a low aggregate content combined with rice straw makes an ideal high-fiber sculpting mix for building out shelves and other horizontal details because stickier clays accept more fiber reinforcing, and the long, flexible rice straw fibers are especially easy to shape and adhere to surfaces. A clay with more sand plus wheat, barley, or oat straw is less versatile as a sculpting material, since both the fiber and the mix are stiffer, less sticky, and less flexible, but this combination is preferred for large-volume cob mixing and rapid wall building since it is less prone to tangle in a machine. For example, a cob recipe containing short, stiff straw and lots of sand can be mixed reasonably well in a concrete mixer; but trying to use the same tool to make a sticky low-sand mix with a lot of long straw is profoundly annoying, if not downright impossible. Rice straw mixes can also resist integration into the wall more than mixes made with other types of straw.

It is important to store straw properly. Wet straw can mildew and start to decompose pretty rapidly, so when bales arrive onsite, it is best to have a fully protected storage spot ready for them. This is one good reason to build your roof before your walls—the bales can be stored underneath as you build. Tarps tend to be ineffective for long-term protection, both because they leak and because they trap moisture inside. It is essential to protect bales not only from the weather but also from moisture in the ground. Store them on pallets rather than directly on the ground, even if under a roof.

Before you purchase bales, make sure that they have been stored properly. If you have one, use a moisture meter to verify that their moisture content is under 20%. Bales that have been previously exposed to moisture, usually have dirty chartreuse or brown discoloration; well-stored bales should be pale gold, light green (rice straw), or beige-ish.

Chapter 5

Tools

Safety Equipment

- Gloves: All kinds of construction can be rough on the hands. Although cob contains no toxic or corrosive ingredients, it often contains abrasive aggregates. The clay dries the skin and can cause it to crack, especially in dry climates. We usually wear tight-fitting nitrile gardening gloves that protect our hands while allowing fine control and sensitivity.
- Dust mask or respirator: Breathing dust and fine particles is not only unpleasant in the short term, but can have severe health consequences over time. Be sure to protect your lungs when handling dry materials such as powdered clay and chopped straw.
- Eye protection: A must when using power tools and when working with lime and cement.
- Ear protection: Manual cob mixing and building can be very quiet, but other parts of the building process (especially those involving power tools) put cumulative stress on the ears.

Tools for Manual Mixing

- Shovels: Rounded or pointed spades are good for harvesting clay soil and moving it around. Square-ended shovels are best for picking up sand and for flattening the bottoms of trenches.
- Buckets: 5-gallon buckets are essential on a cob site, and you can never have too many. They are used for measuring ingredients in a cob mix, storing and transporting water, cob, and other materials. You may be able to source used buckets for free from restaurants or other food businesses. Very durable buckets can be bought at feed stores.

- Garden hose with spray nozzle: Ideally, get enough hose to reach to all of your cob walls and mixing locations.
- Tarps: The best tarps for cob mixing are sturdy, flexible, and about 8' to 10' square. The woven polypropylene tarps available at the hardware store come in a range of grades. The lightest kind start shredding after a few weeks of heavy use, while the most heavy-duty (colored silver to increase UV-resistance) may hold up to a whole season of cobbing, especially if stored in the shade when not in use. Many alternatives can be sourced from the waste stream. Try PVC billboards, RV awnings, reinforced greenhouse plastic, etc. You will need at least one mixing tarp for every two workers—one for each worker is better. Tarps are also useful for protecting materials from the weather, for lining clay-soaking pits, for temporary shade, and for slowing the drying rate at the tops of cob walls.
- Wheelbarrows: Large, sturdy, contractor-grade wheelbarrows are a must for moving heavy materials around the site. Replace the tires with solid rubber wheels that will not deflate.
- Barrels and other large containers: Plastic and steel 55-gallon barrels are useful for storing water and clay slip. If you cut them in half, they are easier to get material out of and make a good sized container for mixing slip and plaster with a drill-type mixer. Larger, flatter vessels including kiddie pools and stock tanks can be used for soaking clay soil prior to mixing cob. You can also make a convenient soaking pit by building a simple wooden frame or placing four or more straw bales in a ring and draping a waterproof tarp over them.

- Sifting screens: Depending on the clay soil and the mixing technique, it is sometimes necessary to remove stones from the soil before mixing. See below in "Tools for Plastering."

Above: *A drum-style mortar mixer is an excellent tool for making large quantities of plaster, including the straw-clay base plaster shown here. It can also be used for mixing cob.* Credit: Laura Sandage

Tools for Mechanical Mixing

- Concrete mixer, mortar mixer, pan mixer: Small- to medium-scale mixing machines. Both electric and gas-powered mixers can mix cob, but in some cases only the clay, sand, and water can be mechanically mixed and the straw has to be added later by foot. See "Other Mechanical Mixing Options" in Chapter 13.
- Tractor, skidsteer, backhoe, excavator: These medium to large machines can all be used effectively to mix large quantities of good-quality cob. A skidsteer is often called a *Bobcat* in the US, and an excavator is called a *JCB* in the UK.
- Rototiller or tractor with a tiller implement: Another way to mix cob mechanically. See "Mixing with a Rototiller" in Chapter 13.

Tools for Cob Wall Building

- Cobber's thumbs: These are sticks you can make yourself about 8" (20 cm) long and 1" (2.5 cm) in diameter, with rounded ends like a thumb. They are used for integrating new cob during construction. You will want one for each builder.

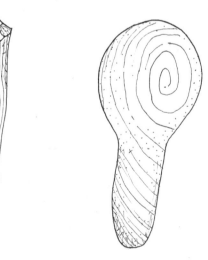

Fig. 5.1: *Cobber's thumbs.*
Credit: Dale Brownson

DRIFTWOOD

WHITTLED BRANCH

TURNED ON A LATHE

- Garden or digging fork: With strong tines and a short, stout handle, these tools are useful for lifting cob onto the wall.
- Spirit level: Levels in different lengths (2', 4', and 6') are useful for checking that the cob wall is going up plumb, as well as for installing window and door frames—and during many other stages of construction.
- Cob saw: An old hand saw works well for trimming moist cob. Eventually, the sand in the mix will wear the teeth away. When that happens, use a bench grinder or angle grinder to cut new, larger teeth into the edge of the saw.
- Machete: Useful for trimming cob after it gets too dry to cut with a saw and for more detailed trimming.
- Hatchet and gardener's adze: Helpful tools for trimming dry cob, especially in detailed areas such as inside niches and around windows.
- Scaffolding: As the wall grows higher, you will need a safe and sturdy surface to stand on while integrating new cob at the top of the wall. Ladders are not ideal for this purpose. You can use commercial scaffolding or build your own.
- Hammer drill: Once cob is dry, it can be very hard to cut or shape with hand tools. If you need to remove a lot of cob or install wiring or plumbing in a dry cob wall, an electric hammer drill with chisel attachments can get the job done quickly.
- Circular saw or angle grinder with masonry blade: Useful for cutting channels into dry cob for the installation of pipes and wires.

Tools for Plastering

- Sifting screens: Screens are used to sift clay soil and other ingredients for making plasters and earthen floors. These can be easily made by attaching hardware cloth or metal screen to a frame made of 2×4s. About 2' by 3' is a good

Cob saws. Note that the lower saw has been modified with an angle grinder to make larger teeth, which cut more efficiently and are less prone to getting clogged with clay and straw.

CREDIT: ELKE COLE

size to fit over a wheelbarrow. Much larger self-supporting screens are good for dry-sifting clay soils. Depending on the fineness of materials desired, you may need screens with mesh sizes of ½", ¼", ⅛", and/or 1⁄16" (window screen). If working with very rocky clay or for bulk screening, heavy-duty expanded metal mesh is a better choice because the thin metal of hardware cloth will quickly shred.
- Chipper-shredder, lawnmower, leaf mulcher, weed whacker: Various machines used for chopping straw. See "Chopping Straw for Plasters" in Chapter 15.
- Drill with mixing paddle or handheld mixer: A large ½" chuck electric drill with a mixing paddle is great for making clay slip and mixing small to medium quantities of plaster. Get a heavy-duty paddle that is made for mixing mortar; paint-mixing paddles aren't designed to move heavy materials and will break quickly. You can make your own paddle by wrapping several lengths of stiff wire around a piece of rebar. A hand-held mortar or plaster mixer is similar to a drill, but better, since it is easier to control the speed.
- Mortar mixer: The typical tool for high-volume production plaster mixing. Keep in

mind that a mortar mixer is different from a concrete mixer in that it has paddles that rotate independently from the drum. Drum mixers are usually gasoline-powered and range in capacity from 4 to 12 ft³. Electric models of both drum and pan-style mixers are also available.

- Mixing tub or box: These are flat-bottomed troughs for mixing mortar, concrete, or plaster. They come in both plastic and metal, and in various sizes. You can also make your own out of plywood. Wheelbarrows are smaller than would be ideal and somewhat awkwardly shaped, but will serve in a pinch.
- Mixing hoe: These are heavy-duty hoes with long handles and holes in the blade, used for mixing mortar and concrete. You will want two of these for mixing large quantities of plaster in a big tub.
- Buckets: Plastic 5-gallon buckets are endlessly useful for measuring and storing materials and for mixing small quantities. Two-gallon buckets are excellent for moving plaster.
- Scoops: For moving plaster from a bucket to a hawk. Don't use your trowel for this, or it will eventually break!
- Wooden floats: Basically, thin hardwood boards with handles, these are ideal for smoothing wet cob and for flattening base coats of clay or lime plaster. They are inexpensive, but you can also make your own.
- Trowels: Metal trowels come in many different shapes and sizes. Standard rectangular plastering trowels will work for clay plasters, especially on very flat walls. On curved walls, pool floats with their oval shapes leave fewer trowel marks. There are many styles of Japanese trowels, made of different materials (iron, stainless steel, wood, and plastic) and in different sizes, each intended for a different purpose.
- Hawk: A flat tray with a handle for carrying plaster and, in combination with a trowel, transferring it to the wall. We like the Japanese design, which is more ergonomic than the typical American variety. You can easily make your own to fit your body size.
- Tile sponges: For cleaning tools and woodwork. Sponges can also be used as a finishing tool for certain finishes such as *alis* clay paints and various kinds of plaster.
- Paint brushes, various sizes: For applying natural paints. Also useful for cleaning fresh plaster off of wood.
- Spray bottle or pump sprayer: For wetting walls prior to plastering and for moistening fresh plaster to extend working time.

A selection of plastering tools, including wooden floats and tile sponge (in crate), spray bottles, small buckets, Japanese-style wooden hawks (upside-down at right) and drills with mixing paddles, awaiting use at a Clay Sand Straw project. CREDIT: JOHN CURRY

Chapter 6

Mix Design and Testing

Variations in Cob Mixes

HISTORICALLY, a wide variety of recipes have been used for cob building around the world because people made use of whatever resources they had at hand. The standard practice in many areas was to start with the local clay soil (which may have contained silt, sand, and/or gravel in varying proportions), add water and mix to a dough-like consistency, then mix in fairly small quantities of whatever type of fiber was most available. Depending on the bioregion, the fiber source could be straw, hay, animal manure, or even bracken, heather, broom, gorse, twigs, reeds, rushes, sedge, moss, wood shavings, pine needles, or animal hair.[1] In the British Isles, where horses and oxen were commonly employed to do the labor of mixing, it can be hard to know whether their manure was deliberately added as a fiber source or whether it was an unintentional byproduct of the mixing process. According to the Devon Historic Buildings Trust, typical cob mixes in the English County of Devon (the heartland of cob in England) contained clay and aggregates in the following proportions: stones and gravel (over 5 mm diameter) 30–40%; fine and coarse sands 25–30%; silt 10–20%, and clay 10–25%.[2] In other parts of the world, the proportions of sand and gravel were often lower and those of silt and clay higher.

Now that it is easy to have truckloads of clean, good-quality sand delivered directly to a building site in many parts of the world, it is common to add imported sand to local clay soil. This makes it possible to work with a much wider range of clay soils than would otherwise be acceptable. Increasing the proportion of sand in the mix and lowering that of clay reduces shrinkage and cracking during drying and improves the compressive strength of the cob (but only up to a certain point—too little clay results in a soft, crumbly mix). High-sand mixes contain less water and therefore dry faster, which can reduce delays and labor costs.

However, there are disadvantages to high-sand mixes which have led some builders to prefer low-sand, high-fiber mixes. Chief among these is their thermal properties. The insulation value of cob mixes goes down as density increases. High-sand mixes—which typically contain only 1–2% fiber by weight—are very dense and therefore insulate poorly. By reducing the proportion of sand in the mix and increasing the amount of straw, the density is reduced, and the insulation value can be increased substantially. For example, thermal testing conducted by the Cob Research Institute (CRI) and Quail Springs in 2021 found that a high-sand cob mix had an R-value of 0.22 per inch of thickness. By eliminating sand from the mix and raising the straw proportion as high as possible (to nearly 7% dry weight), this value could be brought as high as 0.54 per inch. This is a huge practical difference—vastly increasing the geographical range within which cob buildings are thermally efficient. (See "Optimizing Cob's Thermal Performance in Your Climate Zone" in Chapter 3.) High-fiber mixes also produce lighter cob, which is a big plus for engineering reasons, especially in high-seismic zones. High-fiber cob is likely to have lower cost and environmental impact than dense cob for several reasons: 1) lighter materials equates to less energy spent in transportation; 2) the harvesting and processing of sand can have large environmental

53

consequences; and 3) as the straw content increases, so does the amount of carbon that is durably stored and sequestered from the atmosphere.

There are some unresolved concerns about these very-high-fiber cob mixes. They can be more difficult to mix—mechanically *or* manually, especially depending on the aggregate content of the clay soil used. (See descriptions of "The Lasagna Method" and "Mixing with a Rototiller" in Chapter 13.) High-fiber mixes dry slowly and are more prone to mold growth, so they may be most appropriate for dry areas. When fiber content gets extremely high, it may inhibit bonding within the wall, so both compressive and flexural strength may decline. Initial testing of high-fiber cob mixes that Massey and Anthony worked on in collaboration with engineering faculty and students at the University of San Francisco in 2013 seemed reassuring in this regard.

Further testing is necessary to clarify the effects of straw content, length, and type on the physical properties of cob, but it currently appears that the "sweet spot" of density for both thermal and structural performance is about 70–75 pcf of dry cob. Our expectation is that as more research is done, high-fiber mixes will become increasingly common. In the meantime, be sure to test the compressive strength of any high-fiber cob you plan to use in load-bearing applications.

These blocks were made by Sasha Rabin of Earthen Shelter as part of a project to test the thermal and structural properties of various densities of cob. When the blocks were completely dry, they were weighed to calculate their densities. The samples in this photo range from 51 to 113 pcf.
CREDIT: SASHA RABIN

Another way to make a low-density cob mix is to substitute lightweight aggregates such as lava rock or pumice for some or all of the sand. Although we have less experience with this approach, it seems likely that similar densities can be achieved, possibly with greater compressive strength than high-fiber mixes. Ann and Gord Baird chose this approach when building their Eco-Sense home in BC, Canada (see "Thermal Properties of Cob" in Chapter 3 for more details). Their mix consisted of 1 part clay soil to 1 part sand (combined first with water into a slurry using a rototiller) and 1 part of pumice, plus a fairly standard amount of straw. The pumice was sold as "¾-inch minus," meaning that it contained fines and sand-sized particles as well as larger chunks. This recipe yielded R-values as high as 0.39 per inch and an impressive compressive strength of 181 psi (1.25 mPa). The resulting mixture was stiff enough to install in lifts 3–4' (90–120 cm) tall and could be smoothed immediately with a magnesium float, making trimming unnecessary. These labor-saving features may be another strong reason for including pumice in the mix—where it is available.

Recipe Development

The first step toward developing a cob mix recipe is choosing your ingredients. Most often, the base material is locally available clay soil (see "Sourcing Clay Soil" in Chapter 4). Unless your

available clay soil contains a large amount of coarse aggregates (over 50% by volume) or you choose to pursue a high-fiber mix without added sand, you will also need coarse, well-graded sand (see "Sand" in Chapter 4). Finally, you will need a source of clean, dry straw or another fiber. Once you have developed a recipe using your particular set of ingredients, you can't substitute other materials without repeating the recipe development and testing process. Clay soils in particular vary enormously, not only from site to site but even at different depths or locales on the same property. (Field tests such as the "palm test" described in Chapter 4 can help you determine whether your clay source is consistent or variable.) Sand acquired from different sources can also vary in grain size, shape, and gradation, which will change the properties and performance of the resulting cob.

An experienced cob builder can often tell if a particular recipe is going to work just by making and handling a mix, picking up on myriad subtle sensory clues. If the mix has a good ratio of clay, aggregate, water, and straw, it will be sticky—but not too sticky; a lump will hold its shape when tossed in the air but still combine readily with other lumps when placed on a wall. A mix with too much clay (or water) will stick to the mixing tarp, tools, and hands, making it difficult to work with. An excessively sandy mix will be hard to form into a ball and may crumble when tossed or placed on the wall. Water is required to activate the adhesive properties of clay, but too much water makes the mix so runny that it won't hold its shape; not enough, and the mix will be difficult to mix and won't feel cohesive. Acceptable amounts of straw vary widely, but it is important to have at least some straw fibers integrated into every part of the wall. Straw helps

a wet cob mix hold its shape, so adding more straw is a common cure for a too-wet mix.

Another useful field test is to gently squeeze a ball of cob while holding it up to your ear. A good high-density mix will make an audible crunching sound as it is compressed, indicating that the aggregates are touching each other with little clay between. This means there is enough sand in the mix that excessive shrinkage is unlikely.

Once your base ingredients are established, make a series of small mixes with different ratios of clay soil, sand, and straw. The easiest way to measure soil, sand, and water is by volume. Soil and sand can only be accurately weighed if they are bone dry, which is difficult to achieve. Straw, though, is most accurately measured by weight because it is so compressible. In

Recipe

Recipe 1: Sample Recipes for Different Clay Types and Cob Densities

Clay source/type	Recipe for high-density cob (~100–120 pcf)	Recipe for high-fiber cob (~50–75 pcf)
Very sticky clay soil with little aggregate	1 soaked clay soil 2–3 sand ·¼ straw	1 clay slip[1] 1 straw
Moderately sticky clay soil with high silt content	1 dry clay soil 1 sand ⅛ straw	1 clay slip[1] ½–1 straw
Decomposed granite or similar[2]	1 dry clay soil 0–½ sand 0–½ clay slip[1] from other source ⅛ straw	Not appropriate
Bagged mortar or ceramic clay	1 powdered clay 2 sand ¼ straw	1 clay slip[1] 0–1 sand 1–2 straw

Note: All measurements are by volume. Keep in mind that these are only starting ratios. To determine your final recipe, you will need to make sample blocks, let them dry, and test them.

[1] Clay slip should be about the consistency of a thick milkshake or pancake batter.

[2] These soils contain a lot of coarse aggregates and a small amount of clay. They often make good high-density cob with only the addition of water and straw, or they may require extra clay from another source. They are not appropriate for high-fiber cob.

practice, builders usually add straw by eye or by feel. But if you are aiming for a target density to achieve a thermal insulation goal, you will need to carefully weigh the amount of straw that goes into each test mix so that the same ratios can be duplicated later. Expect to make a couple of rounds of samples before you find a recipe for the precise density you want.

Start with a recipe in the range shown in the chart for your soil type. It's always wise to make several different recipes, with ratios varying on each side of your best guess. For example, if you know you are starting with a silty clay soil, you might make five different batches: 2 clay soil:1 sand, 3 clay soil:2 sand, 1:1, 2:3, and 1:2. If you have no idea what kind of soil you are working with, try recipes ranging from 1 clay soil:0 sand, up to 1 clay soil:3 sand, by half-part increments. Small changes in ratios can make surprisingly large differences in the strength and workability of the mix.

Test mixes should be large enough that, at minimum, several test bricks can be made; 2–5 gallons of ingredients is a typical volume. We often make a couple of test bricks out of each mix before adding straw, then add straw and make a couple more. Test bricks can be hand-formed or made in a mold. They should be at least 8" × 4" × 2" (20 × 10 × 5 cm). If you use a mold, make it 4" x 4" x 8" (10 × 10 × 20 cm) tall—the dimensions required by Appendix AU for testing compressive strength. Clearly label each brick to avoid confusion later.

How the ingredients are mixed can affect the properties of the resulting cob. This is especially true when working with high-clay soils. Pre-soaking these soils loosens clay particles and allows them to become more evenly distributed throughout the matrix of aggregates, which tends to result in higher-strength cob but can also cause the mix to shrink and crack more as it dries. Some mechanical mixing systems can

have the same result, so use the same mixing technique for your test samples that you plan to use for your actual project. North Carolina builder Stephen Hren mixed his first round of test samples for one project using a mortar mixer to combine clay soil, water, and sand, then stomped in the straw by foot. When tested at a laboratory, these samples had compressive strengths ranging from 90 to 100 psi. Later, Stephen sent in another round of samples mixed with the concrete mixing attachment for a skidsteer (see "Other Mechanical Mixing Options" in Chapter 13). Although made from the same recipe, these samples achieved higher and more consistent strength results—around 125 psi. Stephen also found that adding small amounts of very strong hemp fiber to his mix in addition to straw significantly improved the compressive strength.

Test bricks should be left to dry completely before being tested. This could take anywhere from 3 days to several weeks depending on drying conditions. It can be hard to tell visually whether a brick is dry all the way through. That's one reason to make several bricks from each recipe. When you break a brick in half, if it is a darker color in the middle, it isn't completely dry yet. Although it is safe to dry cob bricks in an oven at a low temperature, this will affect the results of some tests; don't artificially dry any samples you plan to send to a laboratory.

DIY Mix Testing

To obtain a building permit, you will most likely need to demonstrate the compressive strength of your cob mix (the amount of pressure it can withstand) and sometimes also its bending strength, or *modulus of rupture*. Appendix AU specifies that all cob walls must have a minimum compressive strength of 60 psi (414 kPa), and that cob in shear walls needs a compressive strength of 85 psi (586 kPa) or greater and a minimum modulus of rupture of 50 psi (345 kPa). Even when no permit is required, the amount of load that can be safely carried by a cob wall is calculated using its compressive strength. The standard way to determine these numbers is by submitting test blocks to a materials testing laboratory. This process can be expensive, so it is best to do your own testing first and arrive at a mix you think is suitable.

There are some simple tests and observations you can perform on your dried samples before you send them to a laboratory. First, look for cracking. Severe cracking is the result of shrinkage, which can be addressed by adding more sand and/or straw to the mix. Once a test brick is completely dry, it should be very difficult (if it is 2" [5 cm] thick) or impossible (if it is 4" [10 cm] thick) to break in half with your bare hands without using a fulcrum, such as your knee. Scratch or rub the brick with a metal tool. If it wears away quickly, that is likely an indication that the mix is short on clay, and/or lacking sufficient coarse aggregate for heavier cob or fiber for high-fiber cob. (Although this

test can be used to compare the hardness of several cob samples, even high-quality cob is soft enough to scratch easily with a metal tool.)

Another simple test required by Appendix AU is a shrinkage test. To perform this test, follow the steps as shown in Figure 6.1. Note that

Fig. 6.1:

The shrinkage test. Credit: Dale Brownson

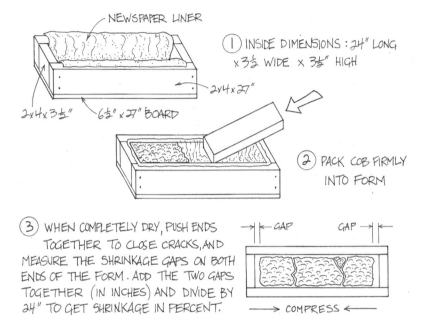

① INSIDE DIMENSIONS: 24" LONG x 3½ WIDE x 3½" HIGH

② PACK COB FIRMLY INTO FORM

③ WHEN COMPLETELY DRY, PUSH ENDS TOGETHER TO CLOSE CRACKS, AND MEASURE THE SHRINKAGE GAPS ON BOTH ENDS OF THE FORM. ADD THE TWO GAPS TOGETHER (IN INCHES) AND DIVIDE BY 24" TO GET SHRINKAGE IN PERCENT.

Table 6.1: Cob Mix Troubleshooting Guide

Symptom	Problem	Solution
Test block cracks while drying or shrinks too much in shrinkage test	Too much clay or	Add more sand and/or fiber
	Mix too wet or	Add less water or allow mix to dry somewhat before forming
	Very expansive clay	Try a different clay source
Test block crumbles easily or breaks when dropped	Too little clay	Add more clay soil or try a different clay source
Compressive strength too low	Too little clay or	Add more clay soil or try a different clay source
	Too little aggregate or	Add more sand
	Improperly mixed or formed	Mix more thoroughly and fill forms carefully
Modulus of rupture too low	Too little clay or	Add more clay soil or try a different clay source
	Too little fiber or	Add more fiber—try fibers of varying lengths and types
	Improperly mixed or formed	Mix more thoroughly and fill forms carefully
Sample erodes easily	Too little clay or	Add more clay soil or
	Unstable clay type	Try a different clay source

the dimensions of the form are specified by the code. In Step 1, a piece of plastic can be used instead of newspaper. The idea is to prevent the cob sample from sticking to the form, allowing it to shrink in one piece as it dries. When you fill the form, press with a small tool to make sure all corners and voids are filled, then smooth the top surface and let the sample dry. Appendix AU specifies that the total shrinkage may not exceed 1", or about 4% shrinkage. The New Zealand Standard NZS 4298-2020:

A cob block being tested for compressive strength. This test was part of a research project conducted by Massey Burke and Hana Mori of the University of San Francisco at the UC Berkeley Structures Laboratory to determine the effects of different amounts and lengths of straw on cob's structural properties. Note that this sample failed in shear compression, as demonstrated by the diagonal cracks.

CREDIT: HANA MORI

Materials and Construction for Earth Buildings allows for a maximum shrinkage of 2%. (This Standard is an excellent source of information for several other simple DIY tests you can perform on your own test blocks.) Either of these figures is a low bar. A good cob mix should probably shrink less than 1%.

Repeat these tests until you find a mix you believe will meet the requirements of the project. If you cannot make a strong test block, you may need to find a different clay soil source or add some pure clay to your mix. Remember that certain clays, such as bentonite, are highly expansive and may cause extreme shrinkage and cracking; these are best avoided.

When you have what you think is a suitable mix, you will need to make a set of test blocks to send to the laboratory if your building is permitted. Or, if you have the mechanical aptitude and a cooperative building inspector, you may be able to do your own testing to determine compressive strength and modulus of rupture. (In which case, see sidebar, "Build Your Own Testing Apparatus.")

Laboratory Testing

To satisfy Appendix AU's requirements for compressive strength, you will have to send five test blocks of your cob mix to a materials testing laboratory. Make these blocks in a form 4" × 4" wide × 8" tall (10 × 10 × 20 cm), and blocks for testing modulus of rupture in a form 6" wide, 12" long, and 6" high (15 × 30 × 15 cm). Pack cob into the form bit by bit, pressing with a small tool to fill voids and prevent seams. For the best results, make sure the samples are built in the same orientation that they will be tested, are handled very carefully, and are fully dry before being tested. Mark the top of each block with the test ratio and wait for the blocks to air-dry completely. Before testing, the top and bottom surfaces of each block should be coated with gypsum plaster to make them flat and parallel. Often, the laboratory will do this for you.

The reason you need to send five samples is that there will likely be variations in the mix and/or construction method. When you get the report from the laboratory, the official result is the fourth number, counting from highest to lowest results. Because this number will be used by your engineer or designer to determine the safety of the design, this process is best completed early in the design process.

Build Your Own Testing Apparatus

by David Wright

Laboratory testing can be expensive. Depending on where in the country it is located, a laboratory might charge as much as $300 per sample, which would mean a minimum outlay of $1,500 for compressive strength testing of one cob mix. Luckily, the code also allows for on-site testing by the builder *if* the building official agrees. Even if your building official insists on laboratory testing, you may still want to pretest your mix to find out whether it will meet the strength requirements. That way, you can adjust your mix if necessary to increase its strength before submitting samples to the lab.

Appendix AU requires all cob to have a minimum compressive strength of 60 psi (414 kPa). Since the standard cob block is 4" x 4" (10 x 10 cm) when viewed from the top, the block has a cross-sectional area of 16 in^2 (103 cm^2) and must support a minimum of 960 lbs (435 kg). ☞

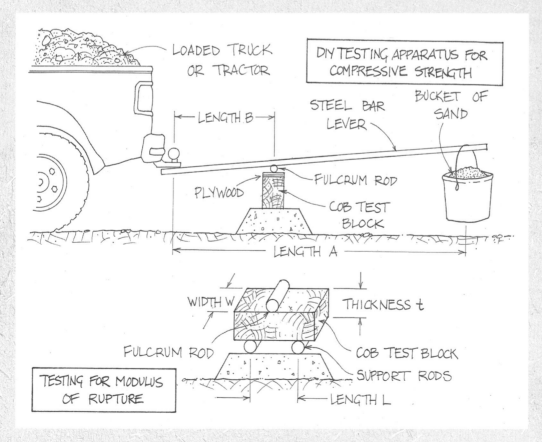

Fig. 6.2: Credit: Dale Brownson

Testing blocks on site presents three main, somewhat challenging, tasks: applying force to the test block, distributing that force evenly across the surface of the block, and measuring how much force is being applied when the block fails. Stacking 960 lbs of weight on top of a small block would be dangerous; all of that weight would fall when the block breaks. A safer approach is to use a long, strong lever such as a steel pipe to apply the force, a piece of plywood to distribute the load to the top of the specimen, and a round bar to accurately measure the length of the lever and apply force to the center of the cob block. Be sure to secure the stationary end of the lever under something very heavy because it will push up on the restraint with over 1,000 pounds of force if the cob meets the Standard. Some objects that could provide sufficient anchoring weight include the bucket or structural beam of a tractor, the tow hitch of a fully loaded truck, or a portion of cob wall specifically designed to withstand and distribute these upward loads.

To measure the force applied to the block, hang a bucket securely from the end of the lever and slowly fill it with sand. Add the sand gradually while another person watches for cracks in the test block. When the block has broken, the bucket will droop or land on the ground. Once the block has cracked or compressed, stop adding more sand. Weigh the sand and bucket to determine the force applied to the lever. The mechanical advantage is found by dividing length A by length B in the figure. The equation for how much force was applied to the test block can be written: $F = \text{Weight} \times A/B$. To calculate the pressure on a standard 4" × 4" test block, the equation would be $PSI = A/B \times \text{Weight}/16$. (Lengths are in inches and weight in pounds.) For example, imagine that the length of the lever bar from the fixed end to where the bucket is attached is 72", the distance between the fixed end of the lever and the fulcrum rod is 3", and the sample breaks when the bucket weighs 60 lbs. In that case, the compressive strength of the sample would be 72/3 × 60/16, or 90 psi. (The formula has been simplified to ignore both the angle of the lever and the weight of the lever itself. To minimize the inaccuracy, use a lightweight lever such as a hollow pipe and set it close to horizontal. Or, if you want a more precise [and complicated] formula, get in touch with CRI.)

The same testing lever described above can be used to test modulus of rupture with only minor changes to how the test block is supported. Rather than placing the block on a flat hard surface, support it on two steel pipes. Measure the distance between these pipes for use in the calculation later. Apply the load to a third pipe on top of the block, placed halfway between the two supporting pipes. The force applied to the test block can be found in the same way as in the compression testing: $F = \text{Weight} \times A/B$. To calculate modulus of rupture, the equation is $MOR = (3 \times F \times L)/(2 \times W \times t^2)$, where t and L are shown in the figure and W is the width of the specimen. For the standard 6" × 6" × 12" test block, this reduces to $MOR = (3 \times F \times L)/432$. (Again, lengths are in inches, and force is measured in pounds.) Imagine that you set up the lever and fulcrum the same as in the previous example, the support rods are placed 10" apart, and the sample once again begins to break when the bucket weighs 60 pounds. Using the equation above, we determine that Force = 60 lbs × 72"/3", or 1,440 lbs. The calculation for the modulus of rupture would be 3 × 1440 × 10/432, or 100 psi.

Setting up your testing apparatus may be a challenge, but no more so than many other stages of the building process. If you want support, feel free to contact the Cob Research Institute. CRI would love to hear about your experience. Sharing your test results, obstacles, and successes will help other cob builders.

Chapter 7

Cob Building Design and Planning

with Martin Hammer, Architect

ALL GOOD BUILDINGS begin with good design, and good design is largely a matter of sound, well-informed decisions. Conversely, a poorly designed building, no matter how well constructed, will always be a flawed building. Careful consideration at the outset, while a building is still a collection of lines on paper or in a computer, will pay large dividends during construction and ultimately when it is lived in.

A good design process is iterative: while many parts of the process can be laid out as a series of steps, expect to cycle through many of these steps more than once, especially in the early stages of design. *Essential Sustainable Home Design* is an excellent resource to help you through these early stages.

Before you begin designing, carry around a tape measure and a camera; record dimensions and take photos of spaces and details you really like. Develop your design as much as possible in three dimensions, not two. Models (physical or computer) are good (include people for scale),

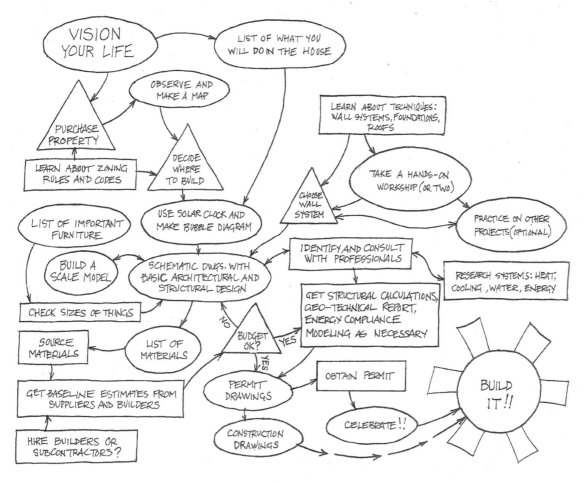

Fig. 7.1:
The process of designing your home.
CREDIT:
DALE BROWNSON,
FROM AN IDEA BY
ELKE COLE

but full-size mock-ups are better still, especially for key elements such as the entry, the kitchen, or a window seat. When you have sketched out a floor plan, test its dimensions at full scale, so your body can provide feedback. If the site is sufficiently flat, chalk out the interior layout on the ground to see how well it fits the site and how your movements and actions fit the plan. Do this on a beach or pavement if you can't use the site. A simple way to mock up a building's volume is to pound stakes into the ground at corners and curves, attach bamboo or wood poles to the stakes, and connect them with strings to indicate the positions of walls, doors, and windows. The goal is to not be surprised by the final result—because you've already *seen it*. *The Hand-Sculpted House* contains many practical suggestions on how to design a small, space-efficient cob home.

Some design considerations apply to all buildings, and others are cob-specific. Let's first look at general concerns, then examine how the particulars of cob affect design decisions. At the time of this book's publication, cob is primarily used for homes and accessory buildings, so this discussion is geared mostly toward home design; but the principles can be applied to other types of buildings as well.

General Design Considerations

Design program and goals: A building's *program* states everything the design intends to achieve. Before you begin working on the floor plan or any other design details, take the time to think deeply about your goals for both the finished building and the building process. What do you want your building to feel like when you live in it? Introspective or extroverted? Austere or colorful? What are the practical purposes of your building, and how do you want it to perform? How might that change in 10 or 20 years? Make a list of your objectives, then rank them in order of importance. Don't be afraid to think ideally at the beginning. Later in the design process your ideals are sure to be tempered by real-world constraints such as budget, building codes, and so on.

One's design philosophy, and the buildings that result, are largely a reflection of one's values. Be clear first with yourself, and then

Architect Féile Butler used models as well as drawings when designing her family home in Ireland. Modeling with clay helped her capture the sculptural potential of cob. For more on the design process of this house, see the color insert.
CREDIT: STEVE ROGERS

with any design professionals you engage, about your prioritized values. These can include the following—and anything else that's important to you:

- Environmental concerns: personal, local, regional, global
- Lifestyle
- Functionality
- Aesthetics: interior and exterior
- Outdoor spaces
- Affordability

Environmental concerns: This is one of the values that most frequently lead people to choose cob as a building material. The concern may be about one's personal environment (e.g., interior air quality, exposure to toxins during the construction process), or it may extend to the local, regional, or global levels. Cob's many environmental benefits are mentioned throughout this book, including its low embodied energy, low carbon footprint, local sourcing (often on-site), thermal mass benefits for energy efficiency, and its graceful return to the earth at the end of a building's life. But remember that cob will only be one component of your building; all the other materials and systems you will use should also be evaluated through an environmental lens. *Making Better Buildings* is an excellent resource to guide you through what is often a complex series of trade-offs.

Lifestyle: The design of a building significantly impacts how its inhabitants live within and around it. Design choices can encourage or discourage physical activity, as well as connections with the outdoors, with other family members, and with neighbors. This is especially true when considering the site design of an entire property. To a certain degree, design decisions should be aspirational; envision your ideal life and design a building that supports and encourages it. But also be realistic about your needs and patterns of living and working.

Functionality: "Form follows function" is a well-known adage of modern architecture that has its roots in the earliest human dwellings. It makes sense that the best forms for shelter are ones that best accommodate the day-to-day functional needs of its inhabitants.

Aesthetics: A building's beauty ideally grows out of and is integrated with its many functions. However, good functional design alone is not a sufficient design goal. Aesthetic beauty is equally important. Beauty can be enhanced by thoughtful use of the primary architectural elements of space and light (natural and artificial), as well as by material, finish, color, and texture choices, and by design elements that are purely decorative. Engendering feelings such as comfort, security, warmth (psychological and thermal), and visual interest in your design will greatly increase the day-to-day enjoyment of your home.

Aesthetics are subjective, so make design decisions and material selections that feel right for *you*. Pay attention to what you like and why. Take photos, or collect online or printed images

Rob Pollacek of California Cob took advantage of cob's sculptural qualities and its fire-resistance when sculpting this fireplace surround and bench.
CREDIT: ROB POLLACEK

of buildings (both exteriors and interiors) with design features that appeal to you. This can be for your own reference, but is especially helpful if you engage a design professional. Bear in mind that aesthetic choices are also an expression of your values, whether explicit or implied. Materials with short lifespans or toxic end-of-use scenarios have very different impacts than durable materials that can return to the earth without harm or be repurposed in the future.

Outdoor spaces: Your building doesn't begin and end at its walls. The transition from outside to inside, the extension of inside to outside through views, doors, and windows, and the outdoor spaces surrounding the building are all important parts of a good design. When choosing the size and placement of windows and doors, consider how they relate to views and privacy as well as their effects on natural lighting and heat gain and loss. Outdoor spaces

should be designed relative to adjacent indoor spaces (kitchen, living room, bedrooms, workspaces, etc.). Consideration should be given to how outdoor spaces will be used seasonally, including addressing issues of sun and shade.

Outdoor spaces can be defined with existing or new trees and other plantings, site walls or fences, benches, steps, roofs, or trellises. The house itself will buffer the microclimate of surrounding areas: a south-facing exterior wall can be shaped to collect the sun's heat in colder climates and seasons, and properly shaped and oriented buildings and their openings can *create* a cooling breeze in hot climates and seasons.

In general, outdoor spaces are much less expensive to construct than indoor ones, and may better meet many of your design objectives. By maximizing the number of activities accommodated outdoors or in unheated spaces, you can reduce both the literal and carbon footprints of the building and its short- and long-term environmental impacts and economic costs.

Outdoor spaces and transitions between the outdoors and indoors can be just as functional and beautiful as rooms inside a home. Here, owners Jean Peters Do and Tu Do enjoy a tree-shaded sitting area outside their cob home Bliss Haven, designed and built by Kindra Welch.

CREDIT: KINDRA WELCH

Affordability: This is an important issue for all building projects and can make or break the success of a project. It can also make the process either stressful or enjoyable. Budgeting is discussed in detail in Chapter 10.

External Influences

Site and climate: Every good building's design is a response to its site and climate. Get to know both, during the day and at night, and through as many seasons and weather conditions as possible, *before* starting your design. If you can, camp out for a day or two (or a year!) on sites you're considering. Bring all your senses to observe, measure, and feel the land, whether rural or urban. Try to avoid getting immediately attached to one location on the property. Small or urban properties offer limited choices in this regard, but whenever possible, consider a few different sites and building orientations. Issues of access (for vehicles, pedestrians, building materials, and equipment), views, privacy, topography, drainage, and solar orientation should all be on your list when evaluating a potential building site.

On a larger scale, both social and environmental factors are important when choosing where to build your home. Ask yourself: what sort of community do I want to be part of? How much commute time is acceptable? What are all the things I envision doing on my property? What climate and environmental features do I find essential? The dream of rural property is often a romantic vision, but good urban sites also have many points in their favor:

- A property that doesn't require driving to and from important activities (such as work, school, shops, socializing, and recreation) conserves time and financial resources and reduces stress.
- Reduced driving may be the single biggest way you can reduce your carbon footprint.

- A degraded property in a walkable urban neighborhood needs the energy and love of your building project to restore it. A beautiful undisturbed parcel out in the country does not.
- Physical distance is not the only way to achieve privacy. You can design privacy into a neighborhood parcel, but you can't design social connection into an isolated one.

General site analysis includes exploration of subsurface geology (Where are the clay layers? How far down to bedrock? How do groundwater and storm runoff travel through the site?) as well as understanding local weather patterns (What direction do storms come from? What is the likelihood of flooding?). In our age of accelerating climate change, do your best to understand climate trends and how they might affect your design. Detrimental examples include increased seasonal temperature extremes, risk of wildfire, flooding potential, and high-wind events. Beneficial examples include reliable wind for generating electricity, seasonal clear skies for solar electricity, passive solar heating or hot water, and access to sources of surface water, groundwater, snow melt, or rainwater.

Before you focus in on the design of any particular building, you should have a general idea of all of the improvements you eventually plan to make on the property and where they will be; this allows you to choose the ideal location for each element. Make a site plan that includes all buildings, roads, parking, gardens, orchards, greywater systems, shade trees, and accessory structures (such as storage sheds), and so on. Make your best guess now, but expect the *master plan* to change over time.

Try to work *with* the building site and not against it. Accept and appreciate what it offers and try not to impose preconceived expectations. If you find yourself fighting the site more than working with it, it's probably not the right site for you. Renew your search until you find a good fit.

Zoning and planning regulations: Local county or municipal zoning regulations dictate what can be built on each property according to its assigned zoning designation. These regulations govern the number and kinds of buildings and other activities (e.g., residential, commercial, industrial, or agricultural) allowed on the property, minimum setbacks of structures from property lines, maximum heights of buildings, maximum lot coverage, amount of required off-street parking, and much more. Research these requirements *before you begin designing,* and even before you purchase a property. These parameters can have a profound effect on your design and may even prevent you from building the project you envision. Check the property deed to determine whether there are any easements (usually for public or private utility use, or through-access to another property) that restrict your use of that portion of your property.

In addition to Planning Approval (which ensures compliance with zoning regulations), some jurisdictions require Design Review as part of project approval. Guidelines for acceptance and a description of the process are usually provided by the local planning department. These are typically limited to the exterior appearance of structures but can also take into account the proposed building's location on the property and potential impacts on neighbors. Adjacent and nearby property owners often have a voice in the Design Review process. If so, connect with them early in the process and hear their concerns. Try to balance their opinions with your project needs and your property rights. Positive neighbor relations can be one of the most important factors affecting how it feels to live in a particular location. Often, with a little forethought and compromise, all parties can be satisfied with a project.

For projects requiring well water or an on-site septic system, the jurisdiction's health department typically determines the requirements and will administer associated reviews, permits, and inspections.

Building codes and the building department: The local building code and building department will impact the design of your building—a fact too often ignored until a problem arises during the permit approval process or an on-site inspection. Simply stated, permitted projects must comply with the building code and all associated codes (including electrical, plumbing, energy, etc.). It's tempting to focus on the process of obtaining a permit for the cob walls specifically, but many other requirements (e.g., minimum room sizes and ceiling heights, natural light and ventilation, attic access, emergency egress and rescue requirements, and energy conservation requirements) will impact your design as much as your choice of wall systems. Understanding key building code provisions is essential to creating a realistic design from the beginning. When developing a design that includes cob, your design choices may also be affected by Appendix AU. For more information on understanding and navigating codes and permits, see Chapter 9.

Resources

Personal resources: All of your experiences and skills may directly or indirectly contribute to a successful project. Building design and construction experience are obviously valuable, but so are many other abilities, such as organizational, planning, budgeting, record-keeping, and communication skills. Personal traits such as patience, perseverance, flexibility, and creative thinking are also very important.

Financial resources: Whether the funds you bring to the project are in hand or borrowed, the budget you establish is likely to be a primary factor shaping your building's design. Try to be realistic with your budget from the beginning, and decide how much, if any, financial stress you're willing and able to endure to achieve your objectives. Simple + Small + Standard = less expensive. Complex + Colossal + Custom = more expensive. For more on project budgeting, see Chapter 10.

Land is the most important resource and will likely be one of the largest project costs if you don't already own or have access to it. Whether you are working on property you already have or selecting a new one specifically for the project, the land will greatly affect the design. Rural properties frequently have substantial development costs—water and septic, electrical hookup, driveway, and required fire truck turnarounds, to name a few. With typically smaller lot sizes, urban properties are especially affected by zoning regulations such as setbacks, building height limits, and off-street parking.

Building materials: How you source materials is both a practical and a philosophical choice. The property may provide some materials. Its soil (often from foundation excavations) can often be used for buildings with earthen walls, floors, and plasters. Stone and lumber can also be site-sourced, especially if trees must be removed to make way for the structure. If using site lumber, be aware that it may need to be air-dried (for several months to several years depending on thickness) and/or graded before use in the building. You may already possess or have access to construction materials, such as lumber, doors, or windows, that can be incorporated into the design. These can also be sourced from salvage yards, Craigslist, and other recycled materials repositories.

Cob-Specific Design: What Does Cob Want to Be?

In order to make the best design decisions, you need to fully understand cob as a material, including its physical characteristics and what it does and doesn't do well. Your design and the construction process will greatly benefit by working *with* the material instead of against it.

To understand what cob wants to be, you can, of course, read books, articles, and online material. It's also valuable to visit cob buildings and to talk with cob designers and builders and people who live in cob homes. Simply observing cob buildings—whether through photographs, drawings, or in person—can tell you a lot, but their context must be understood to draw sound conclusions for your project. The best way to start understanding cob is to work with it! Begin to test clays and experiment with mixes (see Chapters 4 and 6). Attend a cob workshop or volunteer at a cob building project if you can. See Chapter 11 for advice on how to get cob-building experience.

Structural considerations: The subject of structural design is covered extensively in Chapter 8, but the following information is helpful to keep in mind when developing your design. Though low in compressive strength compared with concrete (2,000–4,000 psi [13.8–27.6 mPa]), cob has sufficient compressive strength (60–250+ psi [0.4–1.7+ mPa]) to support the roof, ceiling, and second-floor loads of moderate-sized buildings. This assumes typical wall thicknesses (10–24" [25–61 cm]), and low-to-moderate concentrated loads. Cob has high mass and low tensile and flexural strength. Its mass contributes significantly to seismic loads in earthquakes, and its low

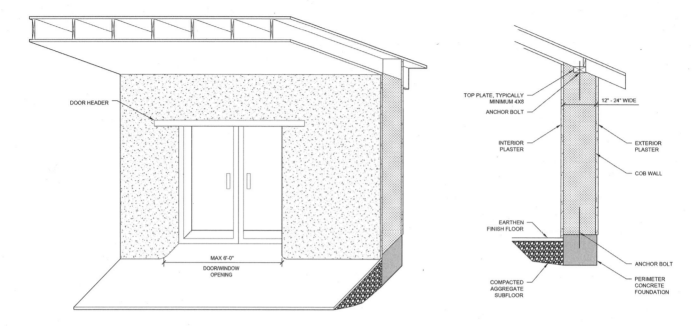

DOOR HEADER

MAX 6'-0"
DOOR/WINDOW
OPENING

TOP PLATE, TYPICALLY
MINIMUM 4X8
ANCHOR BOLT

12" - 24" WIDE

INTERIOR
PLASTER

EXTERIOR
PLASTER

COB WALL

EARTHEN
FINISH FLOOR

ANCHOR BOLT

COMPACTED
AGGREGATE
SUBFLOOR

PERIMETER
CONCRETE
FOUNDATION

Fig. 7.2:

Typical cob wall dimensions and features.

Credit: Rachel Tove-White, Verdant Structural Engineers

tensile and flexural strength demand additional reinforcing beyond its straw in medium-to-high seismic zones. Cob's limited ability to carry concentrated loads encourages narrower door and window openings (generally ≤ 6 ft [1.8 m]), though wider openings are possible per Appendix AU or with engineering.

Geometry of cob walls: Because a wet cob mix is moldable when placed and requires no form-work, many people choose to design cob walls, wall openings, benches, and other cob features with curved or "organic" shapes to take advantage of this fundamental sculptural quality that few other building materials possess. This ability to curve is often evident in a cob building's floor plan; cob walls can exhibit the geometries of circles, semi-circles, or ovals, or they can be highly "free form." Cob benches, other applied cob features, wall openings, and niches can be the most sculptural of all.

However, curves in a building design can increase both complexity and cost, especially when interfacing with rectilinear materials such as lumber, plywood, doors, and windows. Roof framing (and sheathing) geometry can become very complex when bearing on curved walls.

Since cob is often associated with an organic, curvy aesthetic, it is useful to mention that cob doesn't *have* to look this way. Designs with flat, straight walls are also possible for those who desire that look, though guides such as string lines or formwork and careful trimming and finishing are usually needed to achieve this. Many people choose something in between, with walls built on a straight foundation, but allowing the cob wall surfaces to be gently irregular as the successive lifts of cob are placed, or adding sculptural cob elements that are nonstructural. While freely exploring the sculptural potential of cob, keep in mind the practicalities of living inside a building. Be realistic about standard sizes and shapes of appliances and furniture, and make sure to include enough storage area.

The thickness of cob walls creates an intermediate "place" between inside and outside that we're not accustomed to in conventional wood-framed buildings, opening up aesthetic and spatial opportunities for window seats, niches, and built-in storage.

Cob walls are sometimes tapered vertically, becoming thinner with the wall's height, reflecting the wall's greater need and capacity to carry its own weight at the bottom of the wall. Tapered walls reduce material quantity, labor, and drying time and concentrate the material where it is structurally needed most.

Fire resistance: Cob is known to be highly fire-resistant and has a 2-hour fire-resistance rating (resulting from an ASTM E119 test), which makes it appropriate in situations where a fire rating is required by the building code.

However, the fact that cob won't burn does *not* mean that a cob building is fully fire resistant. The parts of a building most vulnerable to fire (wildfire in particular) are often not the walls. Other factors, including roofing materials, eave vents and detailing, window glass type, window shutters, and defensible space around the building, often determine whether a building survives a wildfire.

Wildland-Urban Interface (WUI) zones have been governmentally established in many US states to identify areas of high wildfire risk, and WUI codes contain requirements that significantly increase a home's fire resistance.[1] These codes are derived from the International Wildland-Urban Interface Code (IWUIC) and adopted by individual states and jurisdictions. Even if you aren't building in a designated WUI zone, the strategies they contain can help protect your home from wildfires.

Kiko Denzer built this privacy wall in a residential neighborhood in Corvallis, Oregon. As urban firestorms become more common, cob boundary fences can enhance safety. They also very effectively deaden noise. CREDIT: MICHAEL G. SMITH

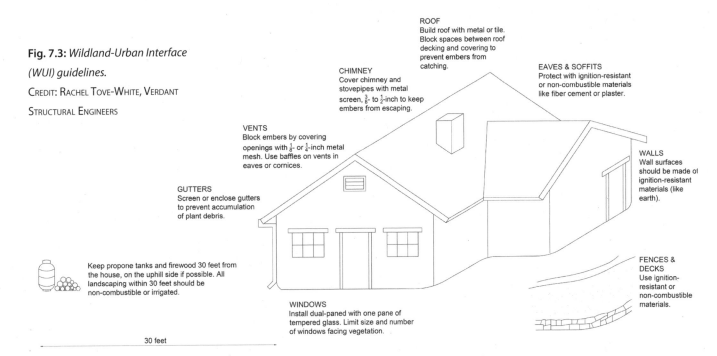

Fig. 7.3: *Wildland-Urban Interface (WUI) guidelines.*
CREDIT: RACHEL TOVE-WHITE, VERDANT STRUCTURAL ENGINEERS

ROOF
Build roof with metal or tile. Block spaces between roof decking and covering to prevent embers from catching.

CHIMNEY
Cover chimney and stovepipes with metal screen, ⅜- to ½-inch to keep embers from escaping.

EAVES & SOFFITS
Protect with ignition-resistant or non-combustible materials like fiber cement or plaster.

VENTS
Block embers by covering openings with ⅜- or ¼-inch metal mesh. Use baffles on vents in eaves or cornices.

WALLS
Wall surfaces should be made of ignition-resistant materials (like earth).

GUTTERS
Screen or enclose gutters to prevent accumulation of plant debris.

Keep propane tanks and firewood 30 feet from the house, on the uphill side if possible. All landscaping within 30 feet should be non-combustible or irrigated.

WINDOWS
Install dual-paned with one pane of tempered glass. Limit size and number of windows facing vegetation.

FENCES & DECKS
Use ignition-resistant or non-combustible materials.

30 feet

Not Just for Walls

Cob's high thermal mass is useful in many applications—not just exterior walls. Built-in furniture, sculptural elements, and interior half-walls made of cob can provide interior thermal mass as part of a passive solar heating and cooling strategy for buildings whose exterior walls are made of other materials. An underexplored use of cob is to retrofit existing wooden structures, which are frequently low in thermal mass. (Wood-framed floors need to be carefully evaluated to make sure they can support the extra weight.)

Because cob is extremely fire-safe, it is a particularly good material for applications relating to fire, including earthen ovens,[2] Rumford fireplaces, certain types of masonry heaters such as Rocket Mass Heaters,[3] and even simple surrounds for wood stoves and fireplaces. In these situations, cob's thermal mass helps extend the heating effect of the fire by storing and slowly releasing heat that would otherwise escape from the building much more quickly.

This cob fireplace surround in a Colorado strawbale home serves multiple functions: as thermal mass heat storage for the wood stove, a safety feature, and a sculptural focal element of the home.
CREDIT: CATHERINE WANEK

This cob bookshelf, sculpted by Greg Allen and Danielle Ackley of the Mud Dauber School of Natural Building, provides function, beauty, and thermal regulation to a strawbale studio. CREDIT: DANIELLE ACKLEY

Thermal attributes: A building's thermal performance is a function of the building envelope's insulation and airtightness as well as the amount and location of its thermal mass, all in the context of the local climate.

Cob has excellent thermal *mass* (ability to store heat and to moderate interior temperature change) but low thermal *resistance* (ability to resist conductive heat flow). It is critical to investigate local code requirements for the thermal resistance of exterior mass walls (cob walls are considered mass walls in the IRC) very early in the design process to ensure that the project won't be disapproved solely on thermal resistance requirements. Mass walls such as cob have lower thermal-resistance requirements in the IRC Table, but compliance depends on the climate zone and the thickness and insulation value of the specific cob mix. The Cob Research Institute has conducted thermal testing on high-fiber cob mixes with higher R-values that will satisfy energy codes in some climate zones. When greater insulation is needed for compliance, it must be added to the exterior of the cob walls and follow the usual rules for combining other materials with cob (for instance, rigid foam is not appropriate because it is vapor impermeable). See "Optimizing Cob's Thermal Performance in Your Climate Zone" in Chapter 3 for more information on cob's thermal performance and designing insulation into a cob wall assembly.

Passive solar design: Passive solar design is a natural fit for cob buildings because of cob's excellent thermal mass. With well-placed glass and mass and proper building orientation, the building itself becomes a solar collector during the heating season. Windows and glass doors let in the sun's heat, and interior mass absorbs and stores the heat, then releases it at night and on cloudy days. Thermal mass is most effective where it directly receives the rays of the winter sun (such as in floors and portions of walls), but mass anywhere in a room heated by the sun will absorb and store the sun's heat to some extent.

For good solar collection in the northern hemisphere,[4] a building's long side and most of its glass should face south or close to south. Southeast- and even east-facing windows serve to introduce warmth in the morning hours when it is needed most. To avoid overheating in the warm season, it's important to provide shading for this same glass (with roof overhangs, trellises, deciduous vines, or trees) from the higher-angle summer sun, and to minimize west-facing glass. In hot climates, skylights should be avoided or used very sparingly to prevent overheating. A good alternative to skylights are small-diameter *solar tubes;* they collect and deliver lots of sunlight but very little of its heat.

Passive *cooling* can be effective in any climate with a summer diurnal (day-night) temperature difference of at least 15°F (8°C). It works by opening air outlets high in the building—such as operable skylights, clerestory windows, or roof or gable vents—at night to allow warm indoor air to escape. At the same time, low windows or doors (preferably on the north side of the building, where the ground has been kept cooler by the building's shade during the day) are opened to let in cool outdoor air. This creates a natural convection current that expels excess daytime heat and cools a building's interior mass—which will then moderate temperature swings the next day. Both inlet and outlet openings should be closed in the morning, before the exterior air temperature exceeds the temperature inside the building.

Edward Mazria's iconic book *The Passive Solar Energy Book* remains an excellent guide to solar design. Another favorite is *Natural Solar Architecture: A Passive Primer* by David Wright.

Foundations, floors, and roofs: Choices for these elements have significant impact on the design and planning of cob buildings. See Chapter 12.

Electrical and plumbing: It's important to understand the cob-specific issues involved. See "Electrical and Plumbing Installation" in Chapter 14.

Appropriate finishes: Various finishing options are appropriate for cob. Finishes should be chosen early in the design process since they impact many other aspects of the design. See Chapter 15.

Cob with Other Wall Systems

Cob walls can be combined with other natural or conventional wall systems for various purposes. Natural builders often use multiple natural materials—straw bales, light straw-clay, rammed earth, hemp-lime, wattle-and-daub, bamboo—deploying each in situations that take advantage of its particular attributes. (Some of these materials share one or more of the constituent materials of cob—clay subsoil, sand, and straw—making for efficient use of materials and equipment on the building site.) For example, external straw-clay or hemp-lime insulation can be added to a cob building to improve its energy efficiency. Or, cob can be used as thermal mass

LCA and Cob Design

It's useful to mention a tool that is becoming increasingly common in building design: a Life Cycle Assessment (LCA). What is an LCA? It's the systematic analysis of the environmental impacts of products or services during their entire life cycle. The purpose of an LCA when applied to building materials is to help designers compare them. Because of the wide range of materials to choose from, and because many common building materials have long and complex supply chains, it has become difficult to truly compare their environmental impacts. An LCA standardizes the metrics of the environmental impacts of a given building product (or whole building, or service), beginning with raw materials excavation and continuing through fabrication, delivery to site, construction, and end-of-life disposal/recycling/composting. LCA categories of environmental impact include climate change, the eutrophication and acidification of bodies of water, smog formation, particulates, and ozone depletion in our atmosphere, and any harmful environmental persistence. An LCA can be structured by different boundaries: "cradle-to-gate" includes all stages from raw materials to the beginning of construction, whereas "cradle-to-grave" extends to end of life.

Several documents are associated with an LCA. An EPD, or Environmental Product Declaration, is essentially a summary of the data collected in the LCA process. The EPD is the document collected into databases for comparison to other materials. Other terms such as LCI (Life Cycle Inventory) and LCIA (Life Cycle Impact Assessment) refer to stages within the LCA development process. A Whole Building LCA (WBLCA) is an LCA for the lifecycle of an entire building.

Using LCA/EPD to compare and assess building materials is becoming a widespread practice. This information was initially available only through proprietary software, but rapidly growing interest in embodied carbon and building materials has helped make it more accessible. There are now several open-source LCA databases and carbon footprint calculators available online: EC3,[6] the BEAM calculator,[7] the Athena EcoCalculator,[8] and others.

To our knowledge, there is no current LCA/EPD for cob, but completing an LCA for cob should be relatively straightforward. Cob has very few ingredients, all of which have short, transparent, uncomplicated supply chains, and minimal processing. They also return directly to the earth without harm. A cob LCA/EPD will help showcase the many environmental benefits of cob, which in turn will help mobilize resources and acceptance for cob as a powerful ecological building tool.

on south-facing walls (e.g., as a *Trombe wall*[5]) in conjunction with more insulating wall systems such as strawbale on the other walls to best utilize the thermal properties of each.

Combining cob with other wall systems can also be aesthetically desirable: for example, since cob walls have structural limits on the size of openings, a design might transition to wood framing with light straw-clay or hemp-lime infill where multiple or large openings are desired. Some wall systems such as wattle-and-daub can be constructed much thinner than cob walls can, making them more appropriate for interior partitions where space is at a premium.

Take extra care when connecting different wall systems to each other—especially in making a weather-tight seal where they meet, and in transferring seismic loads from one system to another. Despite its many advantages, combining wall systems introduces both structural and logistical complexities which can increase engineering and construction costs.

Step-by-Step Checklist

The following list contains the specific steps needed to complete a cob home or other major building project. Some of these steps need to be completed in a particular order, such as foundations before walls. Others are more flexible in their order. We have indicated this by the numbers in brackets, which indicate the range of possible sequences for some steps relative to others. Not all of these steps are necessary for every project. More details about many of them follow.

1. Design program and goals
2. Siting
3. Cob mix design
4. Planning and design (from 4 to 14)
5. Permitting
6. Materials acquisition (from 1 to 24)
7. Sitework: roads, grading, trenching, and utilities
8. Drainage and foundations
9. Plumbing and electrical wiring (from 9 to 15)
10. Subfloor (from 9 to 15)
11. Rough door and window bucks
12. Cob wall construction
13. Roof framing and waterproofing (from 9 to 13)
14. Other major carpentry, such as interior framed walls (from 9 to 14)
15. Base plaster (from 13 through 18)
16. Window and door installation
17. Counters, cabinets, etc.
18. Final electrical outlets, switches, etc.
19. Interior finish plaster
20. Finish floors
21. Painting
22. Finish carpentry/trim (after 19)
23. Install sinks, appliances, electrical fixtures (after 19)
24. Exterior finish plaster (after 16)

- **Design program and goals:** In practice, there is often a fluid back-and-forth between clarifying design goals, choosing a site, and designing a building, which can change the order of operations described above (or lead you to cycle back through several steps more than once).
- **Siting:** Building site selection generally happens in three steps: 1) Determine the region and community for the project, 2) Select a property to build on, and 3) Select the specific site for your building. All three steps are best taken before you start your design process since good designs are tailored to the site.
- **Cob mix design:** If the building is to be permitted, you will need to demonstrate the compressive strength and other properties of your cob mix. This requires knowing what materials and mix recipes you will be using early on, since the results will affect your structural design. Mix design should therefore happen early in the design process. ☞

- **Planning and design:** If the building is to be permitted, it will be difficult to make major design changes after the permit has been issued. With an unpermitted project, you have more flexibility to adjust the design as you build. Either way, there are many design aspects that must be clarified before you begin building. These include structural design of the foundation, walls, and roof, determining interior floor levels relative to exterior grade, and at least a rough plan for plumbing, wiring, and other utilities.
- **Permitting:** Don't underestimate how long this step may take. Contact your planning and building departments early to get an idea of their timeframes for Planning Approval and issuing a permit. Especially if yours is the first cob building to be permitted in your county or city, there may be many months of conversation and unexpected hurdles before the permit is granted. Planning Approval is required by many jurisdictions prior to issuance of building permits.
- **Materials acquisition:** Give yourself as much time as possible to source and gather building materials. This allows you to take advantage of good deals and salvage opportunities. If you start collecting early enough, you can work your special finds into your design. However,

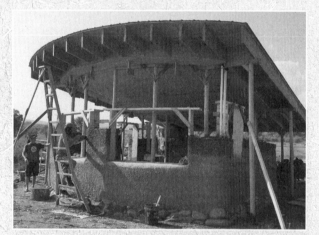

For this project in Colorado, builders Emma Geddes and Clayton Ives chose to erect the roof first to protect the building site from weather during construction. The roof is supported by permanent posts, some of which are surrounded by cob. CREDIT: GEORGE GUMERMAN

if you are planning to get a permit, you might want to wait until after you have the permit in hand before you invest much in materials, in case you have to substantially change your design.

- **Sitework:** Pre-construction sitework can include access roads, driveways and parking (including emergency vehicle turnarounds on rural sites), grading for a building pad, drainage, retaining walls, water supply (sometimes including fire-suppression water tanks on rural sites), water source, electricity source, and wastewater system. Plan to have all major earth moving done at the same time. Unless you own your own heavy equipment or will do all the excavation by hand, it's much cheaper to get equipment onto the building site only once and as early in the process as possible. Also, the grades of access roads and paths, parking areas, and the building site all need to relate to each other. Trenches for drainage and utilities are best dug at the same time so that none of the buried infrastructure is damaged by later excavation. Be sure to set aside topsoil for gardening and landscaping and clay subsoil for construction. Try to schedule sitework for a season with favorable weather (dry, well above freezing).
- **Plumbing and electrical wiring:** As discussed in Chapter 14, wires and pipes may be installed in cob walls at various stages of construction. However, it's important to have a clear plan for these utilities before wall building begins. If any wires or pipes need to pass through the foundation or under a slab floor, they should be installed early in the process.
- **Subfloor:** If you choose a concrete slab subfloor (rather than an earthen floor), it is typically poured along with the foundation. A raised wood subfloor can be installed at various times in the build, depending on the design. If the design includes an earthen floor, it's best to install the first stages (drainage layer and base subfloor) as early in the construction process as possible—usually after foundations, but before walls are begun. This reduces drying time and provides a firm, level surface for the rest of the construction process. If rain is expected during the construction period, wait to install earthen floors until the roof has been built. ☞

- **Roof framing, sheathing, and waterproofing:** In a climate with long periods of reliably dry weather, you may be able to wait to build the roof until after the cob walls are complete. This maximizes air and sun exposure to the walls for faster drying and simplifies construction because the cob walls bear the weight of the roof. However, it is often preferable to construct the roof before the walls, providing both weather protection and shade for the entire building site. In that case, the roof can either be supported on permanent posts and beams or on temporary posts that will be removed once the walls are completed. Sometimes a hybrid approach works best; the roof can be framed ahead of time and covered with removable tarps until wall construction is complete and the permanent sheathing and waterproofing layers are installed.

- **Base plaster:** People often underestimate the time this step will take, especially when you factor in preparing materials, masking to protect wood, and all of the cleanup needed before and after plastering. If the unplastered cob walls are wavy or uneven, it can be helpful to apply base plaster early, since it clarifies the wall planes and makes it easier to see how windows, doors, and other wood elements relate to the cob. All wall shaping should be completed using the base coat before the interior finish plaster, finish floors, or finish carpentry such as baseboards and window trim are installed.

- **Window and door installation:** It's usually best to install windows and doors after base plasters and subfloors are in place. Not only is base plastering messy, but maximum air flow is desirable to speed drying of the floor and plaster.

- **Interior finish plasters:** If you are using clay plaster, you will first need to make samples of your recipes and let them dry completely to test for hardness, color, texture, cracking, and other properties. Leave time for several rounds of samples in case you don't get it right on the first try. Don't start on the final layer of plaster until all significant construction is done in that room, including major carpentry, all base plaster, and the subfloor; finish plasters are somewhat fragile and are

difficult to keep clean when a lot of mud and water is being tossed around. Once you begin finish plastering a wall, keep going until the entire wall is complete, or a visible seam is likely to show.

- **Finish floors:** As with clay finish plasters, it's essential to test earthen floor mixes to make sure they are hard and crack-free, and this will take time. It's usually best to install finish floors (especially earthen floors) after the finish plaster is complete.

- **Exterior finish plasters:** Application of exterior finish plasters is often one of the last tasks in completing a cob building. In fact, because they are not necessary in many cases (unless the walls are exposed to a lot of weather), they are often postponed indefinitely!

The final layer of earthen floor is being installed in this cob house. Note that nearly everything else is already complete, including finish plasters. This was the first permitted cob building in California, built by Katie Jeane in Willits in 1998.
Credit: Michael G. Smith

How Long Will It Take?

It's clear from the discussion above that many steps are involved in designing and constructing a cob building. Some of these steps take significant time; others are done quickly. Some steps involve waiting for other people, such as building officials, engineers, or electricians, to do their parts. Some steps are weather dependent. All of these factors combined make it difficult to predict how long it will take to complete a building.

Most owner-builders are unable to finish even a very small cob house in a single building season. In a temperate location, if you wait to begin excavation until after the soil has dried out and the danger of major rains and freezing is past, you have only a few months before the weather becomes cold and/or wet again. During that time, you need to build your foundations, walls, and roof. By the time those are complete, drying conditions may not be right for the finish plasters and floors. More times than we can count, we've seen an owner-builder move into an unfinished cob house because they needed a warm, dry place to spend the winter. The following spring or summer, they move back out while completing interior floors and plasters.

One good way to get a leg up on this process is to complete the planning, design, permitting, sitework, and foundations *the year before* cobbing will begin. Maybe the roof can go up, too. This allows an early start for cobbing the following spring. Maybe you will be able to finish everything before the second winter. Or maybe you can enclose the space and use heaters and dehumidifiers to allow interior work to continue during the winter.

An experienced builder working with mechanical mixers and a skilled crew can build much faster. Rob Pollacek of California Cob has built many cob structures—large and small—mainly in the Sierra Nevada foothills. He mixes cob with a tractor (see "Mixing with a Tractor or Excavator" in Chapter 13) and uses forms to build the walls (see "Formwork" in Chapter 14). Both of these innovations speed up the mixing and building process. Rob says that no matter how large or small the footprint, it typically takes him 3½ weeks to construct the cob walls for a one-story building. This is because the biggest constraint to how fast he can build a cob wall is the drying time. His forming system allows him to build 15" (38 cm) lifts of cob. He waits two to three days before applying the next lift, resulting in 30" (76 cm) of height per week. This is unusually fast and is made possible in part by the fast drying conditions in California. (In the damper climate of the UK, where cob lifts are commonly between 18" and 24" high [45 and 60 cm], it is common to wait 7 to 14 days before adding the next lift). But remember, wall construction is only one of the many steps required to build a house. When Rob was building a large 3,000-ft^2 cob home (280 m^2), the cob walls still took only 3½ weeks, but the entire house took nearly three years to complete.

Ananth Nagarajan kept track of all the labor hours for one of his cob projects in Hyderabad, India. The building included 100 lineal feet (30 m) of cob walls, which were 9'6" (2.9 m) tall and 14" (36 cm) thick, for a total of 40 cubic yards (30 m^3) of cob. The total labor required to prepare materials, mix the cob, and build and trim the walls was about 600 hours, or 15 hours per cubic yard. The work was done by a combination of experienced cob builders (31% of total hours), semi-skilled laborers (43%), and unskilled volunteers (who may not have been working efficiently much of the time). The cob was mixed very quickly with a backhoe (see photo in Chapter 13). Forty cubic yards of cob were prepared in just 12 hours of machine time, with two or three assistants on the ground adding water and straw.

Some of the most important factors affecting construction time for a cob building are described below. Keep in mind that in many cases, the slower the building goes, the more it will cost. This is clearly the case if you are paying for labor. But even if you're doing all of the building yourself, or with volunteer help, time can still equate to money. For example, while you build your cob home you may be paying rent elsewhere, and you may be unable to work for income because the building is monopolizing your time.

- **Building size:** Obviously, the bigger the building, the more cob needs to be mixed and installed. Tall walls, especially, slow the process. Because the wait time before the next lift is controlled by drying conditions, more wall height corresponds to more building days compared to a shorter building, even with the same volume of cob. Also, building cob walls requires more work as they get higher. More weight needs to be lifted high into the air, scaffolding needs to be built, moved, and raised, and time and energy are spent climbing up and down.
- **Design complexity:** The more complicated the building design, the slower and more expensive it will be to build. An often-overlooked example appears during the plastering process. A skilled plasterer might be able to cover an entire simple wall in a day. But the more complications, such as windows and niches, edges where the plaster meets other materials, and especially sculptural details, the slower the process will be. It might take ten times longer to plaster a wall covered with bas-relief designs compared to a flat wall of the same area. Also, remember that the cob is only one part of the building process.

Non-cob building elements such as the foundation, roof, and finishes may be even more complicated, difficult to permit, or expensive than the cob itself, especially if the floor plan is organic. Think carefully about every part of the design and what it will take in time, materials, labor, skill, and money to complete.

- **Weather:** Plan to do your cob mixing and building during warm, dry weather. Poor

This cob sculpture surrounds a fireplace in an Alaskan birthing center designed and built by Lasse Holmes with many collaborators. The clay finish plaster on the tree is by Sasha Rabin. A well-executed cob sculpture can be the focal point of an entire building, but keep in mind that this level of detail is extremely laborious to plaster. CREDIT: SASHA RABIN

drying conditions limit the height you can add to your walls in one day and may make you wait many days before applying the next lift. Also, you will have to mix your cob stiffer than you would in good drying conditions, which is more difficult. If heavy rains come before the walls are protected by the roof, you may spend a lot of time covering them with tarps, and then uncovering them—maybe many times.

- **Mixing technique:** While it isn't always true that mixing with machines saves time (machines can and do break down, and they sometimes cause accidental damage to the building or surroundings), mechanical mixing often speeds the process considerably. This is especially true of large machines such as excavators, backhoes, and skidsteers, which can quickly produce enormous volumes of cob. See Table 13.1 for a comparison of the efficiency of various mixing approaches.

- **Building crew:** We know individuals who have finished cob homes almost entirely by themselves. We know couples who have built small cob houses with little or no additional help and moved in after six or so months of near-constant work. While there are certainly tasks that are easier to do with a larger crew, there is little that one person can't do given enough motivation and ingenuity. Even though cob mixing and building are fairly simple and easy to learn, the speed and efficiency of different builders varies enormously due to experience, work habits, or motivation level. Just as important as the size of the crew are the individual levels of experience, stamina, and problem-solving ability. The best size for a crew depends on many details of the project: the size of the building, how the cob is to be mixed, and so on. If drying conditions are good, the walls will go up faster with a large crew installing a high lift of cob onto each wall every day. But in poor drying conditions, it might be more efficient to have a smaller crew that takes a week to complete a single lift around the building.

- **Site organization and cleanliness:** There can be a surprising loss of efficiency on a poorly organized building site. Give thought to how to store and organize materials so they will not degrade over time or get in the way of construction. Time will be wasted if materials are not stored in a convenient place or need to be moved repeatedly, or if work has to stop until more materials are delivered. If there isn't enough space for storing and mixing materials close to the walls, time and energy will be lost transporting the cob. Although salvaged materials are readily available in urban areas, storage is a bigger challenge on small urban lots and on steeply sloped land. If there is no vehicle access to the building site, factor in much more time for transporting materials by hand. If the process of planning and design isn't far enough ahead of construction, everyone may have to stand around waiting for decisions to be made. If tools are left dirty or scattered at the end of the day, buckets are left full of solidified cob, or the cob wall isn't trimmed at the appropriate moment, it will take more time the following day to set things right.

Conversely, a well-organized and clean building site is a joy to work in and makes everything about your project easier. A clean building site is also famous for having a positive effect on building inspectors, who are less likely to look for problems if the site is well-managed and feels professional.

Chapter 8

Structural Engineering

THIS CHAPTER is intended for anyone who needs to engage with engineering topics for cob structures—both engineers and non-engineers. It does not provide complete instructions on how to engineer a cob building. Instead, it explains engineering subjects and concepts important for all parties engaging in cob construction to be aware of. At the very least, this chapter will give you a common vocabulary of concepts to make your discussions with building officials, designers, and engineers easier. For assistance with engineering issues that are more complex than those covered in this book, you will likely need to wait for a future textbook on the subject or consult an engineer.

Engineers are a valuable resource in the construction industry, though they often charge high rates that can be problematic for low-budget projects. No matter what the finances of a project are, there are building types and design strategies that do not require the services of a licensed engineer. In particular, Appendix AU provides all of the necessary information for the structural design of cob buildings in Seismic Design Categories A-C (see Figure 8.3) as long as they stay within the prescriptive limits of the Appendix. Chapter 9 covers the distinction between engineered and non-engineered, or prescriptive, designs in more detail.

"Concrete" Materials

There are many different definitions of the word *concrete*.[1,2,3,4] Most of them agree that concrete is an artificial rock made up of aggregates held together by a binder. By this definition, cob is a type of concrete: the binding material is clay; the embedded aggregate is sand, and the

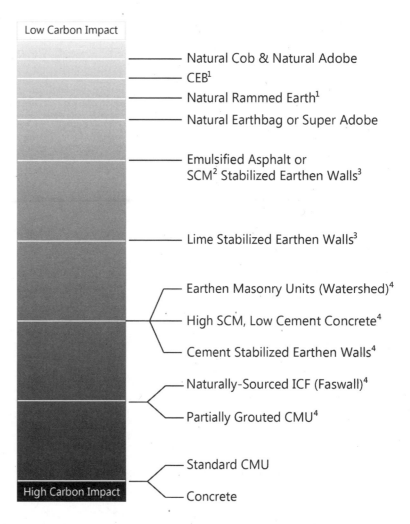

Low Carbon Impact

———— Natural Cob & Natural Adobe
———— CEB[1]
———— Natural Rammed Earth[1]
———— Natural Earthbag or Super Adobe

———— Emulsified Asphalt or SCM[2] Stabilized Earthen Walls[3]

———— Lime Stabilized Earthen Walls[3]

———— Earthen Masonry Units (Watershed)[4]

———— High SCM, Low Cement Concrete[4]

———— Cement Stabilized Earthen Walls[4]

———— Naturally-Sourced ICF (Faswall)[4]

———— Partially Grouted CMU[4]

———— Standard CMU

High Carbon Impact ———— Concrete

[1] Natural Rammed Earth and CEB's are very uncommon due to durability from wetter exposure

[2] SCM = Supplementary Cementitious Materials such as Fly Ash or Slag

[3] Stabilized Earthen Walls typically refer to Rammed Earth, Adobe SCEB, Earthbag, or Super Adobe

[4] These options can easily trade places in order depending on volume of cement and width of wall.

This image is intended as a general aid. Many complex variables such as materials source, reinforcing steel, and non-cement additives are only loosely accounted for here.

Fig. 8.1: *Concrete material carbon impact scale.*

CREDIT: RACHEL TOVE-WHITE, VERDANT STRUCTURAL ENGINEERS

mix contains a reinforcing matrix of straw. The engineering community is mostly accustomed to concrete that contains Portland cement and binds and hardens by curing through chemical reaction rather than through molecular electric charge and shrinkage mediated by wetting and drying—as a cob mix does. Nevertheless, it can be helpful to refer to cob as a type of concrete when speaking to building officials, engineers, and other professionals, rather than as a type of "natural building," a term they may not know or may have negative associations with.

Natural (without cement, lime, or asphalt stabilizers), straw-reinforced cob can be thought of as the simplest, lowest carbon impact example of a range of concrete materials. Cob's closest neighbor on this spectrum is adobe, which is simply the block form of a very similar material mix. The concrete materials with the highest carbon impacts are conventional Portland cement concrete and *concrete masonry units* (CMUs, also known as *concrete blocks* or *cinder blocks*).

Compressive strength is the capacity of a material to support gravity loads without breaking or deforming. Although cob has a lower compressive strength than most other materials shown in Figure 8.1, this does not mean that it is a lower-quality material. This is a common misconception in the conventional building and design paradigm. (A good example is the late David Easton, who was a well-known modern rammed earth pioneer and innovator. While working with him on certifications for his Watershed rammed earth masonry block system,[5] Anthony explained the permitting work we were doing with cob. David was shocked that for nearly his whole career he had been going to great lengths to make rammed earth meet conventional masonry or concrete strengths of over 2000 psi (14 mPa), while we were designing systems with a material that

ranged from 100–300 psi (0.7–2 mPa). If this was the reaction of a very experienced earthen builder, similar skepticism should be expected from building professionals and officials not familiar with natural building.) Material properties like compressive and flexural strength are only one set of variables; other factors such as building size, wall thickness and geometries, geographic location, and reinforcement have just as big a role to play in the success of any structural design.

On the other hand, cob is sometimes portrayed as an all-purpose super-material. It is important to note that all the structural loading considerations relevant to other kinds of buildings pertain to cob as well. The most significant failure modes for cob walls include compression failure (by buckling) due to wall slenderness, and collapse as a result of earthquake forces (mostly lateral) due to being under-reinforced or inadequately anchored to foundations and roof structures.

Forces to Consider

The walls of all buildings are subjected to several kinds of forces. Gravity loads exert downward forces that are divided into two categories: *dead loads* from the weight of the wall itself as well as the roof and any floor bearing on the wall; and *live loads* from occupants, building contents, and temporary snow and ice. The two most likely ways that a cob wall can fail due to gravity loads are crushing under large concentrated loads, like long-spanning ridge beams or lintels, and by buckling due to excessive slenderness and/or a low-strength cob mix. Crushing failure is rarely a human safety concern because it is unlikely to cause the *rapid* collapse of a wall or building. It usually manifests first as cracks in the wall, which grow more severe over time. Eventually, parts of the wall or roof can fall away or shift out of plumb. When this kind of failure

Structural Engineering 81

is observed, it typically requires retrofitting or reconstruction, which can be a major inconvenience and expense.

Shear forces from wind and earthquakes push on walls both horizontally and vertically. Horizontal forces are divided into the two primary ways that they affect and are resisted by a building, *in-plane* and *out-of-plane* forces, as shown in Figure 8.2.

Many cob buildings constructed during the past 30 years are located in seismically active zones in the western US and Canada. This is the result of many factors, including culture and climate. It does not imply that these designs

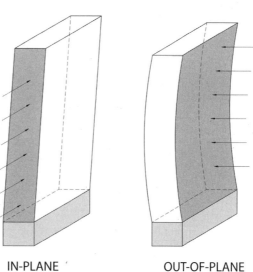

IN-PLANE LOADING   OUT-OF-PLANE LOADING

Fig. 8.2: *Horizontal forces generated by wind or seismic activity may act parallel to a wall (in-plane) and/or perpendicular to a wall (out-of-plane).* CREDIT: RACHEL TOVE-WHITE, VERDANT STRUCTURAL ENGINEERS

Fig. 8.3 (below): *Map of seismic hazard zones relating to cob wall reinforcement options.* CREDIT: RACHEL TOVE-WHITE, VERDANT STRUCTURAL ENGINEERS. SOURCE: US SEISMIC DATA IS FROM THE US GEOLOGICAL SURVEY. HTTPS://WWW.USGS.GOV/MEDIA/IMAGES/2018-LONG-TERM-NATIONAL-SEISMIC-HAZARD-MAP; CANADIAN SEISMIC DATA IS FROM NATURAL RESOURCES CANADA. WWW.SEISMESCANADA.RNCAN.GC.CA/HAZARD-ALEA/SIMPHAZ-EN.PHP

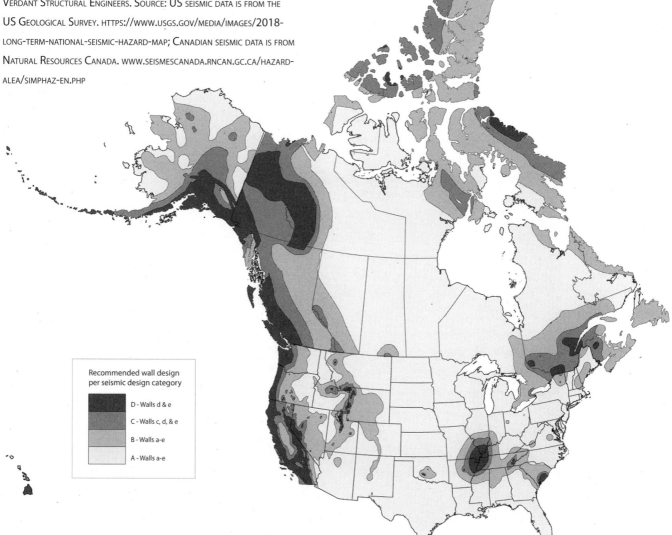

Recommended wall design per seismic design category

- D - Walls d & e
- C - Walls c, d, & e
- B - Walls a-e
- A - Walls a-e

have been proven to be earthquake safe. As of now, most cob buildings in North America are unpermitted, non-engineered buildings and have not experienced anything close to the strongest potential earthquake-induced ground movement in their areas. Just because a cob building that has been standing for two or three decades looks fine so far, that's no guarantee it will perform well in an earthquake. The purpose of a structural design is to ensure that both a building and its inhabitants will make it through extreme events such as hurricanes, storms, and earthquakes intact. Therefore, the recommendations in this book and the requirements of Appendix AU can differ substantially in many respects from common practices, especially those of owner-builders of unpermitted buildings.

The potential destructive force of earthquakes must be carefully considered in the design process. Cob is not inherently earthquake-resistant compared to other building systems, and it certainly is not earthquake-proof. (No building system is!) In an earthquake, the ground acceleration acting on the mass of a building exerts a force called the building's *seismic force*. Because

cob is so massive, its own mass generates most of the earthquake forces that a cob wall (and other structural elements of the building) must resist. This is unlike most lateral-force-resisting wall systems (typically, shear walls) in conventional structures. Conventional shear walls, like plywood over studs, concrete, or CMU walls, typically collect force from a large portion of the building and concentrate it into high-strength panels. Unless the roof is especially heavy (e.g., slate, clay tile, living roof), most of the mass of cob buildings comes from its walls. To resist these forces, strength can be added by making the walls longer or thicker, but longer or thicker walls are heavier and therefore experience more seismic force. The simplest solution to this dilemma is usually to increase reinforcement. The seismic tables in Appendix AU (Tables AU106.11[3], AU106.11[4], and

A straw-reinforced cob wall before and after testing at Santa Clara University in California in 2018. The test performed was an in-plane reverse cyclic test to simulate earthquake forces. Diagonal cracking typically associated with earthquake forces can be observed as well as horizontal separation that is likely the result of repeated cyclic compression forces.

AU106.11[5]) factor in what percentage of the walls are solid cob vs doors or glazing, which are obviously less massive.

If a wall is undersized or under-reinforced for its seismic zone, failure during an earthquake can be quick and catastrophic. Chunks of cob can come loose and fall, or the walls can crack, separate, and shift laterally. This is directly dangerous to occupants, and it can also destabilize wall or roof framing that is supported by the cob walls, leading to building collapse. Both of these failure types were observed in the seismic testing on cob walls described in this book as well as in earthquakes around the globe, including New Zealand earthquakes in 2010,[6] 2011, and 2016.[7]

Many people assume that the safest strategy for cob buildings in seismic areas is to construct a post-and-beam framework to support the roof, then fill in between the posts with cob. They expect that if the cob walls crack, the framework will prevent the roof from falling down. The problem with this plan is that, in the event of an earthquake, the massive cob walls will exert tremendous *lateral* force on the frame. To resist these forces, the frame needs

Cob building damaged in the Kaikoura, New Zealand, earthquake of 2016. CREDIT: GRAEME NORTH

Are Geotechnical Reports Worth the Cost?

To circumvent the prescriptive limitations of Appendix AU, you or your engineer might choose an Alternative Material design using the International Building Code (IBC Section 104.11, see Chapter 9 for more on this). In this case, seismic force on the building will be related to its Soil Site Class, which quantifies how much the soil on the building site will amplify or decrease ground motion during an earthquake. If you do not choose the default "D" option, a geotechnical report is often required to determine your Soil Site Class. Geotechnical reports can be expensive, and most home designers try to avoid them. However, they can be worth the cost due to the massive nature of cob buildings. A geotech's assessment, while expensive, could actually save you money by reducing

design and construction costs of a building that would otherwise be overbuilt. An assignment to a lower Soil Site Class could significantly decrease the seismic design loads on your building for medium and high seismic zones— Seismic Design Category (SDC) C and above. Building sites which may have been classified as SDC D with the default "D" Soil Site Class, may be in SDC C if a geotechnical engineer determines that the site has a "B" Soil Site Class. Conversely, buildings expected to be in SDC B based on mapping, may be in SDC C with the default "D" Soil Site Class. In either case, a geotech's report could save you a lot of money and reduce the environmental impact of your building by reducing the amount of steel and concrete required.

to be as stiff as the cob. If it is not, the cob wall will begin to break before the frame can absorb much of the force. Any system made of wood columns or steel moment frames, which are quite flexible by nature, is likely to be far too flexible to work. Steel braced frames are a stiffer option that can be explored.

Since the mass of cob walls amplifies seismic forces, the walls need to support themselves when subjected to out-of-plane loads. Therefore, any adjacent frame is limited in the ways it can assist. If frames are not specifically designed to help the cob walls resist lateral forces, they may actually weaken the walls by creating voids and disruptions in the straw/clay/reinforcing matrix.

Cob walls typically perform well against moderate to high wind forces. Because wind forces on a building are *independent* of the mass of the structure, the magnitude of potential wind forces on a 1- or 2-story cob building is usually much less than the earthquake force on the same structure in medium or high seismic zones. In this situation, an engineer would say that *seismic forces govern the design.* There is much more flexibility for the design of safe cob buildings in most wind zones in low seismic zones and this is expressed in the *braced wall panel* design tables of Appendix AU (Table AU106.11[2]).

The Structural Design Process

When conducting a structural design, engineers usually start at the top of the building. The reason for this is simple: for the most part, building elements higher in a building transmit forces to those below them until they ultimately reach the foundation; from there, the forces are transmitted through bearing or friction into the ground. For example, roof rafters bear on the walls below them, which bear on the foundation below them. This transfer of forces from element to element is commonly referred to as the *load path.* Two

different load paths need to be considered: *gravity design* and *lateral design,* the latter of which accounts for earthquake and wind forces. The paths of gravity loads vs lateral forces through a building differ even though they are often transferred through many of the same building elements. For example, cob walls both bear gravity loads and resist shear, uplift, and downward loading from earthquake and wind forces.

In order to conduct a structural design, you need a general idea of the shape of the building, and the building elements and their material makeup. This is typically available from the preliminary architectural design. That said, it can be difficult to start any architectural design without some idea of what the structural requirements will be. Therefore, we recommend that you read this chapter, then (re)read Chapter 7, "Cob Building Design and Planning," for guidance on starting your design, and later return to this chapter for assistance with the structural design. You may need to make architectural compromises while conducting the structural design. Therefore, it is a good idea to keep the initial architectural design flexible until the structural analysis has been done.

Gravity Design

Since gravity design starts at the top of the building, the first thing to determine is the weight of all the roofing materials. You also need to know the potential snow load in your region. These loads determine the design of the rafters and beams. These elements are independent of the wall system and are covered in the main body of the building code. The rafters bear on the bond beam or top plate which bears directly on the cob walls. This is the first time in the design process that you will take into account the compressive strength and the density of your cob mix. The design then follows the load though the walls to the foundation.

The following section gives more details on how to approach the structural design of many elements of a cob building. The most critical topics to consider during gravity design are the compressive strength and density of the cob, lintels/headers, and bond beams. Other relevant factors include anchors and connections, cob wall height and thickness, reinforcing, end-of-wall boundary zones, openings, and niches, bottles, and rocks in the cob walls.

Lateral Design

Lateral design covers both earthquake and wind forces. The starting point of the lateral design for most residential buildings is once

again the roof sheathing, referred to in this context as the *roof diaphragm*. Although the connection between sheathing and rafters is important, the most important design feature of the roof diaphragm is the load path between it and the shear walls or *braced wall panels* (in IRC terminology). Shear walls and diaphragm are the primary structural elements that resist lateral forces, as shown in Figure 8.4.

Lateral design typically begins by determining the wind and seismic forces in your area based on the building code. The larger of those forces for each orthogonal direction of the building is used to design its in-plane and out-of-plane lateral-force-resisting system (including the cob walls) in that direction. All

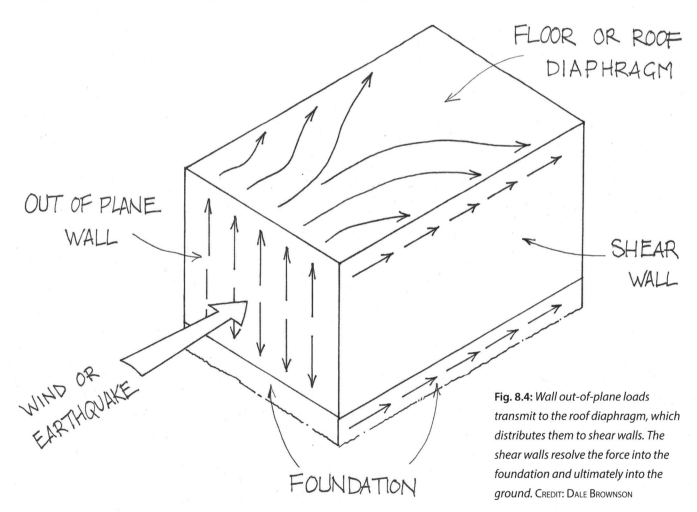

Fig. 8.4: *Wall out-of-plane loads transmit to the roof diaphragm, which distributes them to shear walls. The shear walls resolve the force into the foundation and ultimately into the ground.* CREDIT: DALE BROWNSON

of the topics discussed in the next section are relevant to the lateral design of cob buildings.

Important Considerations for Cob Structural Designs

This section explains what we consider to be the most important considerations for anyone conducting a structural design of a cob building. It should help engineers involved in those projects to quickly focus their attention on the subjects that matter most. For non-engineers planning to perform a prescriptive design based on Appendix AU, it's important to remember that designing a legal, permitted cob building

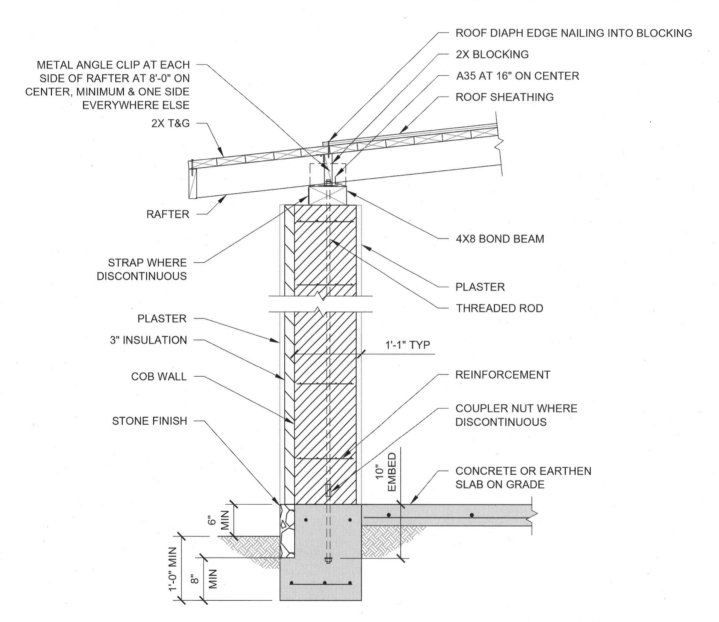

Fig. 8.5: *Structural wall section highlighting connection and reinforcement strategies suitable for all regions, including moderate and high seismic zones.* CREDIT: WILZEN BASSIG, VERDANT STRUCTURAL ENGINEERS

(or any building built to a certain quality standard) is not easy. We recommend dedicating at least 40 hours of your time to work through the process. If you have the means, hiring an engineer and/or architect can make the project go much more smoothly, especially when you hire someone already knowledgeable—or at least enthusiastic—about cob or other types of earthen construction.

Compressive Strength and Buckling

It is unusual for cob wall designs with standard wall thicknesses to be governed by gravity loads. That said, the slenderness limits of thin, tall cob walls have been under-studied. Buckling failure, or the sudden failure of a member under high compressive stress, is possible when cob walls do not meet the thickness and strength requirements in Appendix AU. Tests have been conducted on 10' tall, 12" thick (3 m, 30 cm), steel-reinforced cob walls similar to Wall E in Appendix AU (Figure 8.6) which supported more than 6,000 pounds (2,700 kg) per linear foot. (This test was part of the cob wall ASTM E119 fire-resistance test. A write-up will be published soon and will be available at www.verdantstructural.com.) Wall thickness and steel reinforcing can play a big role in easing buckling concerns.

The engineering equation to calculate buckling capacity, *Euler's Formula*, requires the input of the material's *modulus of elasticity* (MOE), which is a measure of resistance to being deformed. Tests of cob samples collected from around the world suggest that standard mixes can vary widely—from an MOE of 10,000 psi (69 mPa) to 40,000 psi (276 mPa)—which makes it important to act conservatively with an untested mix. These values can be found in the many tests cited in "Review of the Current State of Cob Structural Testing, Structural Design, the Drafting of Code Language, and

Material Based Testing Challenges" by Anthony and Kevin Donahue, SE.[8]

The compressive strength of cob mixes can also vary widely, from 50 psi (0.35 mPa) for low densities (high proportions of straw and low clay content tend to reduce compressive strength) to 350 psi (2.4 mPa) or more for dense mixes (with less straw). Appendix AU requires all cob mixes to have a minimum compressive strength of 60 psi (0.41 mPa, see Section AU106.6.1). Walls resisting in-plane earthquake or wind forces (which will be most walls in typical cob buildings) need to show a compressive strength of 85 psi (0.59 kPa) or greater. To be extra safe, we recommend designing a mix that is between 100–200 psi (0.69–1.38 mPa). (See Chapter 6 for instructions on mix design and sample testing.) However, it is possible to reach even higher strengths. For example, IBC Section 2109 Empirical Design of Adobe Masonry requires 300 psi (2.07 mPa) for adobe blocks. Although cob and adobe are very similar, there are a few reasons why adobe blocks are typically stronger, one being that adobe blocks usually contain much less straw.

You need to know your wall's load-bearing capacity while designing the building. Appendix AU Section AU106.8 contains the equation used to determine the maximum gravity force your cob wall can support; this is based on the compressive strength of your mix and the wall's height, thickness, and density. The two possible ways of determining compressive strength are: 1) be conservative with your design and assume the minimum required compressive strength of 85 psi (0.59 kPa), or 2) test your cob mix prior to design.

Hana Mori, a professor in the Department of Engineering at the University of San Francisco, conducted research (not yet published) into the ideal aspect ratios of cob samples for compressive testing. She determined that the best ratio is around 2 tall:1 wide and thick. Taller, thinner samples test both pure compression and shear-based compression whereas shorter, stouter samples test only pure compression. This is both more conservative and more representative of how cob walls actually fail in reality. As a result of Hana's research, the compressive testing protocol was revised and approved for the 2024 Appendix AU as follows: "Five samples of the proposed cob mix shall be placed moist to completely fill a 4-inch by 4-inch by 8-inch form and dried to ambient moisture conditions. Samples shall be constructed, dried, and tested with the long dimension vertical."

CRI board member David Wright did his doctoral research at the University of Tulsa on the mechanical engineering properties of cob.[9] One important finding was that the moisture content of test samples has a significant effect on their compressive strength. The drier the samples, the greater the compressive strength. A sample may feel or look dry and still not have reached ambient moisture conditions. Care while forming and handling the small test samples also plays a large role in the strength results. Many other factors can affect compression strength, including density, clay content, clay type, aggregate type and content, mixing technique, and more. Unfortunately, we don't yet know how large these effects may be due to the lack of studies that have isolated each variable.

Bending Strength

Bending strength, also known as *flexural strength* or *modulus of rupture*, is a measurement of a material's capacity to withstand bending forces. Technically, it is the maximum force that the tension side of a bending member can withstand. We do not recommend using cob as a horizontal beam without a confident engineering analysis. Cob walls act like a beam vertically when resisting earthquake and wind loads in both out-of-plane and in-plane action as shown in Figure 8.2. Therefore Appendix AU, Section AU106.7 requires a modulus of rupture of 50 psi (0.34 mPa) for all mixes used in braced wall panels.

Anthony conducted full-scale cob wall out-of-plane testing at—and with the enormous help of—Quail Springs Permaculture in 2019. We used a waterbed bladder filled with air to simulate horizontal wind and seismic forces on both a straw-reinforced specimen similar to Wall A (Figure 8.6) and a vertical steel-mesh-reinforced wall similar to Wall 2 (Figure 8.6). The results of these tests influenced the out-of-plane requirements of Appendix AU.

When conducting a prescriptive design based on Appendix AU, you do not need to consider any bending behavior beyond meeting the requirements of Table AU105.3, which specifies the effective reinforcement strategies in different wind and seismic risk categories, and Section AU105.3.4, which mandates reinforcement on either side of window and door openings.

Density

In general, the higher the ratio of sand in a cob mix, the denser it will be. The more straw is in the mix, the less dense it will be. Denser cob mixes are typically stronger in compression while less-dense mixes can be stronger in bending due to increased tensile strength from straw microfiber reinforcing. The cob mixes we have tested had an upper density limit of 110 pcf. The lowest density we've worked with structurally has been approximately 70 pcf. We've seen cob mixes with densities as low as 50 pcf, but these had compressive strengths below what's required by Appendix AU. See "Optimizing Cob's Thermal Performance in Your Climate Zone" in Chapter 3 for more discussion of low-density cob mixes.

Below 50 pcf, a mixture of clay and straw should not be considered cob but rather *light straw-clay* (LSC), also known as *light earth method* (LEM) or *slipstraw*. LSC is a non-structural wall infill material that is placed between posts or studs and governed by IRC Appendix AR. We recommended *Essential Light Straw Clay Construction* by Lydia Doleman for further reading on the subject.

When conducting a prescriptive design based on Appendix AU, there are technically no restrictions on density, though, as discussed above, density and compressive strength are related, and a minimum compressive strength is specified. All sections of Appendix AU, including those concerned with thermal behavior (AU 109.2) and dead loads (AU106.3 and AU106.8) assume the density of cob to be 110 pcf unless proven otherwise.

Height:Thickness Ratio

The height:thickness ratio is the height of the wall from the top of the foundation to the bottom of the bond beam divided by the thickness of the wall from the interior face to the exterior face, not including finishes. Tall, thin cob walls are more subject to failure in earthquakes, which can exert tremendous out-of-plane bending forces. To prevent this, Appendix AU specifies height and thickness limitations for each reinforcement system. Other standards such as the New Zealand Earthen Building Standard, NZS 4299 and the Appendix of ASTM E2392: Design of Earthen Wall Building Systems more generically require that the ratio of a cob wall's height to thickness can be no greater than 8:1 in areas of medium seismic risk, and no more than 6:1 for high seismic risk.

Appendix AU Section 105.2(5) states that wall thickness, excluding finish, shall be not less than 10" (25 cm), not greater than 24" (60 cm) at the top two-thirds, not limited at

the bottom third. Wall taper is permitted in Section AU106.5(1). The reason for limiting the thickness of the upper portion of the wall is that thicker walls exert more lateral force on the building during earthquakes. Ten inches is considered a safe minimum thickness for load-bearing walls, although low non-structural walls such as indoor or outdoor privacy walls can be thinner.

Aspect Ratios and Shear Walls

The ratio of a wall's height to length is referred to as its *aspect ratio* in engineering texts and codes, including Appendix AU. Aspect ratio is especially important for the design of shear walls (called *braced wall panels* in the IRC), which are walls used to resist wind and earthquake forces. Aspect ratios also differentiate the "column behavior" of a wall section that is short in length relative to height from the "shear wall behavior" of one that is longer relative to height. When a wall is short in length compared to its height, seismic and wind forces will induce bending as if it were a column or beam *cantilevering* (extending without other support) out of the ground. This results in high tensile and compression forces at the top and bottom of the wall. One of the practical consequences of this is that the distance between windows or doors in a cob wall should typically be equal to or greater than the height of the opening—or half of that height, depending on the reinforcement system. Appendix AU Section AU105.3.5 Minimum Length of Cob Walls prohibits walls less than 2'6" (76 cm) long between windows and doors. For any wall less than 4'0" (1.2 m) long, it requires vertical reinforcing with ⅝" threaded rod embedded in the cob wall 4" (10 cm) from each end of the wall. Walls less than 2'6" (76 cm) long are considered cob *columns* and are not recommended without an engineered design. Another option is to use a wood, steel, or concrete column between windows that are close together.

A longer wall section, with a length equal to or greater than its height, has more stability to resist horizontal loading through shear. This is much more favorable for the cob material. Shear walls are typically needed on all four sides of a rectangular building, or on each "side" of any curvilinear or non-rectangular building idealized as a rectangle or group of rectangles. Note that cob shear walls can be curved in plan (Appendix AU Section AU106.11.3) as well as straight.

Cob shear walls are wall sections without windows, doors, embedded bottles or stones, niches, or any other cavities. To do their job, they need to be firmly anchored to the bond beam above and, usually, the foundation below. Appendix AU Table AU106.11(1) describes several types of cob shear walls. Depending on the reinforcement system used, the maximum aspect ratio is either 1:1 or 2:1. Tables AU106.11(2–5) list the minimum required length of cob shear walls depending on wind speed and Seismic Design Categories. Also note that the maximum wall height for shear walls is either 7' or 7'6" (2.1 or 2.3 m), depending on the wall type. Appendix AU limits the heights of cob walls (non-shear walls) to a maximum of 8' (2.4 m) without an engineered design. These height limits are based on the heights of the sample walls tested, and may increase in future iterations of the code as more testing is performed. In the meantime, engineered designs could reasonably allow for taller walls with the appropriate analysis. This is especially true for wall types permitted in higher seismic zones that are being used in lower seismic zones.

Curves

Walls curved in plan have out-of-plane structural benefits by virtue of their geometry, meaning that they can be stronger than a flat wall of similar size of the same material. In the case of fully round buildings, it is important to remember that although cylinders are very strong, every time you add a door or a tall window, the walls are broken into partial cylinders, which are weaker. Appendix AU conservatively allows for some wall geometric shape benefits in AU106.11.3. A custom design, especially one performed using a structural modeling

Rocking Shear

Anthony was a primary participant in seismic testing on full-scale cob walls with the California Polytechnic State University and Santa Clara University. The results showed that straw-reinforced cob walls could remain intact even when they were lifted partially off the ground by horizontal, in-plane forces applied to the top of the wall. All the researchers involved were surprised by this result. This means that all cob walls—including those reinforced with straw alone—could theoretically take advantage of *rocking-shear behavior*. Rocking shear allows a wall to resist the uplift demands of wind and seismic loading using the weight of the wall itself. This is an uncommon approach in structural designs using conventional materials, where full foundation anchorage is typically required.

We want to stress that this is a hypothetical solution; it was first proposed by the late Dr. Mark Aschheim, PE, an invaluable strawbale building researcher. Rocking shear was also the resistance type used in the half-scale Stanley Park shake table tests at the University of British Columbia.[10] More research is necessary before we can use rocking shear in the design of cob buildings.

program, would make it possible to take greater advantage of these effects.

Buttresses

Buttresses can assist in bracing a wall against out-of-plane forces. A buttress does not need to be an isolated feature—a wall that is perpendicular to another wall can act like a buttress. In some traditional adobe buildings, the walls that extend beyond the corners of a building act as buttresses. If you and your engineer are able to credit rocking-shear behavior as described above, buttresses could add additional wall weight to counteract lateral forces. Appendix AU permits the use of buttresses, but it does not offer prescriptive design options for these features without an approved engineered design.

Reinforcing

Cob walls are always reinforced with straw. Therefore, cob walls that do not contain steel or other non-straw reinforcing should *not* be referred to as *unreinforced*. The role of straw in cob is similar to that of the steel, glass, and synthetic microfibers commonly used to reinforce conventional concrete. Many laboratory tests[11] have shown increased ductility in samples containing straw microfiber vs those without it. Ductility can be an important seismic design feature, as it describes how flexible a material is and if it is able to dissipate energy by deforming but not breaking.

Where wind and/or seismic demand loads are high, straw microfiber reinforcing alone is often not enough. Additional reinforcing is required, using a material with higher tensile strength or longer unit length than straw. Steel has been most thoroughly tested, though many other materials are possible, including bamboo and polypropylene mesh. Polypropylene mesh is represented in New Zealand's Earth Building Standards.[12] There is some concern that as

bamboo dries and shrinks it could leave weakening cavities between the materials; this is a subject that needs further study.

There are several ways to use steel to reinforce a cob wall. These include: 1) vertical bars running through the wall from foundation to bond beam; 2) a grid of both vertical and horizontal rebar; 3) steel mesh embedded vertically in the middle of the wall; 4) two layers of vertical mesh embedded near each surface of the wall; and 5) strips of mesh laid horizontally. These strategies are represented by Walls A, B, C, D, E and 2 shown in Figure 8.6, which are the only types of steel-reinforced cob walls that have been tested for seismic resistance so far. See Chapter 3 for a discussion of the durability of steel embedded in cob walls.

The wall examples in the illustrations are the only options available to prescriptive designs using Appendix AU. Table AU105.3 specifies which wall types can be used in each Seismic Design Category. None of these options

Sasha Rabin and team placing cob around vertical mesh for a cob ASTM E119 fire test in Bryan, TX. CREDIT: JOHN ORCUTT

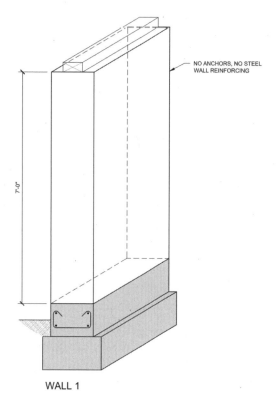

NO ANCHORS, NO STEEL
WALL REINFORCING

7'-0"

WALL 1

include solid horizontal bars; we recommend horizontal mesh instead. In general, meshes are recommended over bars because their smaller diameter and more distributed nature relate more effectively to the lower-strength clay material and straw reinforcing. During an earthquake, embedded bars are more likely to damage a cob wall by cutting through the soft material.

Vertical meshes are one of the strongest seismic and wind reinforcement systems you can use in a cob wall. If you are not planning to install vertical meshes, horizontal meshes are highly recommended. Strips of mesh are easy to install horizontally and add a lot of shear strength, providing a big bang for the buck. Horizontal mesh can be used to *contain* vertical bars, as shown in the photographs on the following pages. The best kinds of mesh to use in order of decreasing strength are steel, polypropylene, and natural fiber fishing nets.

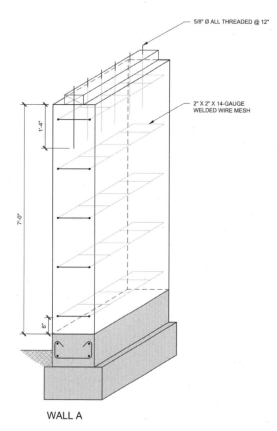

5/8" Ø ALL THREADED @ 12"

2" X 2" X 14-GAUGE
WELDED WIRE MESH

1'-4"

7'-0"

6"

WALL A

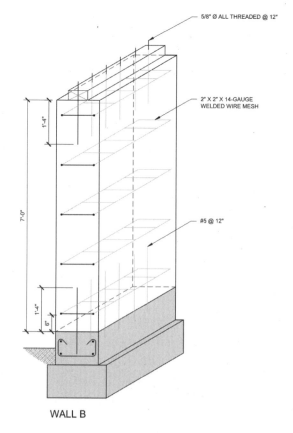

5/8" Ø ALL THREADED @ 12"

2" X 2" X 14-GAUGE
WELDED WIRE MESH

1'-4"

7'-0"

#5 @ 12"

1'-4"

6"

WALL B

Fig. 8.6: *Different wall-reinforcement strategies represented in Appendix AU.*
Credit: Rachel Tove-White, Verdant Structural Engineers

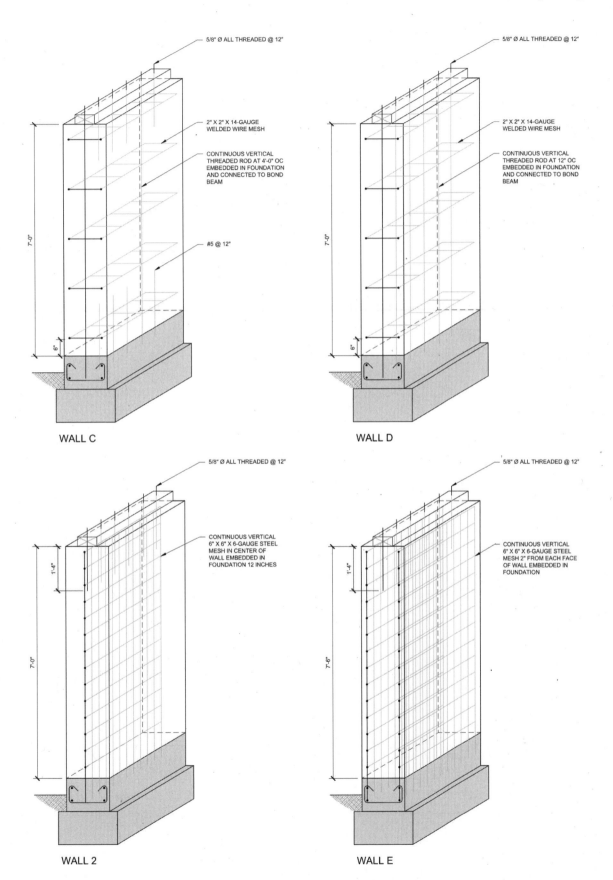

5/8" Ø ALL THREADED @ 12"

2" X 2" X 14-GAUGE
WELDED WIRE MESH

CONTINUOUS VERTICAL
THREADED ROD AT 4'-0" OC
EMBEDDED IN FOUNDATION
AND CONNECTED TO BOND
BEAM

#5 @ 12"

7'-0"

6"

WALL C

5/8" Ø ALL THREADED @ 12"

2" X 2" X 14-GAUGE
WELDED WIRE MESH

CONTINUOUS VERTICAL
THREADED ROD AT 12" OC
EMBEDDED IN FOUNDATION
AND CONNECTED TO BOND
BEAM

7'-0"

6"

WALL D

5/8" Ø ALL THREADED @ 12"

CONTINUOUS VERTICAL
6" X 6" X 6-GAUGE STEEL
MESH IN CENTER OF
WALL EMBEDDED IN
FOUNDATION 12 INCHES

1'-4"

7'-0"

WALL 2

5/8" Ø ALL THREADED @ 12"

CONTINUOUS VERTICAL
6" X 6" X 6-GAUGE STEEL
MESH 2" FROM EACH FACE
OF WALL EMBEDDED IN
FOUNDATION

1'-4"

7'-6"

WALL E

A test wall under construction at Santa Clara University. Horizontal reinforcing mesh can be easily inserted between lifts. CREDIT: ANTHONY DENTE

Little or no testing has been done to determine the strength of lap-spliced bars or bar bonding and development lengths in cob walls. *Lap-splicing* refers to the required length of overlap between pieces of discontinuous reinforcing. *Development length* is the length of embedment required to properly bond reinforcing with the surrounding material and transfer forces from one material to another. Therefore, we do not recommend straying from the tested assemblies shown in the illustrations, which utilize threaded steel rods. These can be spliced where necessary with threaded couplings, although standard lengths are sufficient in most cases to be continuous from foundation to bond beam.

The first image shows a wall similar to Wall D after peak loading in an in-plane reverse cyclic seismic loading simulation. The second is the same wall after the outer layer of cob was chipped away following the test. The third image is of a similar wall in which the horizontal mesh was replaced with horizontal bars. This image is taken shortly after peak loading. You can see from the first two images that the horizontal mesh contained the vertical bar and held on to more cob during testing, whereas the vertical bars sliced more easily through the cob in the assembly with horizontal bars instead of mesh. The vertical bar coupler mechanisms implemented due to threaded rod supply issue are visible. These couplers had an extra-large diameter, but even without this size increase, the same slicing result is expected. CREDIT: JOHN FORDICE

Anchors and Connections

One of the more complicated engineering aspects of a cob design is how to transfer loads in and out of cob walls. This is always a concern at the tops and bottoms of walls, but the transfer can also occur at mid-height if, for example, there is a change of wall systems part way up a wall. Connection strength and anchor shear in clay wall systems is an under-studied subject. The full-scale cob wall tests conducted to date, although not primarily designed for this purpose, all tested the strength of anchors at the top and base of the walls. Appendix AU specifies anchorage options in Tables AU105.3 and AU106.11(1) along with their associated reinforcement systems for wind and earthquake resistance. Where vertical steel bars are used for reinforcing, the strongest design is to extend them from the foundation to the bond beam. If you are using a wood bond beam, the easiest way to anchor the wall to the bond beam is to use threaded bars (instead of rebar) that can receive a washer and nut at the top of the wall. See Chapter 12 for more on roof and floor framing that bears directly on cob walls.

Appendix AU does not include prescriptive design options for anchors used to ledger lofts part of the way up cob walls, or similar side-of-wall applications. There is some strength data on the subject in both IBC Section 2109 on Adobe Masonry and the New Zealand Earthen Building Standard NZS 4297.[13]

End of Wall/Boundary Zone

The boundary zone at the end of a wall is an important area of design because it is responsible for resisting several forces: 1) the compression and tension forces concentrated at wall ends from wind and seismic loading, 2) a larger than usual out-of-plane force if it is next to a window which transmits wind or seismic loads horizontally, and 3) the concentrated gravity load of a window or door header or lintel. Shear wall enhancements intended to address the forces described in item 1 are not included in Appendix AU and are mentioned here as a potential design enhancement for engineers and researchers. This is an under-studied topic, though it is expected that strategies used in concrete wall construction, such as concentrating reinforcing at boundary zones, could also work for cob. The strategy used in Appendix AU to resist out-of-plane forces around windows and doors is to embed threaded rods in the cob wall on either side of openings, as described below.

Lintels/Headers

Lintels, which can also be referred to as *headers,* are beams that span above windows and doors. They support the weight of the cob above the opening, as well as any roof and/or floor loads that bear on that cob, and they transmit those loads into the wall on both sides of the opening. In cob buildings, lintels are most commonly made of wood or concrete. Table AU106.10 contains sizing requirements for wood and concrete lintels based on an array of circumstances (the size of the building, the thickness of the wall, the length of opening, and the height of cob above the opening), and it assumes that the weight of the roof bears on the cob wall above the lintel. In all cases, lintels must extend 1' (30 cm) beyond the opening on both sides. Section AU105.3.4.3 requires that ⅝" threaded rod be embedded in the foundation and run up through the cob walls 4" (10 cm) on both sides of window and door openings. It states that, "the threaded rods shall be embedded in concrete lintels or pass through a drilled hole in wood lintels."

Bond Beams

A *bond beam*, which can also be referred to as a *top plate* or a *collector* when built out of wood, is a structural beam connecting an entire building together at the top of the walls. Bond beams help transfer both compressive and shear forces from the roof into the cob walls. They also act as collectors to transfer earthquake or wind forces from non-shear wall regions, like over windows, to stronger solid cob wall lengths elsewhere in the building. To serve this function they need to be continuous in tension over their entire length. With wooden bond beams, this can be accomplished by connecting across discontinuities with steel strapping. (Where there are discontinuities above a curved section of wall, the strap must be placed on the outer edge of the curve.) Bond beams can also be designed to span directly over windows (with no cob between the bond beam and the window), reducing the demand on the lintel below or replacing the lintel. Appendix AU Section AU106.9 specifies the required dimensions for both wooden and concrete bond beams and how they are to be anchored to the cob walls. It also specifies how roof framing must be connected to bond beams.

For concrete materials like cob, it is common for gable and shed roof *end walls* (the triangular wall sections under the ends of a gable or shed roof) to transition to wood framing. Appendix AU Figure AU106.9.6 shows a recommended detail of this condition. Gable and shed roof end walls that transition to wood framing can create an unstable hinge at the transition. To resist out-of-plane forces, the bond beam should be continuous and straight for the entire length of the gable or shed roof end wall, as described in Section AU106.9.6. Another option is to place the bond beam at the top of the gable end wall, directly under the rafters, and bring cob all the way up to it. This is a cleaner structural detail, with fewer material transitions, but it does introduce some complexities of its own. In this case, the level of the bond beam will change as it follows the top of the walls around the building, and segments may meet each other at complex angles. When designing this type of wood bond beam in medium or high seismic zones, we recommend that you consult an engineer about appropriate bond beam and/or roof diaphragm strapping to ensure bond beam continuity.

Niches, Bottles, and Stones

Don't underestimate the decrease in both shear strength and compressive strength caused by wall cavities such as niches and embedded bottles or stones. Wider walls (24"/60 cm) can likely take more wall reduction than thinner walls (12"/30 cm), but it is difficult to state a

Fig. 8.7: *Top-of-wall connections to roof framing and roof diaphragms suitable for all regions, including moderate and high seismic zones. Note the strap recommended at discontinuous top plates.*
CREDIT: WILZEN BASSIG, VERDANT STRUCTURAL ENGINEERS

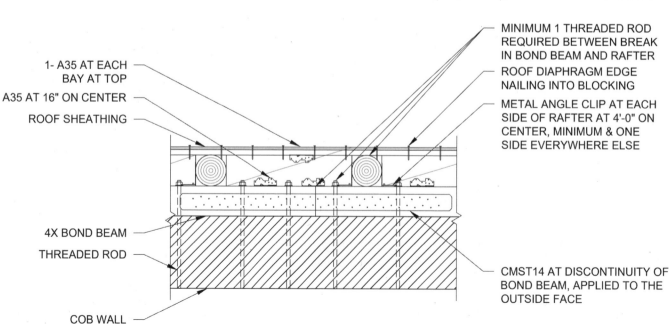

1- A35 AT EACH BAY AT TOP
A35 AT 16" ON CENTER
ROOF SHEATHING

MINIMUM 1 THREADED ROD REQUIRED BETWEEN BREAK IN BOND BEAM AND RAFTER
ROOF DIAPHRAGM EDGE NAILING INTO BLOCKING
METAL ANGLE CLIP AT EACH SIDE OF RAFTER AT 4'-0" ON CENTER, MINIMUM & ONE SIDE EVERYWHERE ELSE

4X BOND BEAM
THREADED ROD

CMST14 AT DISCONTINUITY OF BOND BEAM, APPLIED TO THE OUTSIDE FACE

COB WALL

one-size-fits-all solution. Appendix AU limits niches and other voids to 12" in width and height and 25% of the wall thickness in depth and requires that they be located in the upper two-thirds of the wall. Surface voids that exceed those limits are considered openings, so they need a lintel and boundary reinforcement just like a window opening. Appendix AU also states that surface voids such as niches and bottles are prohibited in braced wall panels. When designing with Appendix AU, it is recommended that spacing between niches be at least 2'6" (76 cm). See "Aspect Ratios and Shear Walls," above, for more on this.

Large stones embedded in cob walls interrupt the otherwise-continuous matrix of clay, small aggregate, and straw microfiber in a manner analogous to a niche. Walls with steel reinforcing rely less on straw reinforcing, especially as seismic risk decreases. For simplicity, it is recommended that embedded stones be treated like niches as described above.

It is likely that a custom-engineered design can justify some amount of force transfer around niches, bottles, and other cavities. One way to strengthen a cob wall with an abundance of bottles is by inserting a vertical steel mesh into the wall, continuous from foundation to top-of-wall, as shown in Wall 2 in Figure 8.6. If the mesh size is large enough, there will be space for bottles to be inserted between the wires. Alternatively, some parts of the mesh can be removed to make space for bottles. Make sure there is at least 2" (5 cm) of cob cover on all sides of the mesh wires to prevent corrosion. Especially in regions with higher wind or seismic forces, engineering of a wall system like this is recommended.

Recent Breakthroughs and Research Needed

There have been significant advances in cob building research during the past 10 years. In 2015, in preparation for developing a cob building code, we reviewed all of the existing testing and research literature on the material properties of cob. There was little to be found, and nearly all of what did exist was small-scale batch testing of compressive and bending strengths. Prior to 2018, no full-scale laboratory testing had been conducted to determine the effects of seismic and wind forces on cob walls. Much more high-quality testing had been done on other earthen wall systems, such as adobe. Though some of these results can be

Cob test wall specimens at Santa Clara University. For this series of tests, materials and experienced builders were brought to the university to take advantage of the equipment and tools available at the lab. Finding space to build and dry the test walls and transportation of the heavy samples was challenging in a lab that typically tests more conventional materials.

Credit: Anthony Dente

generalized to predict the behavior of cob walls, it is important to keep in mind the fundamental differences between the block construction of adobe and the semi-monolithic nature of cob.

With the assistance of many others, Anthony helped propose, design, build, execute, and analyze two rounds of in-plane reverse cyclic testing and one round of out-of-plane testing on full-scale cob sample walls demonstrating a range of reinforcement options. These tests gave us the first reliable data on cob wall performance in earthquakes, which in turn were among the primary drivers for the development of Appendix AU. Other experiments in the last decade have shed light on cob's thermal properties, fire-resistance, and best practices for sample strength testing.

Despite these developments, cob is still very under-tested when compared to conventional building materials. This is probably due to three primary factors. First, since it is difficult to privatize cob, no profit-driven industry has arisen to invest in testing it. Second, cob is lower strength and the general engineering philosophy of our time is geared toward the highest-strength, highest-performing materials. And finally, many of cob's properties are incompatible with current lab testing equipment, protocols, and environments. Can you imagine the logistical hurdles to bringing tons of clay and sand into a laboratory, arranging for builders to construct test walls over a period of a week or more, and then leaving the test walls in place for months while they dry thoroughly? Keep in

Dr. Daniel Jansen, Professor at California Polytechnic University, and Art Ludwig of Oasis Design examine cob walls during in-plane and out-of-plane testing at Quail Springs Permaculture in 2019. These tests were conducted in a rural setting where materials, builders, and drying space were more accessible than at a university, though the testing equipment needed to be trucked in and assembled on site. This caused complications—including a failure of the data equipment during one of the four wall tests—which was a big loss considering the shortage of testing data available. CREDIT: ANTHONY DENTE

mind that these full-scale cob walls can easily weigh over 5,000 pounds (2,250 kg). They are likely to exceed the capacity of the laboratory's forklifts and other equipment, so special rentals are required. To avoid these complications, we have also devised portable lab equipment that can be set up where building is easier, drying faster, and space at less of a premium, but that's complicated too. Logistical issues have caused major headaches in many of the tests we've been involved with—even when we saw the problems coming ahead of time.

The future of cob will be much brighter once we have a complete technical understanding of the complex nature of the clay, sand, and straw matrix. We need to know more about different types of clay and clay soils and their effects on cob mix properties. We want to develop inexpensive, on-site procedures to determine cob's strength and quality. We need to understand best practices for connecting cob walls to other parts of a structure. We need to investigate options for reinforcing cob with materials other than steel. We need to develop more strategies for insulating cob walls. These and many other areas of research need substantial investment in order to overcome conservative estimations, code hurdles, misunderstandings, and misinterpretations.

If you have the opportunity to engage in cob testing, it is important to think hard about all the variables that could be relevant to future practitioners and researchers and to document them carefully. A number of previous researchers failed to keep track of important details such as mix proportions, clay source, moisture content, and so on, requiring practitioners to make conservative assessments and other researchers to duplicate their efforts. If you are a professor or graduate student in materials science or structural engineering, have access to a lab or funding, have technical grant writing experience, or anything else that could assist future cob testing, please reach out to the authors or the Cob Research Institute at cobcode.org.

Recommended Structural Design Strategies for Engineers

This sidebar complements the information in Chapter 9, "Building Codes and Permits."

Appendix AU is a prescriptive, or *non-engineered*, code, which means it has everything you need to conduct a full building design if you are willing to accept the limitations that come with such a code. If you are unable to fit your structural design within the bounds of Appendix AU, and you have the resources to hire an engineer to conduct a custom structural design, we recommend the following approach.

When engineering a cob building that falls outside the scope of Appendix AU, Anthony takes a three-step approach:

Step 1: Generate allowable shear values based on available research (preferably larger-scale research) and factors of safety using industry norms.

Step 2: Use the compressive strength and modulus of rupture determined from tests of your specific cob mix to conduct fundamentals of engineering analysis.

Step 3: Use and cite any codes or standards that apply to or can be used as an analog for cob, even if they are not law in your area.

When looking into existing research, we recommend starting with the paper "Review of the Current State of Cob Structural Testing, Structural Design, the Drafting of Code Language, and Material Based Testing Challenges," that Anthony wrote with his business partner and mentor Kevin Donahue, SE, for the EarthUSA conference in 2019.[14] This paper summarizes most of the cob testing completed at that time. The accompanying poster[15] breaks down an array of compressive and bending strength test results. It also includes a discussion of the safety factors and ☛

seismic parameters such as R-factors used in the development of Appendix AU. Simply stated, one method for running engineering calculations involves applying factors of safety to a material's maximum tested capacity. The resulting forces are referred to as the material's allowable capacity and these values are used in Allowable Stress Design (ASD) analysis. Industry standard safety factor norms for generating allowable capacities can be as high as 10 when using small-scale testing data, while for full-scale wall testing the range is between 2 and 5. CRI's website also contains all of the research cited in the development of Appendix AU.[16] The allowable values you decide on can then be used in a regular IBC-based design approach, or at least as far down that path as you are able to go using another material as an analog. Concrete masonry unit codes are recommended as the closest analog; because cob is constructed in lifts, it is semi-monolithic in a way that is analogous to masonry.

The use of your mix's mechanical properties to conduct fundamentals of engineering analysis on different wall behavior topics can end up looking very similar to step 1. If steps 1 and 2 blend at some point, that's fine.

When earthquake or wind forces are transmitted to the top of a shear wall from the diaphragm, one way the wall resists these forces is through tension on one side of the wall and compression on the opposite side; this is a structural behavior known as a *couple*. There are no accepted standards for modeling these tension/compression coupling forces in a cob wall, so a practitioner must use their judgment on whether their system would take a rectangular shape vs a triangular shape. More research is needed in this area. The best way to use engineering fundamentals to design the most efficient wall system is by letting a finite element analysis program do some of the work for you. If you are comfortable with a program like ETABS, it can be very helpful for curved, tapering shapes.

The development of an engineering material specification is complicated for a highly variable, site-sourced material. We recommend the approach taken by Appendix AU which attempts to be as hands-off as possible on mix design by limiting mixes to those that can pass the shrinkage, compression, and, when needed, bending tests.

There is helpful information in Chapter 9 on the use and citation of existing relevant codes. Keep in mind that New Zealand is an English-speaking, seismically active country with an excellent building code including earthen building systems, but New Zealand's building code differs substantially from those used in the US. IBC 2109 on Adobe Construction is actually part of the IBC, but it's not permitted in high seismic regions and it is slated for removal by The Masonry Society (TMS)—and it is obviously not cob. Anthony is currently on a committee tasked to author a formal adobe chapter in TMS 402, though it is unknown when that will be developed and published. Appendix AU should be your primary reference if you are working in the US; it can potentially be used outside the US as well (if accepted by the relevant code enforcement entities).

Once you complete these three steps, you will likely have three or more outcomes, and it is up to you to determine which one *conservatively*—though not *too* conservatively—fits reality and meets the intent of the building code and the needs of the builder most efficiently. The first time you approach this three-step process, it will not be a light lift—especially if the building is larger or not a simple shape—but the process will get easier over time as you repeat the process.

Chapter 9

Building Codes and Permits

BUILDING *CODES* are laws regulating how one is allowed to build in a particular region. A building *permit* is a license issued by a building department for the construction of an approved building design within a specific timeframe, and (usually) in a specific place. The permitted building is then inspected for compliance with the approved plans during construction. This design approval, permit, and building inspection process requires a fee (from hundreds to tens of thousands of dollars, depending on location and the size and type of building), and carries with it many requirements. Permitted buildings must comply with the local building codes in their design, materials, and construction (although see below for a discussion of "Alternative Materials and Methods Requests," or AMMRs). They also need to meet other health and safety requirements pertaining to water sources and uses, sewers and septic systems, fire and emergency access, means of egress, and (possibly) separate electrical, plumbing, and mechanical codes. In most places, additional planning and zoning regulations govern what can be built where, as discussed below.

Keep in mind that just because it might be difficult for you to build a legal cob building, that doesn't mean cob building is illegal. You will need to research the relevant laws, procedures, and accepted practices wherever you intend to build. The entire regulatory system can vary greatly from jurisdiction to jurisdiction. The information in this chapter pertains primarily to the United States, and secondarily to Canada, and it may be helpful in other regions with similar building regulation systems.

Obtaining a permit for a cob building can be challenging, but is made easier by first understanding all that is involved. Revisiting the purpose of building codes can offset some of the frustration with the process if it arises.

Building Codes

The primary purpose of building codes is to establish minimum requirements that provide a reasonable level of safety, health, and welfare. However, there is still much work to be done for the codes to fully achieve this goal, given that buildings have a wide range of impacts beyond the traditional hazards addressed by codes, especially in the age of climate change and new public health awareness. Many of these impacts are external to the safety of the building itself—in the extraction of resources, manufacture and transportation of materials, energy sources, and the generation of waste during construction and a building's eventual disuse.

In the US, anyone can propose changes to the building code, though that can be challenging for someone new to the process. It is common for professional organizations (like the National Association of Home Builders, building materials industries, and trade associations) to play a major role in code development. Building officials, whose public-interest charge is to enforce the building code locally, also play a large role in US building code development—including having the final vote on code change proposals. Design and building professionals and other interested groups such as universities and researchers also play a part, and public-minded individuals and nonprofits are increasingly stepping up to participate in this important process.

The International Code Council (ICC) publishes and oversees the development of model codes (the *I-Codes*), which for the purposes of our discussion are primarily the *International Residential Code (IRC)* and the *International Building Code (IBC)*; these are the basis for nearly all building codes in the US. The word "International" in these titles is mostly aspirational, though several of the I-Codes are currently used in Mexico, Ghana, and some Caribbean countries. In Canada, the equivalent is the National Building Code of Canada (NBC). In the US, states and/or local jurisdictions adopt model building codes, often with modifications, into law. Certain states mandate some or all aspects of building codes statewide; in others, individual municipalities and counties have authority over what is adopted, often adding unique allowances or requirements of their own. This process is similar in Canada for each province and territory.

The IRC is primarily a *prescriptive* code, describing what must be done to meet minimum requirements, mainly for widely accepted and commonly used materials and systems (though alternatives such as strawbale, light straw-clay, and cob have in recent years been added as Appendices). This means it isn't necessary to have a licensed design professional create a building's plans as long as there aren't significant departures from what is prescribed.

The IBC, on the other hand, is mainly a *strength-based* code for structural design, and contains performance and/or prescriptive language for all non-structural requirements. It is geared toward design professionals and commercial, institutional, and large-scale multifamily residential buildings, allowing custom designs within established parameters.

Some countries (including Canada, Australia, New Zealand, and Scotland) have mostly *performance-* or *objective*-based codes, which provide detailed processes to determine whether something meets the intent of the code. These typically also provide prescriptive paths for widely accepted building materials and systems, like Part 9: Housing and Small Buildings in the NBC. A similar code in the US is the ICC's International Performance Code for Buildings and Facilities.

Though the IRC is not perfect, its intention is fundamentally sound—a stand-alone building code with all the requirements for one- and two-family dwellings and townhouses, including electrical, plumbing, HVAC, and other systems. It allows homeowners and builders who understand the code to design their own buildings and submit them for permits without the help of design professionals. However, the technical code language can be difficult for non-professionals to understand. This is partly because the code must be unambiguous and enforceable during plan review and inspection. Many people therefore choose to work with a design professional if they have the means to do so. This can vary from guidance as needed, to full responsibility for the design, plans approval, bidding, and assistance during construction (see Chapter 11 for more on working with professionals).

Both the IRC and the IBC are revised on a three-year cycle, as are most state and local building codes. The NBC of Canada is revised every five years. The 2021 IRC is the first edition with provisions for cob, *Appendix AU: Cob Construction (Monolithic Adobe)*. Appendices to the IRC are expressly optional, meaning a state or local jurisdiction must specifically adopt them to be enforceable. If IRC Appendix AU has not yet been adopted by your state or local jurisdiction, you can advocate for its adoption. You can also propose its use for your project to your local building official. With good presentation, this well-developed appendix published by the ICC will likely be accepted. Earthen building codes and standards exist in several parts of the world including New Zealand and Germany. In most places, for now, cob will be treated as an alternative material, which we will discuss later in this chapter.

Most I-Codes, including the IRC and IBC, have accompanying commentary that explains code requirements and their intent, along with compliance examples. The commentary can help building officials understand and approve designs not explicitly allowed in the code, especially when an alternative meets the intent stated in the commentary. This commentary is not included with most printed or PDF versions of the code, and it is not contained in the free version at codes.iccsafe.org. When reviewing codes during your design process, it's best to purchase a copy with commentary from the ICC. The 2021 IRC Appendix AU with commentary is included in this book and offered for download on CRI's website, cobcode. org; IRC Appendices AR: Light Straw-Clay Construction and AS: Strawbale Construction are available from the California Straw Building Association (CASBA) at https://www.strawbuilding.org. You'll need the edition of the code that is (or will be) in force in your jurisdiction on the date or expected date of your permit application. Your local building department likely has the code and commentary, and you can ask to view it. Importantly, the 2024 update to Appendix AU includes fire-resistance rated cob walls. Even if your jurisdiction is still using the 2021 version of Appendix AU, you can propose the 2024 version for the purpose of a fire-rated cob wall if one is required for your project.

Building Permits

In the US and Canada, the permitting process can vary significantly from region to region. Building permit applications are submitted to either a county or municipal building department, or, in some states, to a state agency; each entity usually has its own rules. Some jurisdictions do not require permits (usually rural areas with sparse populations), while most moderately and densely populated areas do. Some jurisdictions that require permits require either a licensed engineer's or architect's stamp on the building plans, particularly in high-seismic, high-wind, and flood- or fire-prone areas. Some only require an engineer's stamp, and some require both.

Most places in the US require a *planning submittal* prior to review of the *building submittal*. It's important to obtain planning approval first, before spending the time and money to fully develop the design into a set of construction/ permit drawings. The planning department enforces regulations related to siting, density, building size and height limits, setbacks from property lines, allowed uses, and other requirements. Most building and planning departments have a list of submittal requirements on their websites that includes the size of drawings, the types of drawings (like site plans, floor plans, etc.), as well as support documents, such as stamped and signed structural calculations and energy code documentation. It is very important to read and understand each requirement before beginning the application process.

To apply for a permit in most jurisdictions, you assemble the required documents along with a permit application. Prior to the 2020 pandemic, the plans and any structural calculations were typically paper copies, with *wet stamps* (inked stamps with the design professional's license information and signed in pen) when required. Since then, most jurisdictions accept or even require electronic submittals.

The more organized and complete your application is, the more likely you are to obtain a permit in a reasonable time and without problems.

We have found that if you approach the building department as an ally rather than an adversary, you will have much more success. Employees of the building department are typically good public servants, trying to do their jobs well. The chief building official in each jurisdiction has significant authority over what projects will be approved. This can be either a help or a hindrance depending on their stance on the issue at hand. The department's other building officials and staff have little ability to step outside the lines of the governing code. It is our task as cob practitioners to work within those lines—or as close to them as possible—with appropriate interpretations and justifications. The better your relationship with your building department, the more likely your interpretations and justifications will be accepted. Also, you never know the personal experiences or interests of someone at the building department unless they tell you. Our (Anthony and Massey's) first cob building permit submittal meeting with a senior building official in 2013 began with him sharing that he grew up in a beloved adobe house in Mexico. Our project was in the City of Berkeley, CA, known to be one of the most strict building departments in the county, but we were ultimately successful—in part, we believe, due to the human relationships we built in the process. That, and very committed clients. Other clients have chosen a more antagonistic route. Some were successful, too, though their process wasn't easy, efficient, or supportive of the long-term success of cob in their regions.

The process of applying for a permit may feel expensive and daunting, and it may force certain choices you would rather not make for budgetary or environmental reasons—for

example, you might be required to build a conventional septic system instead of a composting toilet, or a concrete foundation instead of a gravel trench foundation. Certain types of structures can be legally built in some jurisdictions without a permit. For example, in some rural regions, agricultural buildings don't require permits. Most places also exclude accessory buildings not exceeding a certain size (typically either 120 or 200 ft^2 [11 or 19 m^2]) from permitting requirements. However, nearly everywhere, buildings that are intended for human occupancy require permits.

In some cases, the choice to build without a building permit carries relatively low risk. These situations include small buildings hidden from sight in areas with a culture of unpermitted construction. In many rural areas, the primary enforcement mechanism for building regulations is neighbor complaints. We know of several examples of unpermitted cob buildings that were reported by neighbors to the local building department, which was then legally required to open an investigation. In affluent and more densely populated areas, you may not get far with construction before you are flagged for building without a permit.

No matter how you assess your risk level, keep in mind that the actions required after being cited for building without a permit are never easy, inexpensive, or fun. The most extreme outcome is being forced to tear down the building if it can't be brought into code compliance. Steep fines may be assessed daily until one of those two ends is achieved. Building departments commonly refuse to be helpful or lenient after you are caught building without a permit. You usually are offered the opportunity to prove that your building is designed and constructed to meet the intent of the code. With a cob building, it is very likely that you and your building department won't agree on some of its design elements. Once a building has been built, it can be impossible to make satisfactory retrofits. Anthony has been called out to inspect a number of red-tagged cob buildings. Luckily, none of these were torn down, but the process was an ordeal for everyone involved.

If you build without a permit (or even with one), document the construction photographically—especially structural, electrical, plumbing, and any other building elements that will be covered by finishes. This can be useful for future repair or other work. These elements ordinarily must also be inspected by the building inspector before they are covered.

There are other good reasons to obtain a permit besides the risk of enforcement. A permitted building raises your property value (though that can also increase property taxes), while unpermitted ones generally do not. Permits can make a huge difference when dealing with other institutions; it is usually impossible to get a bank loan, fire or homeowner insurance, or an electrical hookup for a non-permitted building. Another positive result of choosing to get a permit is that you will be contributing to the growing number of permitted cob buildings. Each new cob permit helps building departments become more familiar and comfortable with cob and the cob code. The more cob permits are acquired, and the more the cob community shares and addresses the issues they encounter with the permitting process, the easier that process will become for future cob designers and builders.

Alternative Materials and Methods Requests

Almost all building codes include a statement that the provisions in the code are not intended to exclude materials, designs, or methods of construction that are not specifically included. There is usually a general description of what

criteria alternatives must meet to be approved. This is often the most valuable code section for those interested in pushing boundaries in natural and low-carbon building systems.

This is the relevant section of the 2021 IRC:

R104.11 Alternative materials, design and methods of construction and equipment: The provisions of this code are not intended to prevent the installation of any material or to prohibit any design or method of construction not specifically prescribed by this code. The *building official* shall have the authority to approve an alternative material, design or method of construction upon application of the *owner* or the owner's authorized agent. The *building official* shall first find that the proposed design is satisfactory and complies with the intent of the provisions of this code, and that the material, method or work offered is, for the purpose intended, not less than the equivalent of that prescribed in this code in quality, strength, effectiveness, fire resistance, durability and safety. Compliance with the specific performance-based provisions of the International Codes shall be an alternative to the specific requirements of this code. Where the alternative material, design or method of construction is not *approved*, the *building official* shall respond in writing, stating the reasons why the alternative was not *approved*.

R104.11.1 Tests: Where there is insufficient evidence of compliance with the provisions of this code, or evidence that a material or method does not conform to the requirements of this code, or in order to substantiate claims for alternative materials or methods, the *building official* shall have the authority to require tests as evidence of compliance to be made at no expense to the *jurisdiction*. Test methods shall be as specified in this code or by other recognized test standards. In the absence of recognized and accepted test methods, the *building official* shall approve the testing procedures. Tests shall be performed by an *approved* agency. Reports of such tests shall be retained by the *building official* for the period required for retention of public records.

These provisions allow building officials to accept systems and materials that are not explicitly in the code, if they can be convinced of the safety and equivalency of the alternative. In practice, this generally requires the stamp of a structural engineer or architect. By stamping your plans, the engineer or architect assumes responsibility for the safety of the design, relieving the building department of legal liability.

Anthony's company, Verdant Structural Engineers, has submitted plans for over 200 custom natural building projects. The vast majority have been under the Alternative Materials and Methods section of the code, but only a few jurisdictions have required what is sometimes called an Alternative Materials and Methods Request, or AMMR. (It is up to the jurisdiction whether to require submittal of a form stating what your proposal is an alternative to.) This allows use of not only the IRC, but many other code sources (some international) and testing documents to justify proposed designs. For strawbale construction, we rarely need justification beyond IRC Appendix AS on Strawbale Construction, though we often use the testing that code was based on to design a more architecturally flexible building than Appendix AS allows prescriptively. For every Alternative Materials submittal that Anthony has been a part of, we have, often with compromise, obtained a permit.[1] The alternative materials and methods approval process in Canada—called an Alternative Solutions Proposal in the NBC—is very similar.

It's helpful to remember that submitting a project proposal with only one alternative non-codified material is much easier and more likely to succeed than a project with many. On projects Anthony has engineered, the cob or other natural wall system is often not the most

Existing Natural Building Codes and Standards

The current natural building codes available or in development in the US are:

IRC Appendices:

- AR: Light Straw-Clay Construction (2015–2024 IRC)
- AS: Strawbale Construction (2015–2024 IRC)
- AU: Cob Construction (Monolithic Adobe) (2021, 2024 IRC)
- BA: Hemp-Lime (Hempcrete) Construction (2024 IRC)

IBC Sections:

- Adobe Masonry: Section 2109

US State and County Codes:

- Adobe Masonry: New Mexico, Pima County, AZ
- Compressed Earth Block (CEB): New Mexico, Pima County, AZ

Related Codes:

- Crushed Stone (Rubble/Gravel Trench) Footings: 2024 IRC Section R403.5

- Low-Carbon Concrete: Marin County, CA[2]
- IRC Appendix AQ: Tiny Houses[3]

ASTM Standards:

- ASTM E2392: Standard Guide for Design of Earthen Wall Building Systems

Non-US Standards (New Zealand):

- NZS4297-2020: Engineering Design of Earth Buildings
- NZS4298-2020: Materials and Construction of Earth Buildings
- NZS4299-2020: Non-Specific Design of Earth Buildings

Codes and Standards Currently in Planning or Development:

- Earthen Floors: ASTM Standard in progress
- Earthbag/Superadobe: IRC Appendix proposal in development
- Adobe Masonry and CEB: TMS section in development[4]

difficult aspect of the design to permit—or it may be just one among many challenges. If you have the funds, the persistence, and the time, you may be able to get a permit for a radical design. But keep in mind that a building with cob walls that curve in plan, taper in thickness, and change density as they rise, plus a concrete-free foundation, an alternative wastewater system, a heavy, vegetated reciprocal roof, and an earthen floor is not going to be easy to design, to permit, or to build—and is likely not going to be low-cost when all is said and done.

Revising the Building Code

Building codes are developed and refined through collaboration among design, research, and building practitioners over many years or decades. The IRC Appendices listed in the sidebar were generated largely through the efforts of architect Martin Hammer, who was the lead or co-author of all of them. David Eisenberg has also been a foundational force in code development for natural building systems; he wrote, with Matts Myhrman, the first load-bearing strawbale code in the US, was lead author for ASTM E2392-05, and collaborated on all the IRC Appendices listed above. Other invaluable contributors include, but are not

limited to: Bruce King, PE; Mark Aschheim, SE; Kevin Donahue, SE; Paula Baker-Laporte, FAIA (Appendix AR); and Andrew Morrison (Appendix AQ). Anthony was lead engineer or a collaborator in the development or advancement of nine of the 13 US codes listed above.

Changing building codes takes time. Even though these IRC Appendices have been successfully added to the I-Codes, most have not yet been officially adopted in many jurisdictions. In the case of Appendix AU, as of this writing it has yet to be adopted by *any* state or local jurisdiction. Even so, these model codes can be of significant value to your permit submission because they provide a fully vetted and robust basis for an Alternative Materials and Methods Request. They give a building department all the information needed to review and inspect a project that conforms to the Appendix.

In the US, the IRC was the easiest first step for code acceptance of natural building systems. The Strawbale Appendix approved for the 2015 IRC, with the support of the California Straw Building Association (CASBA) and others, opened the door for Appendix approval of other natural materials. Now, a nationwide, state-by-state effort to promote adoption of

Martin Hammer, lead author of Appendix AU, testifying to the IRC Code Committee in 2019 with longtime collaborator David Eisenberg at his side. CREDIT: ANTHONY DENTE

all of these IRC Appendices is needed. If you feel you are the right person—or know the right person—to lead this effort in your region, please contact us. Expect a long-term process of education, advocacy, and relationship-building, but with gratifying and profound effects.

The approval of Appendix AU was a huge step toward broader acceptance of cob in the US and beyond. The development of this code was the foundational purpose of CRI and its founding president, John Fordice, architect. However, Appendix AU, like all building codes, is not perfect and requires continued development. Many provisions of the code must be supported by data—in particular structural, thermal-resistance, and fire-resistance. The testing Appendix AU was based on was limited by CRI's available resources at the time. As time and funding allow for additional testing, CRI plans to expand the range of options allowed by the code.

One major advancement for Appendix AU that we hope to make in the next few years is the use of cob-braced wall panels (shear walls) in higher seismic zones by utilizing *rocking shear* behavior enhanced by buttressing and cross-wall support. (See "Aspect Ratios and Shear Walls" in Chapter 8.) We also hope to test alternative reinforcement strategies using materials other than steel. Continued research, work, and input from every discipline involved in cob building is needed to improve Appendix AU in future code cycles. If you have ideas on how to make the code better, please contact CRI.

Quality Control

Quality control is a mainstay of conventional building. Wider cob success, which is simply the cumulative success of many individuals, including you, will be difficult without similar, quantifiable quality control practices. Conventional materials are held to strict quality control standards, and it is important to demonstrate that cob can be built with the same consistency and reliability.

The importance of quality control increases as buildings become larger and as life safety concerns increase. Some of these concerns include whether the building is residential or public, the length of roof spans, the height of walls, the number of stories (only one is explicitly allowed in Appendix AU, though the commentary opens the door to multi-story buildings), and the location's expected level of seismic activity. As a cob builder, you may find yourself playing the role of raw material excavator, cob recipe developer, product manufacturer, builder, and end-user. This is significant responsibility and should not be underestimated.

Due to the high variability of cob mixes, quality control is a greater challenge for cob than for more uniform, conventional materials. Don't skimp on testing! If your building permit requires testing for shrinkage, compression or bending strength, you will have to do new tests for any cob mix that contains different material proportions or sources. The tests required by Appendix AU can be conducted either by an approved laboratory or, if your building official agrees, on-site. Due to the cost, you will probably want to limit lab testing to what is mandatory. We recommend setting up your own testing operation on the building site (see "Build Your Own Testing Apparatus" in Chapter 6) to confirm that new mixes are in line with any lab tests or to provide the required data if your on-site testing is approved. This will require planning and work before construction begins to allow time for samples to dry. Documenting your testing procedures and results will help both the consistency and quality of your materials and may help establish trust with the building department that you are serious about doing quality work. It might also further the cause of cob building if you share your documentation with designers and builders you know, as well as with CRI.

Chapter 10

Budgeting

There is a persistent myth that the cost of building with cob is much lower than conventional construction. Though it's true that natural materials such as earth are usually inexpensive or even free, those materials are generally used only for the wall system (and sometimes floors and finishes), and their installation tends to be labor intensive. Who is doing the work of mixing and building, and what they are getting paid, become major factors in the cost of the building. (See "Volunteers, Work Parties, and Workshops" in Chapter 11 for a discussion of how to keep these costs low.) But even if the materials and labor for the cob walls were entirely free, you wouldn't save that much; a building's walls typically account for only 10–15% of total construction costs. The biggest expense categories usually include: design, engineering, and permitting; earthwork and foundation; roof; services and mechanical systems; and finishes—not to mention the building site itself, which is often the single largest expense.

For a cob home that will be built by paid professionals, start with the assumption that the costs will be as high or higher than conventional construction in the same region. (And remember that more expensive finishes and fixtures will cause costs to go up even more.) For preliminary budgeting, a cost per square foot of floor area is typically used. Inquire with local builders or design professionals for the number or range that they use.

The ongoing costs of operation are also important to consider. These include energy costs (for lighting, heating, cooling, refrigeration, appliances, electric vehicles, and other electrical devices) as well as maintenance. Good design can drastically reduce the operational costs of a building. For example, a building designed for passive solar heating in the winter, shading from the summer sun, and solar hot water and solar electricity will be dramatically cheaper to live in than a neighboring home with none of these elements. Some of these design features (like

Bliss Haven (also shown on the cover of this book) was designed and built by Kindra Welch of Clay Sand Straw. It earned a 5-star rating for sustainability from Austin Energy Green Building. The cost of this professionally-built house was on par with other custom homes in Austin, Texas, but its energy performance is much higher than most, so operational costs are low. Most importantly, the clients love it!
Credit: Kindra Welch

solar hot water and electricity) add construction costs but can pay for themselves over time. Others, like optimally-oriented windows and shading, and well-placed insulation and thermal mass, add little or no cost, but can save large amounts of energy and expense. Likewise, smart design decisions, high-quality construction materials, and careful detailing can all increase durability and decrease maintenance costs.

Compared to labor expenses, the cost of materials is relatively easy to estimate. But these costs range so widely from place to place that it is difficult to generalize. For example, many builders are able to harvest clay soil from their building site or get it delivered for free, while others have to pay for the material, or delivery, or both. In Austin, Texas, John Curry and Kindra Welch of Clay Sand Straw like to use powdered *caliche* from the local landfill site as their base material for cob. The clay costs only a dollar a yard—but a single delivery with a 12-yard dump truck costs $500, even over a short distance. To develop a budget for your building, you will need to research the availability and cost of materials in your region. You also need to calculate the quantities of each material you require.

Calculating Volumes of Materials

Let's say you are planning to build a cob home with 100 linear feet of cob walls 10' high by 15" (or 1.25') thick. We multiply length times height times thickness to get a total volume of 1,250 ft^3. Dividing by 27 converts that number into 46.3 cubic yards of cob.

Let's assume that through your mix design process you have chosen a clay soil-to-sand ratio of 1:2 by volume. You might understandably expect that if you mix one ft^3 of soil with two ft^3 of sand you will end up with three ft^3 of cob, but this is not correct. Because there is a large volume of air in both a bucket of soil and a bucket of sand, and because in the mixing process much of the clay soil fills what were previously air spaces between sand grains, the resulting volume of mix might be 2.5 ft^3 or even less. So to make 25 cubic yards of cob you will need at least 10 yards of soil and 20 yards of sand. By doing the math (46.3 divided by 25 times 10) you determine that to make all the cob for this house you will need 18.5 yards of soil and 37 yards of sand. To be on the safe side, round up; in this case, we would recommend having 20 yards of clay soil and 40 yards of sand available.

For a standard cob mix, you will need enough straw bales to equal about 10% of the volume of the wall—but the straw won't actually contribute significant volume to the cob itself. That means you will need about 4.6 cubic yards, or 125 ft^3 of straw bales. Three-string bales of straw measure 15" high by 22" wide by about 46" long, or about 5 ft^3 each. (To convert cubic inches into cubic feet, divide by 12^3, or 1,728.) So you will need to buy 125 ÷ 5, or 25 straw bales. This translates to approximately one half bale per cubic yard of cob. You may need straw for plasters and for an earthen floor, and bales come in handy as scaffolding and temporary seating. Why not get 40?

Pricing Materials

Next, you will need to know how much each of these ingredients will cost per unit of volume. The prices of materials vary so much from place to place and are changing so rapidly that we won't attempt to estimate the costs for the materials in the previous example. You will have to research the best prices in your region. Make sure that the materials you get prices for are actually the ones you need. See Chapter 4 for a discussion of desired qualities of clay soil, sand, and straw.

Don't forget to account for delivery costs. Sand and gravel yards will often quote you a price per yard including delivery. If you only need a small quantity, you can save money by taking your own pickup truck to the yard. In other situations, the material itself may be free, but transportation needs to be arranged. Remember that clay soil is often available as a waste material from construction sites. If you are lucky, a local excavation contractor will offload extra soil in exchange for a case of his favorite beer!

Other Expense Categories

When making a budget for a cob project, first break the project down into a series of smaller stages—the more, the better. Don't forget the many expenses that will be necessary even before construction starts, including land purchase, site work, design and engineering, planning and building permits.

Design and engineering costs range widely, depending on the size and complexity of the project and whether you design the building yourself or hire a designer, architect, and/or engineer. Complying with prescriptive codes such as Appendix AU, as well as the prescriptive requirements for foundations, roofs, and other building elements in the IRC, will place limitations on your design. Stepping outside of these prescriptive limits is often possible, but will likely require a licensed design professional. The fees charged by these professionals vary depending on project location and the nature of the project. Expect to pay between 5–15% of construction costs for combined architectural and engineering design (including between 2–3% for an engineer, if necessary). This will get you through permit approval, with limited involvement during construction. Small projects tend to require a higher percentage.

Planning and/or building permit costs also vary widely. In many parts of the US, planning and building permit fees are based on a percentage between 1–3% of the *valuation* of the project determined at the time of permit application. The applicant's valuation is usually accepted for this purpose, but some jurisdictions establish their own valuations based on floor area or other metrics. They may also add fees for schools, parks, or other community infrastructure, often linking them to the number of bedrooms added by the project. When required, Design Review usually has an additional fee. Inquire about all fees associated with your proposed building project early in the process.

The costs of materials and labor are difficult to estimate before the design process is complete and construction drawings are available. At that time, material costs can be calculated fairly accurately. The cost of labor is much more difficult to estimate. Labor costs are highly variable by region. They also depend on how much labor will be done by the homeowner or volunteers, who is managing the project (a general contractor typically adds 10–20% to expenses for overhead and profit), as well as site organization and efficiency. It is very difficult to predict the cost of any building system until you have experience to draw from. Once you have completed a cob project with a certain set of materials, in a particular region, then you can begin to make accurate estimates of time and expenses. Even then, unexpected weather, materials supply and price volatility or other setbacks can easily upset your calculations.

For each stage of construction, figure out what materials will be needed and what quantity of each. Next, determine where the materials will come from and how much they will cost. Then estimate the number of hours or days of labor that will be required and multiply that by how much each worker is getting paid. When you have completed this process for each stage of the building, add up all of the expenses including a *contingency* to try to account for expenses you haven't foreseen and the many unanticipated complications and delays. This contingency factor could be as low as 10% for materials, which are relatively easy to budget. But add more if material costs are likely to increase during the course of the project. For labor, the contingency should be at least 25% and possibly as high as 100% if the project leader is inexperienced, the design seems likely to change, or other conditions seem unpredictable.

A Sample Budget

The example below is a simplified budget for a very small cob building (about 100 interior ft^2 [9 m^2] of space) that Michael designed and helped build for a client in California in 2019—before the pandemic and supply-chain issues

Michael G. Smith designed and built the Earth Chapel with the help of Marisol Lopez, the client Laura Sandage, and many volunteers. It is used for meditation, music, and healing work.
CREDIT: LAURA SANDAGE

sent prices skyward. The building was intended as a healing and meditation space. It was small enough to qualify as a "playhouse" in the local city planning regulations, which exempted the project from needing a permit. (The building has no electricity or plumbing and is not intended as a sleeping space.) To make the cob, clay soil, sand, and water were premixed using a mortar mixer, then straw was stomped in by foot on tarps. The mixing and wall building were mostly completed during three weekend work parties. Two experienced leaders were hired to coordinate the work parties, and the other attendees were given an excellent lunch. The rest of the work, including foundations, roofing, and plasters, was done by builders who were paid between $20–45 per hour depending on the activity and the experience level of the builder.

A few more details about the building: the foundations were made of recycled *urbanite*, or concrete chunks, mortared with cement/sand mortar on top of a rubble trench. The floor is poured earth over a lava rock drainage layer, with a beautiful spiral pattern of inlaid river stones. The roof is a roughly conical living roof with 2×6 tongue-and-groove decking exposed inside as the finished ceiling. On top of this went a layer of rigid foam insulation, then a one-piece EPDM pond liner, then a layer of recycled carpet to protect the liner, then a drainage layer of lava rock, then several inches of specially formulated lightweight soil which was later planted with wildflower seeds and bulbs. Interior finish plaster is partly clay but mostly lime with fine straw (a local version of the Japanese *tosa shikkui*), and the exterior plaster was made with site soil, sand, starch paste, and chopped straw.

A couple of main points stand out from this example. First, even though much of the simpler work such as mixing and building cob was done by volunteers during work parties, labor costs still account for two-thirds of the entire budget. Second, the cob walls themselves (labor and materials) make up only 15% of the total cost. (You can see how both of these numbers would have changed significantly if either a) the owners had done more of the work themselves and hired out less; or b) no work parties had been planned, and all of the labor had been paid.) In contrast to the walls, the roof accounted for 42% of the total budget.

When a contingency of 10% was added for materials and 25% for labor, the total budget was $24,065. The total cost of the building ended up at $24,625. This may seem like very good budgeting, but a lot of luck was involved. If you look at the individual expense categories, you will see that some were far over the projected budget and others way under. As an example, the actual labor cost for plastering was 165% of what had been budgeted. This was partly because there hadn't been a clear plan for the finishes when construction began. The client ended up selecting a difficult plaster system that the crew was unfamiliar with. So, there was extra learning time required, and a

Table 10.1: Earth Chapel Budget[1]

STAGES OF CONSTRUCTION	Materials Budget	Materials Actual	Labor Budget	Labor Actual
PLANNING				
Finalize design[2]	0	0	350	315
Budgeting, ordering, shopping	0	0	490	460
TOTAL PLANNING	**0**	**0**	**840**	**775**
TOTAL PLUS CONTINGENCY	**0**		**1,050**	
SITE PREP/FOUNDATION				
Site leveling & trenches	0	0	270	435
Landscape fabric and drain pipe	100	190	165	110
Fill trench with lava rock	100	40	80	100
Build foundation	50	105	550	1,230
TOTAL FOUNDATION	**250**	**335**	**1,065**	**1,875**
TOTAL PLUS CONTINGENCY	**275**		**1,330**	
COB WALLS				
Cob mixing and building	550	320	1,480	1,290
Work parties	500	405	2,600	1,390
TOTAL COB	**1,050**	**725**	**4,080**	**2,680**
TOTAL PLUS CONTINGENCY	**1,160**		**5,100**	
ROOF, WINDOWS, DOORS				
Windows and doors	900	1,625	210	110
Roof framing & decking	2,100	2,290	1,540	2,820
Roof insulation	200	115	220	135
Liner, drainage, gutters	830	1,290	440	530
Roof soil medium	500	1,070	240	375
TOTAL ROOF	**4,530**	**6,390**	**2,650**	**3,970**
TOTAL PLUS CONTINGENCY	**4,985**		**3,315**	

[1] All costs in 2019 US dollars.

[2] Design work was mostly finished when this budget was prepared, so the total design cost is not reflected here.

Table 10.1: Earth Chapel Budget (continued)

STAGES OF CONSTRUCTION	Materials Budget	Materials Actual	Labor Budget	Labor Actual
PLASTER				
Wall trimming & plaster prep	0	0	275	540
Base coat plaster	30	35	600	1,035
Finish plaster tests	0	0	140	150
Finish plaster	100	315	1,380	3,240
TOTAL PLASTER	**130**	**350**	**2,395**	**4,965**
TOTAL PLUS CONTINGENCY	**145**		**2,995**	
FLOOR				
Lava rock drainage layer	100	40	240	110
Base coats earthen floor	100	70	1,200	495
Stone inlay	0	20	330	440
Finish coat	100	95	600	760
Sealant	300	460	80	70
TOTAL FLOOR	**600**	**685**	**2,450**	**1,875**
TOTAL PLUS CONTINGENCY	**660**		**3,065**	
TOTALS	**6,560**	**8,485**	**13,480**	**16,140**
Contingency	10%		25%	
Total with contingency	**7,215**		**16,850**	

larger-than-expected crew was needed for that phase of the project.

Conversely, the cost of the cob walls came in at just over half of budget. This was because the work parties had an excellent turnout of energetic helpers, so the building went faster than anticipated. Nearly all the cob for this tiny building was completed in just five days. The work parties were so fun and successful that we ended up scheduling two more for later stages of the project: one for the drainage layer and first pour of the earthen floor, and another to get the soil medium onto the roof. This volunteer labor contributed to the floor coming in under budget overall. The roof still went considerably over budget, but not by as much as it otherwise would have.

What is the main takeaway from this example? Budgeting is hard! Even with more than 25 years of cob-building experience, and on a very small building, the actual costs fell within plus or minus 25% of budget in only two of the six project stages. The more completely and accurately the building is planned in advance (including sourcing materials, deciding on finish materials, etc.), the more accurate the budget will be. Changing decisions during the building process usually increases the cost.

Typical Costs per Square Foot

Why was this little building so expensive? $240 per ft^2 may seem outrageous, but it is actually low for a custom building in California. (Other natural builders in California report that their cob and similar buildings typically cost between $250 and $325 per ft^2 at the time of this writing.) This little chapel is sculptural and beautifully finished, and it was built mainly by experienced professionals. The complicated and heavy roof was a major contributor to the cost. And very small buildings often cost more per square foot than larger ones, since one still has to go through all of the same steps on a small building but there is less time to get into the swing of things before it's time to clean up and get out a whole new set of tools for the next task. On the other hand, it's worth noting that this building contained no services or mechanical equipment, and there were no permit fees.

John Curry of Clay Straw Sand in Texas says that before the pandemic their company was able to build a no-frills natural house (using the cheapest available materials, inexpensive premade cabinets, and so on) for $150 per ft^2. This lower price is partly a reflection of the difference in labor prices between Texas and California. But the cost of materials has been so volatile in the last couple of years that they no longer offer binding bids on their projects because they have no confidence that prices will remain stable.

Other builders who specialize in subcontracting only cob walls rather than entire structures have developed standard costs per area of wall built. For example, Adam Weismann and Katy Bryce of Clayworks Ltd. in Cornwall, England, charge their clients a rate of 180–200 pounds sterling per square meter (about US $18–22 per ft^2) for a 2'-thick cob wall, including materials and mixing.

Sculptural details can add an enormous amount of time and cost to a project. Because of the amount of detail work involved, this building—the "Bottle House" in Bastrop, Texas—was more expensive per area ($300 per ft^2 in 2019) than any other Clay Sand Straw project at that time. CREDIT: JOHN CURRY

Chapter 11

Getting Help and Training

Builder Education

THE MORE EXPERIENCE you have, both with cob building and with construction in general, the better able you will be to plan and budget effectively, work quickly and efficiently, and keep costs down. So how do you get that experience?

Reading books like this one and watching videos will help you understand what you want to build and how to go about it, but there is no substitute for getting your hands and feet muddy. A common way for people to get their first cob experience is to take a hands-on workshop. There are many individuals and groups offering cob workshops. One good resource is the website cobworkshops.org, where you will find a calendar of upcoming workshops around the world, with descriptions, prices, and contact information. The scope of these workshops varies widely, from introductory weekend workshops to "start-to-finish" programs lasting several weeks. There is no certifying body that approves curricula or vets instructors; subject matter is determined by the instructors and/or organizations offering the course, and it often depends on the specifics of the building project. Cob courses are commonly light on "non-cob" aspects of building. Even so, a workshop can be an invaluable experience. In just a couple of days you can learn the feel of a good cob mix (which is very difficult to convey any other way) and how to reliably make high-quality cob and build a strong wall. Longer courses help you build up stamina for this physically demanding work and refine your mixing and building skills for greater efficiency and control. They usually also introduce other useful skills

such as plastering, roundwood carpentry, and so on.

Building a home is a complex activity requiring many different skill sets. Although the basics of mixing and building cob walls can be learned fairly quickly, there are many other pieces to the puzzle. Michael and Massey used to lead extended 10- to 12-week workshops with the goal of training participants to become professional natural builders. Although we had some notable successes, we found that the biggest determining factor for someone's skill level at the end of the program was their skill level coming in. Becoming a good builder takes many years of practice and exposure.

We strongly advise against taking on your "dream house" as your first major building project. Start small—with an oven, playhouse, or shed—to get a feel for what it takes to plan a project and carry it to completion. Then move up to a tiny cottage, studio, or guest house. If you can, spend time working with more experienced builders on their own projects; you'll build skills at low risk to yourself while observing someone else's approach to project management. Perhaps you can get hired onto a cob crew as a laborer or spend weekends volunteering on a local project. Every builder has a unique approach and skill set; working with several will give you a more complete education.

Volunteers, Work Parties, and Workshops

People often expect cob buildings to be extremely inexpensive, because "what could be cheaper than the earth under your feet?" But, this is only the case when the people doing the hard work of building are not paid. When labor is hired at the local going rate, cob buildings typically end up as expensive or more expensive than those constructed from conventional materials. So how can you get access to free labor without being exploitative? The main options are: 1) do it yourself; 2) work with volunteers or work parties; or 3) host a workshop.

Volunteer assistance comes in many forms. It could mean your neighbor or your sister pitching in for an afternoon or a whole summer because they want to support you. Or someone with a serious interest in natural building might agree to work on your project in exchange for experience, food, and a place to camp. As your confidence grows, you might want to publicize

Students at a cob workshop at Quail Springs Permaculture in Southern California. This is a common way to learn the basics of cob construction. CREDIT: JOHN ORCUTT

work parties, at which you offer instruction in cob building in exchange for other people's labor, for anywhere from a day to a week. It requires good planning and organizational skills to make sure there are enough materials, tools, space, and productive learning opportunities for everyone who shows up. How many people can you effectively supervise and teach at one time? Make sure to include good nourishing meals and time for everyone to socialize and to ask questions about the project. Keep in mind that the opportunity to learn is a valuable resource that many people are interested in.

Workshops where participants pay to learn skills require a much higher level of organization and commitment. It's the responsibility of both organizers and instructors to ensure that participants get good value in exchange for their money, time, and labor. It's almost impossible for one or even two people to effectively fill all of the necessary roles: instructor; site supervisor (in charge of tools, materials, and site organization); host (in charge of facilities such as camping, parking, and meals); communications (publicity, registration, and payment); and cook. You will likely need to hire help for some of these jobs. Workshops are not usually a money-maker for the host, but they can be a great way to infuse knowledge, experience, and labor into your project. Keep in mind that the primary purpose of a workshop is education, rather than production. It is often possible to link that goal with building tasks that need to get done, but the emphasis may not be on speed and efficiency. A significant amount of everyone's time will be spent on theoretical learning, skill-building, and practice not necessary for the building project to advance. If your intention is to get as much work done as possible while minimizing expenses, a work party is probably a better choice.

Hiring Professional Help

Few people have all the necessary skills, experience, and time to personally carry out every step of a major cob building project. While getting help from friends, family, and other volunteers is often a great option, there are also times when it makes sense to pay for specialized assistance. Below is a brief discussion of when you might want to seek out particular kinds of design and building professionals.

Designer/architect: Many of the most important decisions made on any building project involve siting and design. Small buildings are relatively easy to design, but the larger the building is, the more complex the process becomes. If you feel in over your head (or even if you feel fairly confident with your decisions but realize a second opinion could be invaluable) you may want to hire a professional designer. Licensed architects are building designers who have received professional training followed by working experience and then passed a test administered by the state where they operate. They typically have broad design skills and experience navigating the bureaucracy of building departments. But most of them aren't knowledgeable about cob or other natural building systems. It can be difficult to find a designer who is familiar with your local building department and local climate and also has cob experience. You may need to put together a design team in order to cover all of these important angles.

Engineer: If you are taking on a large or complex cob project, especially in a seismic area, you may want to consult a structural engineer. Appendix AU requires an engineer's stamp in Seismic Design Categories D and E, which include most of California, Alaska, Hawaii, and Puerto Rico and parts of Washington State,

Oregon, Nevada, Idaho, Montana, Wyoming, Utah, Arkansas, Missouri, Illinois, Kentucky, Tennessee, and South Carolina (see Figure 8.3). The engineer's job is to ensure that the building design is structurally sound. There are many ways to accomplish this goal. There is a small but growing number of structural engineers who really understand the physical properties of cob and other natural building systems. These sought-after individuals are much more likely to know how to use cob's inherent strengths as part of the structural design, rather than relying entirely on other materials such as concrete, steel, and wood to do the work. For that reason, you may actually save money by hiring a more expensive engineer who knows the materials well, compared to a less expensive one who knows only conventional building systems.

Natural building consultant: Because it can be hard to find an architect or contractor who is familiar with these materials, you may want to hire someone specifically for their knowledge of cob, plasters, and other natural building systems. This information is important in the design phase. A natural building consultant can help you determine whether cob is the best choice for your building project, or if you might be better off considering strawbale, light straw-clay, rammed earth, or some other option. They can also test your soils, design your cob mix, and think through the many details and connections between cob and other building components. You may want to hire someone to train your building crew in natural building techniques, to supervise work parties, or to teach a workshop as part of the construction process. And even if you intend to do most of the work yourself, it's great to have someone on call to answer questions as they come up—which they will! A few hours per month of an experienced consultant's time can save you a huge amount of time and money.

General contractor: A good general contractor has a wide set of building and management skills and carries a license (from the state) and a bond, which is a sort of insurance policy to ensure that the project is completed within the agreed-upon time frame and terms. A general contractor typically takes responsibility for all stages of a building project, from purchasing materials to hiring sub-contractors. Of course, there is a price for this service. The GC will tack a fee onto all of the building costs, usually 10–25%. By serving as your own general contractor, you trade your time (and probably some extra stress and gray hairs) for cost savings and more control of the project. In any case, you may have a hard time finding a GC who has experience with cob and other natural building systems. The natural building community needs more licensed contractors with passion and skills in these areas—maybe you could be one!

Subcontractors: Whether you hire a general contractor to manage the building project or choose to do that yourself, you may want to bring in people with specific skill sets from time to time. Some of the most commonly needed specialties include heavy equipment operators, concrete finishers, framers, electricians, plumbers, plasterers, drywall installers (for ceilings), painters, and finish carpenters. If you are serving as the general contractor for your project, you will need to get clear with each subcontractor or specialist about what exactly they are committing to (cleaning up afterwards? buying materials?), what and how they expect to be paid, and who is responsible if something goes wrong. Some sub-contractors carry their own licenses and bonds; others don't. Some will bid for the entire job, letting you know how much it will cost before they start, while others will charge you for their time and materials. Whenever possible, get recommendations from people you know (or at least from the internet) before hiring subs.

Chapter 12

Foundations, Floors, and Roofs

WALL SYSTEMS tend to get most of the attention. For example, it is common to hear someone say, "Come look at this cob building!" even though cob walls make up only half of the building envelope, and much less than half of the total cost and complexity of the project. Foundations, floors, and roofs are just as important as walls, but they are rarely thoroughly addressed in natural building courses. Because there are many high-quality design and building resources available elsewhere for the conventional materials commonly used for these building elements, we cover them only briefly in this book. On the other hand, there are plenty of alternative, low carbon, and natural options to explore when designing foundations, floors, and roofs; we will discuss some of those here.

Right off the bat, it is worth remembering that although wood is a natural and biogenic material, paying attention to the source is very important. Products certified by the Forest Stewardship Council (FSC) and/or locally milled lumber is recommended, though these

sources are still not guaranteed to be environmentally neutral or beneficial.

Foundations

A good foundation is of utmost importance for many reasons:

- It prevents settlement that can lead to cracked walls, sticky doors, and other issues.
- It transfers both vertical (gravity) and horizontal (wind and earthquake) forces to the ground.
- It can receive anchorage of the wall above.
- It protects the base of the wall from water damage.

There are many possible materials and techniques for building foundations; some of these can be more easily permitted and built by a conventional contractor, and some have significantly more embodied carbon than others. Some systems are also better suited to use less embodied-carbon materials to reach deeper depths below weak soils or frost lines when required.

Table 12.1: Foundation Types and Their Characteristics

	Easily permitted or built by conventional contractor	Good for medium or high seismic zones	Embodied carbon efficient for deeper foundations	Typical embodied carbon
Concrete or CMU	x	x		med-high
Concrete w/ voids	x	x	x	low-high
Concrete w/ crushed stone	x	x	x	low-high
Stone or non-CMU masonry			x	low-medium
Stabilized earth or earthbags		possibly	possibly	low-high
Gabions			x	low
Gravel bags			x	low

In most climate zones, it is *essential* to lift the base of a cob wall above ground level to protect it from water damage. Common sources of water intrusion into cob walls include moist earth or other wet materials in contact with the base of the wall, flooding, splash from roof runoff, and inadvertent irrigation spray. Keeping all but the surface of a cob wall dry is critical; if cob gets saturated, it can become soft and lose most of its compressive strength (see "Moisture Management and Control" in Chapter 3).

In the UK, it is common for old cob buildings to have stone foundations that elevate the base of the cob wall 3' (90 cm) or more above the ground. Appendix AU requires that the bottom of cob walls be at least 8" (20 cm) above finished exterior grade, although it allows for an exception in dry climate zones down to 4" (10 cm) above grade. It is possible that more research on the topic may conclude that it is safe to build cob even closer to grade in very dry climates. However, for the sake of moisture protection and release, *natural cob* (without cement, lime, or asphalt stabilizers) should never be placed directly against ground.

There are also possibilities of moisture intrusion into cob walls from *inside* the building

This cob home in Cornwall, UK, was built by Adam Weismann and Katy Bryce of Clayworks Ltd. High stone foundations like this one were traditional on many historic English cob buildings, which helped them survive in good condition for centuries despite the rainy climate. CREDIT: RAY MAIN

due to severe plumbing leaks or fire sprinklers, for example. For that reason, we recommend elevating the base of the walls above interior floors as well as above exterior grade.

Concrete wicks moisture through capillary action, so where cob meets concrete, a pathway is available for ground moisture to enter your cob wall. Appendix AU Section 105.4.3 requires the separation of cob from foundation with a liquid-applied or bituminous Class II vapor retarder. Exceptions are allowed where local climate, site conditions, and foundation design limit ground moisture migration into the base of the cob wall; in such cases, options include the use of a moisture barrier or capillary break between the supporting concrete or masonry and the surrounding earth.

Especially in cold climates, it is recommended to insulate your *stem wall* (called a *plinth* in the UK). This is the above-ground portion of your foundation, and it needs to be considered as part of the building envelope through which unwanted heat loss (or gain) can occur. A common way to achieve this is to use rigid foam insulation against the inside or outside surface of the foundation. Foam insulation has negative environmental impacts, and many cob builders have sought more natural alternatives. The UK CobBauge prototype used terracotta bricks, foam glass insulation, and damp-proof membranes to provide a 16"-high (40 cm) water-resistant plinth.

Concrete and CMU

Concrete and concrete masonry unit (CMU) foundations are the most common in conventional construction and the easiest by far to permit. But a conventional concrete foundation for a heavy and wide cob wall requires a great deal of Portland cement—a high-embodied-carbon material. Most designers and builders using a concrete foundation design it as wide as the cob wall it supports. This has been the historical practice, and it is required by Appendix AU (see Section AU106.4). Depending on your wall width, this has the potential to use significantly more concrete (and, therefore, cement) than a foundation for a typical wood-framed wall. Global cement production has doubled in the last 20 years and is currently responsible for 7–8% of global CO_2 emissions. This troubling trend has led many builders to seek lower-cement alternatives.

Low-carbon mixes can be ordered from concrete suppliers if you know what to ask for. These mixes can contain as little as 100 lbs (45 kg) of cement per cubic yard of concrete (typical mixes can easily have above 400 or even 500 lbs [180 or 225 kg] of cement per cubic yard). The reduction in Portland cement is made possible primarily by the addition of *supplementary cementitious materials* (SCMs) like fly ash, a coal-fired plant byproduct, and slag, a steel manufacturing byproduct. The good news about these toxic-sounding products is that their heavy metal content is locked in a harmless state following the chemical process of cementitious conversion. There are other ways to achieve a low-carbon concrete mix, including adding larger aggregate. (But increased aggregate size requires larger hoses for pumping.) Cure times beyond 28 days can also reduce cement use. This is strongly recommended for SCM mixes because of their naturally slow cure times, though increased cure times allow cement reduction of standard mixes too. We recommend referencing the low-carbon concrete code originally commissioned by Marin County, California.[1]

Another cement-saving technique that has been used by some natural builders is to add large stones to the foundation in order to displace concrete. This is only recommended if the stones avoid the structurally essential parts of the concrete—the areas shown around the void

in Figure 12.3. Large aggregate in structurally essential parts of the concrete makes the concrete harder to properly consolidate through vibration, especially around reinforcing steel. The governing building codes for concrete require increased rebar cover for mixes using large aggregate, and those requirements assume stones far smaller than the ones often proposed on natural building sites.

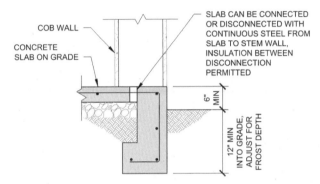

Fig. 12.1: *If you are using a concrete slab-on-grade floor with reinforcing steel that is tied into your foundation, one way to save concrete is to reduce the width of the concrete stem wall. In this detail, the cob wall bears partly on the foundation and partly on the slab. The stem wall should be a minimum of 8" wide (20 cm), and the footing should be a minimum of 12" (30 cm) wide.*
Credit: Mia Zohdi, Verdant Structural Engineers

To achieve this decorative effect, Kiko Denzer smeared a layer of mud on the inside of form boards and made a pattern in the wet mud with the point of a trowel. After the concrete was poured, the mud was washed off and the concrete was stained. Credit: Kiko Denzer

The welcome hut at Shasta Hot Springs in California was designed by John Fordice, engineered by Verdant Structural Engineers, and built by Rob Pollacek. This photo shows the decorative stone facing that was added outside the structural concrete foundation after the cob wall was built. This allows the full weight of the cob wall to bear on the concrete stem wall.
CREDIT: SEABROOK MUNKO

One common objection to concrete foundations is that they look ugly and institutional, especially in contrast to cob and natural plasters. Oregon artist and builder Kiko Denzer has come up with a great way to mitigate this. When forms are in place for the concrete pour, he smears a layer of mud on the inside surfaces of the forms and then carves relief patterns into the mud. When the concrete is poured and the forms removed, the concrete has a much more organic appearance—which can be further enhanced by staining the concrete. Another option is to face the surface of a concrete stem wall with stone. The photo shows one way to accomplish that aesthetic while making sure the full base of the cob wall is properly and evenly supported. Some people find board-form concrete a relatively organic aesthetic option.

Deeper Foundations

Some climate and soil types require larger and deeper foundation systems. In cold winter climates where the soil can freeze to the *frost line*, freezing and expanding soil water exerts significant upward pressure on a foundation. Some sites have unstable soils with reduced bearing capacity, especially soft clay (which has the further problem of expanding and contracting with changing moisture levels) or

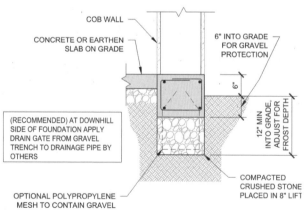

COB WALL

CONCRETE OR EARTHEN
SLAB ON GRADE

6" INTO GRADE
FOR GRAVEL
PROTECTION

6"

12" MIN
INTO GRADE,
ADJUST FOR
FROST DEPTH

(RECOMMENDED) AT DOWNHILL
SIDE OF FOUNDATION APPLY
DRAIN GATE FROM GRAVEL
TRENCH TO DRAINAGE PIPE BY
OTHERS

OPTIONAL POLYPROPYLENE
MESH TO CONTAIN GRAVEL

COMPACTED
CRUSHED STONE
PLACED IN 8" LIFT

Fig. 12.2: *One option when you need a stem wall that is the full width of your wall is to use a crushed stone, or rubble trench footing. This is an approved footing option for cast-in-place concrete foundations in the 2024 IRC and can save a great deal of concrete where deep footings are needed.*
CREDIT: MIA ZOHDI, VERDANT STRUCTURAL ENGINEERS

A crushed stone footing in Nicasio, CA. In this photo, formwork is being assembled in which to pour a concrete stem wall directly on top of the crushed stone. Architect: Arkin Tilt Architects; Engineer: Verdant Structural Engineers. CREDIT: HOYT DINGWALL

loose sand. The conventional solution in these cases is to dig a deep trench (which may need to be 6' deep (2 m)—or deeper—to reach frost line or to a layer of mineral soil or bedrock with sufficient bearing capacity) and pour a tall underground concrete wall into the trench. The main problem with this approach is that it uses very large quantities of Portland cement.

For material efficiency, we recommend two foundation strategies in these cases. The first is known as a *rubble trench, gravel trench,* or *crushed stone footing.* All three terms mean the same thing: a trench filled with small stones, appropriately compacted, used as the footing system for a wall. For simplicity, we will refer to this system as a crushed stone footing, because this is the term codified in the IRC. The bottom of the trench is as deep as the footing requires, whether due to frost line or soil capacity. It is most common for crushed stone footings to support a foundation stem wall or grade beam made of concrete or one of the alternative materials discussed below. The above-grade portion of the foundation elevates and protects the base of the

wall and also covers the compacted rock to protect it from erosion. The crushed stone provides an excellent medium for foundation drainage if the bottom of the trench is gently sloped to one or more drainage outlets. We recommend lining the trench with filter fabric before filling with stone, so that over time soil doesn't enter and fill the voids within the crushed stone. If you are building in a climate and with soil conditions that allow for a shallow foundation (e.g., 12" [30 cm] deep), it is probably not worth the extra effort to build a crushed stone footing. See IRC Section R403.5 for more guidance on the design of crushed stone footings.

Another concrete-saving foundation design involves creating voids in the foundation as illustrated in Figure 12.3. The voids need to be carefully placed to maintain the minimum concrete thicknesses required by anchorage and structural integrity. They can be made up of many kinds of material, though simple pipes are the most obvious choice.

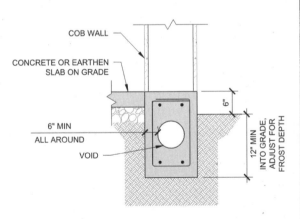

Fig. 12.3: *Another way to reduce cement use when building foundations for very wide walls, (especially if you need a stem wall that is the full width of your wall—which is recommended when using an earth floor slab), is to leave a void in the non-structural part of the foundation. The void could be either an airspace or filled with non-structural materials and must be surrounded by at least 6" (15 cm) of concrete on all sides.* CREDIT: MIA ZOHDI, VERDANT STRUCTURAL ENGINEERS

This project in Joshua Tree, California, took a unique approach to the void-in-concrete foundation option by using waste materials—including old toasters from a local dump—to displace non-essential concrete. Owner: Kevin Cain; Architect: Leger Wanaselja Architecture; Engineer: Verdant Structural Engineers; Builder: Nicholas Holmes. CREDIT: KEVIN CAIN

Stone and Other Alternative Foundations

Stone masonry was historically the foundation material of choice for earthen buildings in many parts of the world. Stones have the advantage of being low-carbon, locally available in many places, beautiful, and extremely durable, especially with regard to moisture. The most significant disadvantage is that stone masonry is not as good at resisting seismic forces. Stone and other non-CMU masonry foundations such as *urbanite* (recycled concrete chunks) can be appropriate where mechanical anchors are not required between the cob wall and foundation, especially in the lowest seismic zones like Seismic Design Category A in US building codes. This zone covers large areas of the US and Canada east of the Continental Divide, but almost nowhere to the west of it (see Figure 8.3).

Other alternative foundation systems include stabilized earth, earthbags, gravel bags, and gabions. We will briefly discuss each of these options below, but keep in mind that no US building code contains prescriptive or performance design parameters for any of these systems. If you choose to pursue a permit using one of these systems, an Alternative Material submittal with supporting documentation will likely be required (see Chapter 9). Alternative foundation systems are among many cob-related topics in need of more research.

Stabilized Earth

If you want to use earthen materials near and below grade, stabilized earthen systems are an option. Stabilization is common in adobe and many other earthen wall systems, though it is uncommon for cob mixes. If done correctly, the addition of stabilizing elements such as Portland cement (using lesser quantities than in concrete), lime, or asphalt emulsion to earthen materials makes them stronger and much more water-resistant. Of these stabilizing additives, asphalt emulsion is considered to have the least embodied carbon because it is a waste product of the petroleum industry, followed by lime—which has the ability to carbonize, taking additional carbon from the atmosphere as it cures—and then Portland cement, though formal Environmental Product Declarations (EPDs) are needed for lime and emulsified asphalt. (More about EPDs can be found in "LCA and Cob Design" in Chapter 7.) Lime-stabilized earth also typically has a lower compressive strength than earth stabilized with emulsified asphalt or cement. Read UK lime experts Stafford Holmes and Bee Rowan's book, *Building with Lime-Stabilized Soil*, if you want more information on this subject. It is important to remember that stabilized soil materials of any kind (cob, rammed earth, adobe) can be even *more* carbon intensive than concrete due to the width of the wall system; it depends on which stabilizer is used and at what volume.

Earthbags, Superadobe, and Gravel Bags

Bags such as polypropylene feed sacks can be filled with damp earth and tamped in place to create a building system known as *earthbags*. Barbed wire is often placed between courses to lock the bags together, and sometimes long continuous tubes of woven material are substituted for shorter bags in a system called *superadobe*. Both techniques were developed and popularized by Persian architect Nader Khalili in the late 20th century and are still promoted at the California Institute of Earth Art and Architecture (CalEarth), which he founded. Whole buildings—often including domes and vaults—can be built out of earthbags, and even more so with superadobe. Some natural builders have adopted the system as a foundation for wall systems including cob and strawbale.

Advantages of earthbag foundations include very low cost and possibly low embodied carbon (depending on the type of stabilization used, if any), use of largely local and waste stream materials, simplicity of construction, and the ability to easily make curved shapes that align with many cob designs.

If the bags are filled with unstabilized earth and that earth gets wet, it can expand, cracking plaster and creating instability. This can be addressed by using either stabilized earth or gravel as a filler. Gravel has the advantages of being relatively low-carbon and non-expansive and serving as a capillary break to prevent moisture from rising into the wall above. However, especially if gravel is used, the durability of the bags themselves becomes a major concern. The woven polypropylene typically used as bag fabric (similar to what blue tarps are made of) degrades rapidly when exposed to the sun. Some builders have wrapped gravel bag foundations with stucco netting and covered them with lime- or Portland cement-based stucco for long-term protection.

Anchoring walls to earthbag foundations can be complicated, and they do not have the bending and shear transfer strength of a concrete base, which is often needed in moderate and high seismic zones. Although we believe this approach has many merits, it needs further testing and development before it will be accepted into the building code. For more information on earthbag construction, we recommend *Essential Earthbag Construction* by Kelly Hart and *Earthbag Building* by Donald Kiffmeyer and Kaki Hunter.

Gabions

Gabions are wire cages, usually made from heavy-gauge galvanized wire (like that used for chain link fences) and filled with large stones. They are commonly used as retaining walls along road embankments, but some builders have been experimenting with gabion foundations. Advantages: it's a relatively quick and easy system that requires little or no cement. As long as the wire mesh remains in good condition, gabions are very strong in both tension and in compression. Air space between the stones serves as a capillary break to prevent moisture wicking up into walls above and as a self-draining foundation. The exterior surface can be covered with a lime or cement plaster to prevent the passage of air and animals through the gabion. Plaster should also slow down corrosion of the metal wire, which is an essential structural component of the system. Even so, it is difficult to know how long the wire will last before corroding. Also, some means is necessary to secure the cob wall and any other structural elements of the building to the gabions. To date there have been few, if any, examples of gabion foundations for cob buildings, so these and other details still need to be refined.

Gabion retaining wall behind a residence in Petaluma, CA. Gabions like these can also be used as a foundation system. Architect: Nathan Pundt; Engineer: Verdant Structural Engineers.
CREDIT: ANTHONY DENTE

Floors

In most cases, cob walls bear directly on the foundation, which is as high or higher than the interior floor level. Typically, cob buildings have a slab floor. The two most common slab floor options are a *concrete slab-on-grade* (SOG) or an *earthen floor.* (Cob-like mixes can be used for floors, but other earthen mixes can be used too, so we use the more inclusive term of *earthen floors* here.) Earthen floors have many positive attributes. The best available resource for their design and installation is *Earthen Floors: A Modern Approach to an Ancient Practice.* Anthony and Massey have both consulted on the development of an earthen floor ASTM Standard (in process, as of publication date of this book), which will be a big step forward in permitting acceptance of that system.

Lime-stabilized earthen floors may have greater strength and durability than natural earthen floors. Projects that Anthony engineered for liquor distilleries in California and New York successfully used this system.

Loft or second story floors are not uncommon in cob buildings (though they are not explicitly allowed in Appendix AU, and therefore require an approved engineered design). Most flooring materials and systems can be used for these raised floor elements if integrated properly with the walls. There should be a wood ledger anchored into a concrete bond beam that separates the upper and lower cob wall, or the ledger can be anchored into the cob wall itself at the floor level (refer to "Anchors and Connections" and "Bond Beams" in Chapter 8). Floor joists can be connected with hardware commonly known as *joist hangers* from the side of the bond beam or ledger. It's best to avoid building cob walls on top of wooden joists and flooring, both for structural and maintenance reasons. One common detail is to reduce the thickness of the cob wall at the level of the second floor, leaving the wall plumb on the exterior but creating a step on the inside where the ledger is supported. This allows wall construction to continue but leaves the ledger exposed for joists to be attached to after the wall has finished settling. The ledger should still be anchored horizontally into the wall.

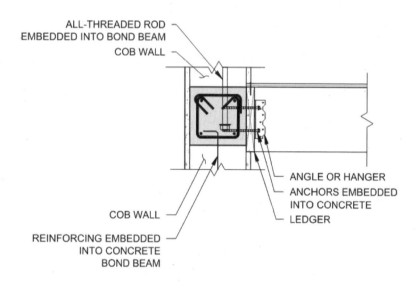

ALL-THREADED ROD EMBEDDED INTO BOND BEAM
COB WALL
ANGLE OR HANGER
ANCHORS EMBEDDED INTO CONCRETE
LEDGER
COB WALL
REINFORCING EMBEDDED INTO CONCRETE BOND BEAM

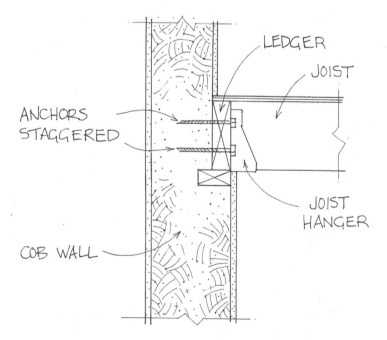

ANCHORS STAGGERED
LEDGER
JOIST
JOIST HANGER
COB WALL

Fig. 12.5: *Second-story ledger connection with wall thickness adjustment.*
CREDIT: DALE BROWNSON

Roofs

Roofs serve three primary functions:

- They keep weather out of the building and allow the temperature and humidity inside to be controlled.
- They protect the walls from weather and unwanted solar exposure by means of appropriate overhangs.
- They resist and collect out-of-plane wind and seismic forces exerted on the walls below and transmit those forces to shear walls.

Dimensional lumber and round poles are the two most common framing materials for roofs on cob buildings. As many readers know, letting rafters bear directly on the wall is a time-tested approach in both the cob and adobe traditions. This requires careful assessment of the wall material's capacity to support concentrated loads from ridge beams or rafters, especially when the rafters are spaced far apart. Rafters must also be effectively embedded or anchored into the cob wall to resist wind and seismic uplift forces. Where framing members are embedded in cob, we recommend driving nails partway into the wood as anchors for the cob. (See "Connecting Cob to Wood" in Chapter 14.)

It is common (especially in permitted or engineered buildings) for the rafters to bear on a wooden bond beam, or *top plate*. When a concrete bond beam is used instead, it is standard practice to install a flat wooden member, such as a 2×4, anchored into the concrete as a convenient connection material for the rafters. A wooden top plate or a flat wooden member over a concrete bond beam can be a very practical surface to attach any type of wood framing members (rafters, joists, or trusses) because the same convenient wood-to-wood connections that are used in conventional wood-framed buildings can be employed. When framing

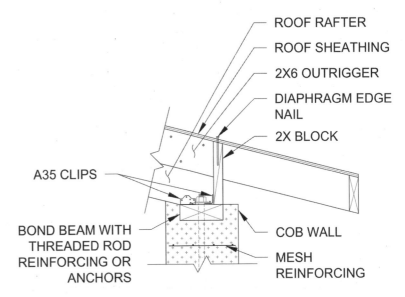

OUT-OF-PLANE WALL CONNECTION

Fig. 12.6: *Top-of-wall connections to roof framing and roof diaphragms suitable for all regions, including moderate and high seismic zones.* CREDIT: WILZEN BASSIG, VERDANT STRUCTURAL ENGINEERS

Wood blocking between rafters above the cob walls of the welcome hut at Shasta Hot Springs in California. CREDIT: SEABROOK MUNKO

rafters or joists above a wooden top plate or bond beam, it is helpful and often required to place blocking between the rafters as shown in Figure 12.6 (see also Figure AU106.9.5 in Appendix AU). Along with the hardware shown in the diagram, this blocking can be an essential part of the lateral (earthquake and wind) force-resisting load path.

Round Roofs

The design of a building's roof framing will depend greatly on the shape of the building. Round and curvilinear floor plans are much more common with cob buildings than with most other wall systems, but roofs that match these shapes can be considerably more complicated to design and build than rectilinear roofs.

Most round roofs require a small compression ring at the top and a wide tension ring at the bottom to resist the outward thrust of the rafters. This technique is most famous in the framing of the *yurt*, a traditional Mongolian round tent, so this roofing system is commonly referred to as a *yurt roof*, even when it's on a cob building. Many different materials can serve as compression and tension rings. For example, Anthony helped the late Bill Coperthwaite, a very innovative yurt building pioneer from Maine, design a roof that used an actual wagon wheel for the compression ring. The main point is not to forget that these two structural rings are required. Anthony has performed inspections for red-tagged round cob buildings that

Bill Coperthwaite-style wooden yurts at Experience Learning in West Virginia. These were designed by Michael Lacona, engineered by Verdant Structural Engineers, and raised with a volunteer crew. They contained a full circle tension ring at the base of the rafters. CREDIT: EVAN BLUMENSTEIN

did not have them, and they can be extremely difficult to retrofit.

Because of its popularity, we would like to highlight the relative difficulty of the *reciprocal roof* framing system. This is a unique way to frame a conical roof on a round building with no need for central support. Like other yurt roofs, a reciprocal roof requires tension resistance at the bottom, but in this case the rafters themselves overlap each other near the top to *create* a central compression ring. Verdant Structural Engineers, Anthony's company, receives a disproportionate number of requests for reciprocal roofs from clients who want natural wall systems. This is not surprising because cob and similar natural building systems are conducive to round shapes; cob builders and designers are often interested in non-conventional forms; and reciprocal roofs can be beautiful. Their self-supporting geometry has a magical quality that belies logic. Unfortunately, reciprocal roofs are one of the most difficult kinds of roofs to frame because of how the rafters bear on one another to form an internal compression ring. It can also be difficult to install the roof sheathing and water-resistive barrier on this multi-faceted, nonlinear roof

This reciprocal roof is being framed with round poles. Each pole rests on an adjacent rafter, leaving a central opening called an oculus *that serves as a compression ring.* CREDIT: SCOTT HOWARD

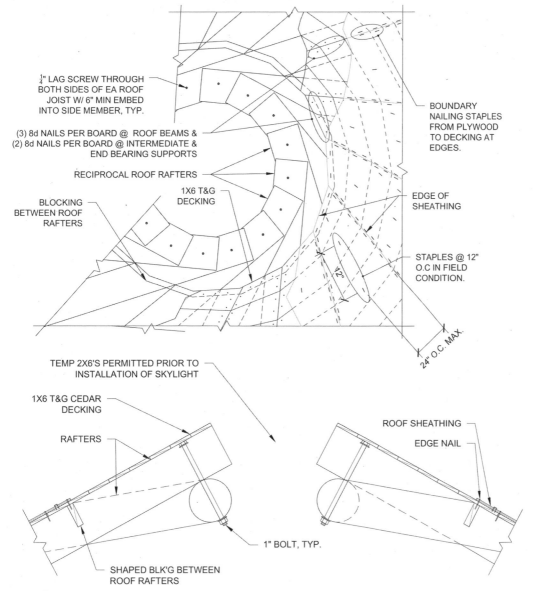

$\frac{1}{4}$" LAG SCREW THROUGH BOTH SIDES OF EA ROOF JOIST W/ 6" MIN EMBED INTO SIDE MEMBER, TYP.

(3) 8d NAILS PER BOARD @ ROOF BEAMS & (2) 8d NAILS PER BOARD @ INTERMEDIATE & END BEARING SUPPORTS

RECIPROCAL ROOF RAFTERS

1X6 T&G DECKING

BLOCKING BETWEEN ROOF RAFTERS

BOUNDARY NAILING STAPLES FROM PLYWOOD TO DECKING AT EDGES.

EDGE OF SHEATHING

STAPLES @ 12" O.C IN FIELD CONDITION.

12"

24" O.C. MAX.

TEMP 2X6'S PERMITTED PRIOR TO INSTALLATION OF SKYLIGHT

1X6 T&G CEDAR DECKING

RAFTERS

1" BOLT, TYP.

SHAPED BLK'G BETWEEN ROOF RAFTERS

ROOF SHEATHING

EDGE NAIL

Fig. 12.7: *Plan (above) and section (below) views of the oculus of a reciprocal roof framed with round pole rafter (similar to the one in the photo above). These design details highlight the difficulty of articulating not only reciprocal roof designs, but round elements in general—especially ones that curve in three dimensions—on 2D plans and details.* CREDIT ELLI TERWIEL, VERDANT STRUCTURAL ENGINEERS

Architect Ya-yin Lin designed and built this cob building in Taiwan. Local bamboo was used to frame the roof, with ample overhangs to shade the walls and protect them from typhoons. CREDIT: YA-YIN LIN

system. If you are an owner-builder or a less experienced builder, we recommend considering a simpler design or else bringing in someone with reciprocal roof-building experience to assist.

Overhangs

As historic adobe buildings in the Southwestern US have exhibited for hundreds of years, earthen parapet walls (with no roof overhang at all) *can* perform well in certain dry climate zones. In wetter climates, though, ample roof overhangs are strongly recommended. At the very least, roof overhangs cut down on maintenance by keeping the erosive force of running water off the walls. They also help protect the bottom of the cob wall, which is typically the most vulnerable to water damage. Wide overhangs, effective gutters, and high foundations combine to keep wall bottoms dry even in the rainiest climates.

Depending on your climate and building design, an ample roof overhang can be anywhere from 1'6"–5'0" (40–150 cm). Design considerations include the height of the wall and the direction and quantity of wind-driven rain in your region (or on your particular site, as surrounding buildings and vegetation can be effective at reducing wind speeds). Where storms come predictably from one particular direction, a good solution is to have a wide porch on that side of the building. In severe climates with harsh weather exposures, wrap-around porches or verandas can completely shield a building's walls from weather. Obviously, the wider the overhang, the more complicated the structural support of that feature, especially on wall ends that are perpendicular to the main rafter framing, such as gable end walls.

There is a critical interaction between roof overhangs and solar heating and cooling. For an effective passive solar design, the roof overhang should fully shade south-facing windows during the hottest time of the year, but allow sun to penetrate the building for winter heating. Incorporating passive solar design principles will help keep your building comfortable year-round with much lower energy inputs and costs (see "Cob-Specific Design" in Chapter 7).

Above the Framing

On top of the framing goes the sheathing (if any), which is typically plywood, oriented strand board (OSB), or wooden boards. On top of this is the roofing material tasked with keeping rain out of the building. Nearly any roofing material can be used on a cob building, including, but not limited to metal, green or vegetated roofs, wood or asphalt shingles, ceramic tiles, slate, and thatch. Roofing systems made of discrete overlapping units such as shingles, tiles, and thatch require a fairly steep roof pitch (3:12 or greater) in order not to leak.

Keep in mind that in seismic zones, weight is directly correlated to seismic force on the building. The higher up in the building that weight is located, the more complicated it can be to resist the seismic force it generates. So, use extra caution when building roofs made of heavy materials such as earth, tile, or slate in earthquake-prone regions. The roof structure has to be able to support not only the materials you put up there, but any water they may absorb, as well as snow loads. When saturated, soil can weigh as much as 60% more than when it is dry. The weight of vegetated roofs can be reduced somewhat by using a very light, easily drained soil medium. Specially formulated soils for living roofs are increasingly available from landscape materials suppliers. We recommend *Essential Green Roof Construction* by Leslie Doyle for more information on vegetated roofs.

A building's *diaphragm* is the system of structural materials in a floor or roof that acts like a wide, thin beam to resist, collect, and distribute horizontal loads from earthquake and wind forces into shear walls or other vertical lateral-force-resisting elements. In most residential construction in the US, this is typically made of plywood or OSB over wood framing. Traditionally, though, board sheathing has served this function in wooden roofs and floors. The governing wood design code in the US, the National Design Specification for Wood Construction (NDS) still contains design values and specifications for board diaphragms, though they are a fraction of the strength of plywood. (If you are interested in a building with solid walls on three sides with the fourth side open for views and/or solar gain, you can employ a three-sided, or *open-front diaphragm*, approach, which accounts for the torsional effects of this building type and allows for architectural flexibility. These designs require a licensed engineer.)

Out-of-plane forces, which are forces from wind and seismic activity that act perpendicular to the wall (see Figures 8.2 and 8.4), are resisted primarily by the foundation and the roof diaphragm. The wall acts like a beam standing straight up, connecting the foundation to the roof assembly. Because cob walls are many times more massive than wood-framed walls, in seismic zones they require greater out-of-plane resistance and load transfer by the roof diaphragm. Therefore, the detailing for a diaphragm on a cob building may appear overbuilt to someone used to wood-framed buildings. It is important to follow the top-of-wall diaphragm bracing and support requirements in Appendix AU or the requirement for concrete and masonry wall anchorage in ASCE 7 (The American Society of Civil Engineers structural design manual) because these could affect roof framing and diaphragm design, particularly in seismic zones.

Fire-Safe Roofs

We recommend referencing Wildland-Urban Interface (WUI) codes to ensure you protect wooden soffits and roof overhangs from fire.[2]

A Nubian vault under construction on Michael's farm in California. This millennia-old technique is being revived today as a practical shelter solution in areas with little available wood. CREDIT: MICHAEL G. SMITH

The vault is built out of many small adobe blocks stuck together with clay-sand mortar. Here, workshop leader Stevan De La Rosa places the final key adobe to complete an arch. Nubian vaults are built as a series of inclined arches leaning against the end walls. After the vault is complete, gravity forces transfer the weight of the vaulted roof into the supporting side walls. CREDIT: MICHAEL G. SMITH

Fire-safe roofing options include non-combustible roofing materials such as metal (with corrugated sheets or standing seam systems), terra cotta or cement tile, or vaulted or domed earthen roofs.

There are many ways to build earthen roofs: cob, adobe blocks, compressed earth blocks, sprayed earth over metal mesh, or wattle-and-daub. Although it may seem ideal to make an entire building out of locally available, fire-resistant material, there are significant downsides to structural earthen roofs. The most important is that, except in the very driest of desert climates, vaults and domes made of earthen materials have the potential to absorb rainfall and collapse. It is extremely difficult to design and install a reliable waterproofing system that will protect the cob but not deteriorate or crack over time and still allow water vapor to escape from the earthen material. The safest solution to this dilemma is to construct a secondary roof structure to protect the earthen vault or dome.

There are also many structural issues when using such heavy materials overhead, especially in seismic areas. Monolithic cob vaults and domes require complicated formwork, which typically negates most benefits of using the system. A more appealing alternative is the *Nubian vault*, which is a formless vault system made of small adobe blocks. The Nubian Vault Association[3] is a good resource for more information on the revival of this ancient technique; it is worth investigating the excellent work they are doing in Africa. Domes are also fairly easy to construct out of adobe or compressed earth blocks, as are wattle-and-daub domes (for small structures). However, capable engineering and care in construction must be employed for vaults or domes made of earthen masonry in areas of seismic risk, in order to prevent collapse during an earthquake.

Chapter 13

Mixing

N
O MATTER WHICH METHOD you choose, the goals of cob mixing are the same: the base ingredients (usually clay subsoil and sand) must be thoroughly combined with the right amount of water and mixed to a stiff, but still plastic, dough-like consistency. In a process similar to kneading dough or wedging clay, the lumps of clay soil are broken down, hydrated, and activated by a combination of moisture and percussion. If you are making a classic high-density mix, the texture at this point will resemble cookie dough. If the goal is a low-density, high-fiber mix, then the consistency might be as wet as pancake batter. Then straw is mixed in little by little until the desired consistency and proportion of straw have been reached.

The mixing system needs to allow you to alternately turn and compress a large volume of heavy materials. Historically, this was done either by groups of people stomping in a large cob pit, or by draft animals, such as oxen. In Oregon in the 1990s, cob instructor Becky Bee discovered that placing the materials on a flexible, water-resistant sheet like a tarp allowed for better control and faster, more efficient mixing. This is still the preferred method on many cob building sites around the world, as tarps are inexpensive, accessible, and reliable.

Some machines can make excellent cob. In the hands of a skilled operator, an excavator, a front loader, or a tractor with a bucket makes quick work of a small mountain of cob. This is a common technique in Britain and is becoming more common in North America as well. Some builders have had good success with rototillers. See below for other mechanical mixing possibilities.

Once you have determined your mix recipe (see "Recipe Development" in Chapter 6), it is important to maintain the same sources and ratios of clay soil and sand—unless you want to go through the mix testing process all over again. However, the proportions of water and straw are less critical. In fact, you may choose to deliberately vary the amounts of these ingredients. Adding more water to a batch makes the mixing process easier and more thorough. However, if the mix is too wet, it will limit the height of wall you can build at one time. In hot, dry weather, you can mix wet and then leave your mix to dry partially before placing it on the wall—or put it on the wall wet and let it dry there. But if drying conditions are poor, you are better off making a stiffer mix with less added water. If your mix does get too wet, you can stiffen it up by adding extra straw.

More straw in a mix increases its strength up to a certain point; the main difficulty with very high-straw mixes is that they are more difficult to produce using typical cob mixing methods (see below for more on high-fiber cob mixing methods). You may choose to use extra-strong, straw-rich cob in the places where the reinforcement will do the most good: around and above windows and doors, for instance. But remember that adding a lot of straw will eventually decrease the cob's compressive strength. If you plan to vary the amount of straw throughout your building, you may have to test two different recipes—a low-straw mix and a high-straw mix—to make sure that both fall within the required range for compressive and flexural strength.

Preparing Clay Soil

Mixing will generally be faster and the quality of the resulting cob better if the clay soil is hydrated in advance. This is especially important if the soil has a high clay content. Dry lumps of high-purity clay are very hard, and they often absorb water slowly. Some soils, though, such as silty clay soils and decomposed granites, are easy to mix dry. If you are blessed with a soil like this, take advantage of it and skip the extra step of soaking because working with dry clay is a lot easier than working with heavy, wet clay.

The best way to hydrate clay soil is to let it soak in water. Large quantities can be soaked in a stock trough or kiddie swimming pool, but it can then be difficult to get the heavy, sticky material out. 55-gallon-drums cut in half make convenient soaking tubs (plastic drums are easier to cut than steel ones, and their edges aren't

as sharp). Another good solution is to make a temporary pool by draping a waterproof tarp or sheet of plastic over a ring of straw bales. Some of the straw bales that form the edge of the pool can later be rolled away to make it easier to extract the clay. Another simple soaking pit design is a wooden frame draped with a tarp; this is sometimes simpler than a bale soaking pit, especially on tight building sites.

Always put water in the soaking vessel first, then add the clay soil broken down as small as is practical. The smaller the clumps are, the more quickly they will hydrate. Barely cover the clay soil with water and leave it to soak for a day or more. It should absorb most of the water as it soaks, resulting in mud about the consistency of pudding. Soils that are very high in clay may hydrate extremely slowly. One solution in this situation is to smash the soil into small clods, soak it in buckets of water for one or more days, then beat it into a creamy slip with a handheld mixer. Similar results can often be achieved with much less work simply by letting the soil soak for a longer time.

Sometimes, cob builders remove stones from their clay soil. Although gravel and small stones will not compromise the strength of the cob, they can impede mixing—especially when it's done barefoot! The easiest way to remove stones is to pre-sift dry soil through a coarse screen (½" mesh is typical) made of hardware cloth or expanded metal lath. Alternatively, one can convert the clay soil to clay slip as described above, then pour it through a screen to remove stones and gravel. Depending on the recipe, clay slip may be too wet to use as the basis for a cob mix. In that case, you will need to let it settle for some time, pour off the water, and then wait for more water to evaporate before you can make cob from it. This adds up to extra work and time, which is why we recommend dry sifting when possible.

A temporary pit made of straw bales and a tarp is often the most convenient way to soak large volumes of clay soil.
CREDIT: SCOTT HOWARD

Soil being sifted to remove rocks at a cob project in Oaxaca, Mexico.
CREDIT: VALENTINA MENDEZ MARQUEZ

Mixing on a Tarp

Mixing cob with your feet may seem like a throw-back to a pre-industrial era, and in many ways it is. To produce enough cob for a large structure in this way will require many hours, days, and weeks of demanding aerobic physical exercise. But it need not be unpleasant or even inefficient. Turning off the machines makes a building site more enjoyable and safer. It helps everyone to relax and have fun. And tarps are available almost everywhere in the world for low cost.

The standard mixing tarp is a piece of flexible, water-resistant material about 8′ square (2.5 m) (see Chapter 5 for recommended tarp materials). Tarp mixing is often done barefoot, partly because cob tends to adhere to the soles of shoes and boots, making the mixing more difficult and the mixer more tired. Although the soles of your feet may feel tender at first, they will soon get tougher. However, if there are sharp rocks or detritus such as broken glass and metal in the soil, it is best to keep shoes on. This is often a concern when working with urban soils. Conrad Rogue at House Alive teaches a very efficient mixing system he calls *the Canadian method*, in which two shod mixers alternately stomp on the mix through the tarp, in a kind of high-energy choreographed dance. See his book *House of Earth*. Other great guides to tarp mixing include *The Hand-Sculpted House* and *Building Green*, which contains an excellent photo essay on the topic.

Whether barefoot or shod, the first step is to pour measured quantities of clay soil and sand onto the tarp. A common volume is about five 5-gallon buckets of dry ingredients, which will yield around 3 ft³ of cob. (This amount can be doubled or tripled by the use of a straw bale pit: see below.) Lay down a layer of sand first so that the mix doesn't stick to the tarp. Then alternately dump buckets of clay soil and sand until you reach your predetermined recipe.

The sand and soil should be pre-mixed by rolling the tarp back and forth several times. It is helpful to have two people working together at this stage. Each grasps two corners of the tarp, leans back slightly and rocks from side to side, tumbling the materials back and forth across the tarp. Be careful not to hurt yourself moving so much heavy material around. Keep your back straight; use the strong muscles of your legs to lift as you rock back and forth, but keep most of the weight of the cob mix on the ground at all times. After rolling the materials across the tarp a couple of times, rotate your positions on the tarp by 90° and roll a few more times in that direction. One person can also do this job by standing on the tarp, crouching

Tarp mixing can be an enjoyable team-building activity or an aerobic solo exercise. CREDIT: MARK MAZZIOTTI

down, grasping one corner, and walking backward.

When the dry ingredients are thoroughly combined, roll them into a pile in the center of the tarp, make a crater in the middle with your foot, and pour in some water. It is easy to go overboard on the water; add it slowly or you may end up with an excessively wet mix that needs to be left to dry before it can be used on the wall. Push dry material into the crater with your foot to absorb some of the water, roll the tarp once or twice as before, and then begin dancing on the mix. Alternate stomping the mix out flat under your feet and rolling it back into a mound, adding more water as necessary. Make sure that the whole pile is flipped each time so that nothing is left unmixed at the bottom.

Once the clay soil, sand, and water have been thoroughly blended into a soft paste, it is time to add the straw. This should be done little by little to avoid clumping, but it can be helpful to measure out the straw ahead of time so that you know how much you are putting in. Roll the mix into a mound, sprinkle a little straw on top and stomp it into the mix. Repeat until the mix reaches the desired consistency. Make sure to roll the tarp far enough that no unmixed portion is left in the middle and pull from different directions so that straw is uniformly blended in everywhere. Each mix should take about 5 or 10 minutes for one or two people to complete. Picking your feet up and jumping or dancing vigorously makes the process faster … and more fun, especially when there is music to help you keep the beat.

How much straw is the right amount? It's difficult to be precise. Sometimes (to meet energy code, for example), you may need a specific amount added to each mix. In that case, since measuring straw by stuffing it into a bucket is very inaccurate, weigh the straw out instead, or use a certain number of *flakes* per mix. (Flakes are the subunits into which bales naturally separate when cut open.) If you don't have a target density, you can be more lax about the amount of straw in each mix—but make sure you add enough to achieve the many functions that straw plays in the cob, including immediate stiffening and permanent strengthening of the wall. A mix containing sufficient straw should hold together as a unit when the tarp is rolled. It should also resist your weight. If you can stand on the mix without sinking in, that's a good indication that it will support the weight of more cob added on top once it has been placed on the wall. Reach down and grab a few handfuls of cob. It should take a little effort to pull them away from the rest of the mix, and you should see plenty of straw sticking out in every direction. You should be able to easily form it into a ball 6" (15 cm) across that can be tossed 10' (3 m) and caught without breaking apart. If not, it may need more mixing.

It is sometimes more convenient to make your mixes very wet—too wet to hold your weight when you stand on them. As long as drying conditions are good (fairly dry and warm) the mix should dry out enough to use by the following day. Certain mixing methods such as straw bale pits, mortar mixers, and concrete mixers lend themselves to making wetter mixes. And even with normal tarp mixing, the amount of effort is reduced by adding more water.

The Lasagna Method and Pit Mixing

Very high-fiber cob is difficult to make using the standard tarp method. When the whole mix is tied together with straw into a single unified mass, it becomes almost impossible to flip, roll, and stomp flat. Our preferred method of mixing small to medium quantities of high-fiber cob is the *lasagna method*, which can be done in a wheelbarrow, mixing tub, or constructed pit.

Start by pouring 2–3" (5–8 cm) of thick clay slip (about the consistency of pudding) into the mixing vessel, then press in straw until the moisture has been absorbed and the whole mixture moves as a single mass. Repeat this process until the vessel is full or the desired amount of material is reached. This kind of mix can be made wet—which is faster—then left to sit overnight until it reaches a nice workable consistency.

A larger lasagna mix can be made in a temporary pit constructed of four or more straw bales placed corner to corner, leaving an empty space in the middle which is lined with a tarp. A pit like this will accommodate a larger volume of wet ingredients than a flat tarp and is much better at containing slip. Follow the instructions above, using your feet to stomp straw into each layer of slip. The hardest areas to mix are in the corners of the pit. Using five or six bales instead of four makes the corners easier to access.

Pit mixing is also effective for regular lower-fiber cob. It allows a much larger batch to be made, although the mixing may not be as thorough as with regular tarp mixing. Add water to the pit before your dry ingredients. You may want to incorporate sand and soil bit by bit to make it easier to stir them into the water with your feet. If possible, have two or three people pull the tarp from side to side a few times to dislodge unmixed material from the corners. Except for this last part, pit mixing is a great job for children who are not strong enough to flip a tarp full of cob. *Building Green* contains an excellent photo essay on pit mixing.

High-fiber cob mixes are best suited to sticky clay soils with low aggregate content. Moderately sandy or silty clays also work fairly well. Using a soil with a lot of coarse sand or gravel for high-fiber cob is technically possible, but it is harder to mix and to build with, so some clays that work fine for standard cob mixes prove difficult for high-fiber cob. It's easiest to use long, flexible straw such as rice straw, though any straw will work. When using stiffer straw (such as wheat), it is a good idea to let the

Making a lasagna mix *in a wheelbarrow is a relatively quick and effective way to mix high-fiber cob.* CREDIT: MASSEY BURKE

Ananth Nagarajan hired a backhoe operator to mix cob for a project in India. It took about 4 hours for the machine (and two assistants on the ground) to make 14 cubic yards (11 m³) of cob. The ingredients were spread over a mixing area about 100′ × 50′ (30 × 15 m). CREDIT: ANANTH NAGARAJAN

high-fiber mix sit wet overnight before building; this makes it much more workable.

Mixing with a Tractor or Excavator

One way to make larger batches of cob with reasonable quality is to use a tractor with a blade or (preferably) a front loader. Suitable machines range in size from a small skidsteer or mini-excavator to a full-size excavator or backhoe. Larger machines mix bigger batches but require a much larger level mixing area.

Rob Pollacek of California Cob has used a tractor to mix cob for over a dozen buildings. He likes to start by laying down a level pad of prepared cob about 12′ wide, 20′ long (4 × 6 m), and 6–9″ (15–23 cm) thick. This is allowed to dry, and then all further mixing is done on top to ensure that no unmixed soil will be picked up during mixing, which would change the mix proportions. Other options are to mix on a driveway, a road, or a concrete slab. Mixing on bare ground is possible, but it's hard to maintain your proportions and to avoid incorporating unmixed clumps of earth. Kindra Welch and John Curry of Clay Sand Straw in Texas mix cob for large projects with a small Bobcat skidsteer. They set aside an unsurfaced circular mixing area about 30′ across as close to the building as possible, usually in what will later become a driveway or parking area.

As with tarp mixing, soil and sand should be mixed well before adding any water. Rob combines soil and sand using a rotary hoe (aka rototiller) on the back of his tractor. If you don't have a rototiller, you can also push the materials around with the bucket until they are thoroughly mixed. Rob recommends a clamshell loader bucket, which can be opened into a flat blade for pushing with or closed around a pile of cob to pick it up.[1]

While the tractor operator continues to mix with blade or rototiller, another worker on the ground sprays the pile with a hose. After the mix is thoroughly and evenly wetted, the assistant begins to add straw while the operator drives the tractor through the mix. Rob says: "Driving through the cob is really important, as it doesn't quite become cob until it has been driven through and smashed together." Using this method, experienced mixers like Kindra and Rob can mix 2 yd³ of high-quality cob in about half an hour. A 2-yard mix requires about a bale of straw.

A common approach in the UK is to make cob with an excavator (called a *JCB* in Britain) in the same pit from which the raw material is harvested. The book *Building with Cob* by Adam Weismann and Katy Bryce contains good directions. They start by removing the topsoil from an area about 10' wide by 20' (3 × 6 m) long, depending on the size of the machine. Then they break up the subsoil with the bucket of the excavator, to a depth of 8–10' (2.5–3 m). Then they fill the pit with water and let it soak. When the clay soil is hydrated, the sand is mixed in, followed by straw. In this way, it takes 4 to 5 hours to mix about 13 tons of cob using only the bucket of an excavator. This is about the same speed as the tractor method and has the advantage of eliminating the steps of excavating and transporting clay soil which precede any other mixing technique. The downside is that the quality of the mix may not be as high.

We know builders who have mixed as much as 17 yd³ (over 20 tons) of cob in a single day. If the batch size is too large, it can be a challenge to keep the proportions accurate and the mix consistent. With practice, a skilled operator will find the most effective batch size and mixing method for their machine, materials, and conditions. It is also important to make sure that your building method can keep up with large-volume mixing. See "Staging, Transport and Scaffolding" in Chapter 14 for more on this.

Mixing with a Rototiller

Cob can be mixed using either a small garden-scale rototiller or a large tiller attachment on a tractor. Ed Raduazo of Washington, DC, recommends a 5-horsepower garden tiller with tines in the front rather than the rear to improve maneuverability. He likes to mix around a ton of ingredients at a time (10 full wheelbarrows) on top of a concrete slab. He first layers the sand and clay soil about 6" (15 cm) thick and then begins driving the tiller around the pile. It's helpful to have an assistant with a shovel to pick up material from the edges of the pile and throw it into the center. Once the sand and clay are uniformly mixed, Ed uses the tiller to dig a hole in the middle of the pile and fills it

with water. Ed says, "I find it easiest to turn the center into a batch of gooey slop and then have my assistant throw in shovels of dry stuff from the edges. Then I add more water from the hose until I get what I want."[2]

Instead of a concrete slab, Graeme North of New Zealand prefers to construct a trough out of old scaffolding planks, a little wider than the tires of the rototiller. The trough is the length of whatever large boards he has handy, usually

Mixing a large batch of cob in situ with a tracked excavator.
CREDIT: JOHN BUTLER

Karen MacDonald mixing cob for an earthen floor with a rototiller. Both the rotating tines and the wheels contributed to thorough mixing. Underneath the cob is a flat surface of ¾" plywood salvaged from concrete forms, which ensured a clean mix and made shoveling the finished cob easier. Each half-cubic yard batch took about 20 minutes. CREDIT: ROB WEST

10–12' (3–4 m) long, and the sides slope slightly outward. It can have a wooden bottom or be built on a concrete slab to prevent the rototiller from grabbing extra soil from underneath. Graeme likes to make a fairly wet mix in his trough, leave it overnight to absorb the water, and then mix it up again the following day, adjusting the consistency by adding more water or more straw as needed. Both Ed and Graeme find that long straw tangles around the tiller blades, so they pre-chop all of their straw first. (For instructions, see "Chopping Straw for Plasters" in Chapter 15.) This may impact the strength of the wall; we aren't yet certain how straw length affects cob strength.

Rob Pollacek has found the rototiller attachment on his tractor to be an effective

way to make high-fiber cob. When driven by a 50-horsepower tractor, the tiller has no trouble incorporating plenty of full-length straw.

Other Mechanical Mixing Options

A heavy-duty electric drill with a mortar-mixing paddle (or a hand-held mortar-mixer) is capable of producing very small batches of cob in a bucket. More practically, the same tool can be used to pre-mix clay soil, water, and sand into a homogenous paste inside a round plastic storage container or trash can. This is then dumped onto a tarp or into a pit where the straw can be added by foot.

Another machine capable of making good-quality cob is a *mortar mixer*. Not to be confused with the much-more-common cement mixer, a mortar mixer has a stationary drum like a large barrel on its side, inside of which a horizontal axle turns several paddles. Designed to churn out stiff mixes of mortar and stucco, they are usually powered by gasoline or diesel engines. If you can find one, an electric model is much more pleasant to use because it is quiet and produces no fumes. There are two ways to use these machines to make cob. One is to mix clay soil, sand, and water in the machine, then dump it out on a tarp to add straw by foot. This allows you to make a larger mix (6–10 ft³ [.2–.3 m³], depending on the capacity of the machine) without unduly straining the motor. The other option is to make a much smaller mix (say 2 ft³ [.06 m³]), in which case one can add straw in the machine to a ready-to-use consistency.

A *pan-style* mortar mixer is shaped like a giant cake mixer; its blades rotate around a central vertical shaft. These machines are capable of mixing small batches of high-quality cob, and are better at incorporating straw than the horizontal drum-type mortar mixers.

Greg Allen of the Mud Dauber School of Natural Building in North Carolina mixing cob with a mortar mixer. He can make up to 5 small batches (2 ft³ each) in 20 minutes by himself. CREDIT: DANIELLE ACKLEY

A regular home-scale cement mixer, consisting of a rotating drum with attached paddles, can be effective at combining clay soil, sand, and water, but it may struggle to incorporate long straw. When straw is added, the entire mix will roll around in a ball inside the machine but not really blend. In a pinch, since they are much more commonly available than mortar mixers, they can be used to pre-mix some of the ingredients, with the straw being added by foot.

Stephen Hren, a builder in North Carolina, experimented with most of the mixing techniques described in this chapter before coming up with a new one. He bought a concrete mixing implement for a skidsteer. The attachment consists of a horizontal auger that turns inside a loader bucket. To make cob, Stephen starts by soaking clay soil in 5-gallon buckets, then pouring it into the mixer. The powerful hydraulic auger smashes any remaining lumps of soil and rapidly produces a smooth clay slip. Sand is added next, and when that is fully incorporated he begins adding straw gradually. The total yield is 5–6 ft³ in about 10 minutes. This mixer has no trouble incorporating large amounts of straw or making a stiff mix ready to build with. The bucket opens to dump the finished mix onto the ground or a prepared platform (see photo in Chapter 14, under "Staging, Transport, and Scaffolding"). Compared with a mortar mixer or a tractor, Stephen feels that this machine produces higher-quality cob. It also has the advantage of not requiring a level mixing area, since the mixing takes place inside the bucket of the machine. This makes it ideal for challenging building sites with slopes, trees, and other obstacles.

The perfect cob-mixing machine would probably have to be custom built. The closest approximation we've seen is a large Kirby silage mixer. Designed to blend hay with grains for feeding cattle, it was repurposed to mix clay plaster by a natural building company called Vital Systems in the early 2000s and was dubbed the *cobasaurus*. Running the length of a large open wagon are several helical augurs that rotate in different directions and are powered by the PTO of a tractor. Each augur resembles a corkscrew wrapped around a tube. The cob materials are forced between and around the augurs as they turn. The cobasaurus produced about 5 yd^3 (4 m^3) of extremely consistent cob or clay plaster in 30 minutes or less—a much higher rate of production than any other known mixing method.[3] There are probably other pieces of so-far-unexplored agricultural equipment that would be well suited to large-scale cob mixing. We got excited about this potential after discovering that a flail mower on a tractor is a fabulous way to chop large amounts of straw for rough plasters and floors, and a hammermill made for grinding animal feed is the most efficient way to chop very fine straw for finish plasters.

The inside of Stephen Hren's concrete mixing implement for his skidsteer. In this photo, the hydraulically powered auger has reduced clay soil and water to slip. This machine produces fairly small batches of high-quality cob.

CREDIT: STEPHEN HREN

Table 13.1: Comparing the Efficiency and Quality of Cob Mixing Techniques

Mixing technique	Volume of batch	Time per batch	Mixing rate (cu yds/hr)	No of workers	Mix quality
Tarp	2–4 cu ft	5–15 mins	0.4–0.9	1–2	med-high
Straw bale pit	6–10 cu ft	15–30 mins	0.6–0.9	2	medium
Mortar mixer	2–10 cu ft	4–20 mins	0.6–1.2	1–2	medium
Skidsteer with concrete mixer	5–6 cu ft	10 mins	1.1–1.3	2	high
Rototiller	14–16 cu ft	20 mins	1.5–1.8	2	low-med
Backhoe	14 cu yds	4 hours	3.5	2–3	medium
Tractor or skidsteer	0.5 cu yds	30–40 mins	3–4	2	medium
Excavator *in situ*	17 cu yds	4–5 hrs	3.4–4.2	2	low
"Cobasaurus"	5 cu yds	30 mins	10	3	high

3D Printing and Automated Construction

One approach to make earthen building more affordable in areas with high labor costs is 3D printing. With recent technological advancements, 3D printing of buildings is rising in popularity. Printers designed for this purpose consist of a computer-automated tube that extrudes a wet material mixture while moving through the programmed shape of a building's walls (and sometimes roof, in the case of domes and vaults). Most building-scale printers currently use Portland cement-based products. Though the technology is touted for being materials-efficient, the mixes required by these systems—including plasticizers—are typically higher carbon than cast-in-place concrete. One reason is that most use significantly more cement than is required for standard concrete. Because of the many benefits of earthen construction, a number of researchers, including Dr. Mohamed Gomaa of RMIT in Australia, are encouraging the growing 3D building industry to develop the use of clay-based materials.

In addition to labor savings, another advantage of the technology is its flexibility to create an expanded range of geometric shapes. This allows for the highly efficient use of materials while maximizing thermal, acoustical, and structural performance. Optimizing for all of these factors often results in walls printed from thin clay lines with interior air spaces and undulating exterior forms. Most printed clay materials being produced to date have not been tested for moderate to high structural-loading scenarios.

This technology is not cheap. Some leading manufacturers include WASP, based in Bologna, Italy, and 3D Potter, based in Florida. WASP offers their crane-mounted 3D clay printer/extruder for around $140,000. Each printer is limited to about 550 square feet and 10' of height, although multiple printers can work together to produce larger buildings.

It is debatable whether these systems should be appropriately referred to as *cob* since both the techniques and the mixes differ so greatly from traditional building. Different companies have taken contrasting approaches to materials. Many start with site-harvested clay soils as the base material in order to keep both economic and environmental costs low and to make the technology applicable in less industrialized areas. In order to meet the exacting specifications of the printers, these soils often have to be carefully sifted, moistened, and mixed to a precise consistency. Other projects start with refined clay, which causes many fewer problems with the machines, but is more expensive, less available, and more prone to cracking. To increase strength and reduce cracking, both sand and fine fiber such as rice hulls or chopped straw are often added. Other additives may include hydraulic lime or enzymes intended to improve water resistance.

In addition to continuous on-site printing, other automated earthen building technologies are under development. One is a printed block system developed ☛

One of the many internal wall geometries possible with 3D printed earth. This is an experimental test wall built by Sandy Curth, a PhD candidate at MIT. Credit: Sandy Curth

This prototype printed earthen house, called Tecla, was designed by Mario Cucinella Architects and printed on WASP printers in Italy in 2020. The walls contain internal insulation cavities and ventilation channels for passive cooling.
CREDIT: WASP

by Yelda Gin and her colleagues at Cardiff and Cambridge universities in the UK.[4] These lightweight, interlocking blocks can be produced in a factory and shipped to the building site, which circumvents the difficulties of transporting and assembling a large, heavy printer on the site and keeping it operating smoothly in demanding outdoor conditions.

A company called Terran Robotics in Indiana originally developed an AI-piloted drone and computer visioning technology to place cob in a similar way to hand-building techniques. They then adapted the AI control system to a *cable-driven robot*—a wire-operated, lightweight crane. Terran Robotics has also automated interior and exterior finishing with machine-operated impulse hammers, among other tools, that can be carried by the robots and operated by their AI. These tools shape and smooth the base wall mix as it is erected, making plastering unnecessary. The company claims that this approach can be more easily scaled to large building sites than common gantry and arm printers, and that their AI has the potential to reduce labor costs even further.

A growing number of small-scale—and some full-scale—prototype buildings are being constructed in different parts of the world. Although these automated systems are still in the research and development phase, some enthusiasts see 3D printing as the future of cob construction. In an expensive labor market, this technology is likely to offer big cost savings at a larger scale of production due to reduced labor, shape efficiencies, and speed of construction. These new systems may also help to overcome negative associations about earthen building as a backward-looking technology appropriate only for impoverished populations with no other alternatives.

Left: *Higher Boden Farm in Cornwall, UK, was listed in the Domesday Book in 1066 AD. These cob buildings, which are still inhabited today, are believed to have been constructed incrementally over the last thousand years. Cob has been a common building technique in many parts of the UK for millennia.* CREDIT: KATY BRYCE

Below: *The Great Mosque in Djenné, Mali, was reconstructed in its current form in 1907. Approximately 250 feet (75 m) long on each side, it is considered to be the largest mud building in the world.* CREDIT: JON ARNOLD/ALAMY STOCK PHOTO

The walled city of Shibam in Yemen was built starting in the 16th century using a combination of cob and adobe blocks. Known as the "Manhattan of the Desert," it contains over 500 high-rise buildings, some of which are 11 stories tall. CREDIT: JON ARNOLD/ALAMY STOCK PHOTO

Horsecollar Ruins in Natural Bridges National Monument, Utah, was made by ancestral Puebloan people over 700 years ago. The detail shows the rounded profile of two broken cob lifts. Indigenous people of the American Southwest commonly built with cob (known in the Southwest as coursed adobe) until the Spanish introduced the practice of making adobe blocks in wooden forms in the 17th century. CREDIT: NPS PHOTO/ALAMY STOCK PHOTO.

Above: *Dingle Dell, a cob home complex built in Devon, England, by Kevin McCabe, contains 13,500 square feet (1,250 sq m) of space enclosed by a quarter mile of cob walls 3' thick and up to 29' high. It was built in the 2010s to PassivHaus thermal standards and reached the highest level of UK certification for sustainable homes.* CREDIT: KEVIN MCCABE

Right: *A cob mezcal bar in Ensenada, Baja California, Mexico. The architect was Claudia Turrent, and Jess Shockley was cob consultant.* CREDIT: JESS SHOCKLEY

Below left: *Detail of a sculptural cob cabin at the Mud Dauber School of Natural Building in North Carolina.* CREDIT: DANIELLE ACKLEY

Below right: *One of seven 17th-century cob barns in Cornwall, UK, renovated by Adam Weismann and Katy Bryce of Clayworks Ltd. The buildings now house an art gallery and artist-in-residence retreat.* CREDIT: RAY MAIN

Left: *Sukita Reay Crimmel and Brandon Smyton built this healing retreat in a Portland, OR, backyard. Columns are cob and earthen plaster around wood posts.*
CREDIT: MICHAEL O'BRIEN

Below: *Sculptural cob bench and wall inside a cob addition to a 1920s farmhouse in Austin, Texas. The project was designed by Kindra Welch and built by Lakshmi Jackson and Clay Sand Straw.* CREDIT: JOHN CURRY

Left and above: *Details of cob saunas built by Rob Pollacek of California Cob. Both buildings feature granite stone foundations and round pole roof structures.* CREDIT: ROB POLLACEK

Right: *CobBauge prototype building under construction at Plymouth University in the UK. The purpose of the project was to develop a cob wall that meets the structural and thermal codes in the UK and France. Note the forming system which contains both a layer of structural cob on the interior of the building and a layer of insulating light earth on the exterior.*
CREDIT: LLOYD RUSSELL/PLYMOUTH UNIVERSITY

Center right: *The finished CobBauge prototype. The designers chose a modern aesthetic which looks at home on a university campus.*
CREDIT: LLOYD RUSSELL/PLYMOUTH UNIVERSITY

Above and lower right: *Authors Massey Burke and Anthony Dente, along with designer Jess Tong (shown here), guided this small cob studio through the City of Berkeley (California) Building Department permitting process. This pioneering effort was not easy or fast and required patience and support on the part of the clients, the Tong family. The structural design relies on a rebar grid to resist seismic forces.* CREDIT: MASSEY BURKE

Top left and right: *Full-scale cob walls being subjected to in-plane reverse cyclic testing to simulate seismic forces. These tests provided critical data for writing IRC Appendix AU. The wall above is in the materials testing lab at Santa Clara University and the one at left is at Quail Springs Permaculture in collaboration with Oasis Design and others. Professor Daniel Jansen of the California Polytechnic State University and graduate students Dezire Perez-Barbante and Julia Sargent are shown supervising the testing.*
CREDIT: SASHA RABIN (LEFT) AND ANTHONY DENTE (ABOVE)

Center left: *This half-scale cob building was built on a shake table at the University of British Columbia to model earthquake forces.*
CREDIT: CARLOS VENTURA

Below right: *The Gaia prototype, built in Italy in 2018, was the first demonstration building constructed using the WASP crane-operated 3D printer. The mix consisted of site soil, rice straw, rice hulls, and hydraulic lime. Wall cavities were insulated with rice hulls and include built-in ventilation ducts for cooling.* CREDIT: PHOTO COURTESY OF WASP

Wall under construction for laboratory fire resistance test. This cob wall is reinforced with two layers of vertical wire mesh, similar to wall E, the strongest reinforcement system yet tested. CREDIT: JOHN ORCUTT

Right: *This cob home in southwestern Ireland has been protected by its lime plaster from the force of Atlantic gales for centuries. Limewash is applied annually to seal and repair the plaster.* CREDIT: FÉILE BUTLER

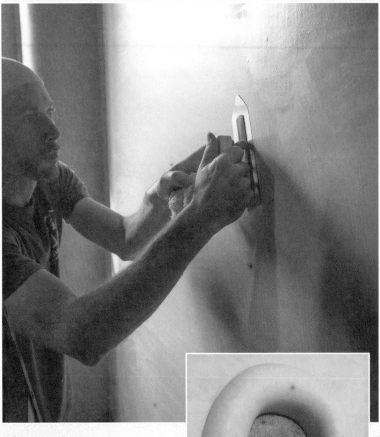

Above: *Exterior clay plasters can also be surprisingly durable, especially when the building is designed to keep the walls protected from rain and snow. This cob cottage in New Mexico was designed and built by Kindra Welch.* CREDIT: KINDRA WELCH

Above and insert: *Kyle Holzheuter burnishes a Japanese-style otsu lime-clay plaster in the cob meeting room at Emerald Earth Sanctuary in California. Although difficult to master, clay-lime finishes can be beautiful, durable, and relatively water-resistant.* CREDIT: JOHN HUTTON (ABOVE) AND KYLE HOLZHEUTER

Right: *In the cob library at the Mud Dauber School of Natural Building in North Carolina, the mushroom design was sculpted into the finish plaster and then painted with clay paint.* CREDIT: DANIELLE ACKLEY

Left: *Carole Crews sculpted these decorative shelves from straw-clay plaster and then finished them with alis, a clay paint containing mica. When leather-hard, the alis is burnished with a damp sponge to reveal the mica and increase the sparkle.* CREDIT: MICHAEL G. SMITH

Center left: *Athena Steen applying straw-clay plaster to a cob sculpture designed by Indigenous artist Nora Naranjo-Morse as part of an installation at the National Museum of the American Indian in Washington, DC.* CREDIT: ATHENA STEEN

Center right: *Earthen floors make a natural complement to cob walls and clay plasters, as in this project by Clay Sand Straw in Texas.* CREDIT: KINDRA WELCH

Left: *Burnishing the tadelakt finish in a cob shower at Quail Springs Permaculture, California. When done correctly, tadelakt is a truly waterproof finish. Note the decorative use of bottles.* CREDIT: JOEL CALDWELL

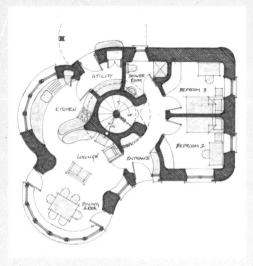

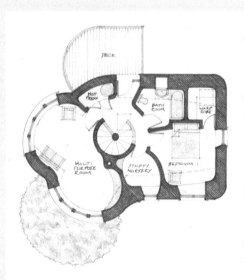

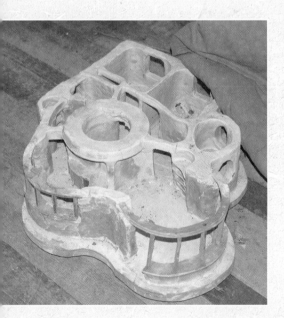

Irish architect Féile Butler took her time when designing a cob home for her family.
She made a series of clay models (lower left) and many drawings to help her visualize
the space. Her husband Colin Ritchie was the primary builder, and the family
lived on site in a trailer during construction in order to eliminate commute time,
save on rent, and maximize everyone's involvement with the project. The sleeping
areas—with their rectangular beds—are located on the more conventional-looking,
rectilinear side of the house that faces the road. The more expressive, curvilinear
side accommodates the living spaces with amazing views of mountains and ocean.
The design also includes many outdoors spaces such as gardens, greenhouses, and
outdoor living areas.

Chapter 14

Building Cob Walls

Staging, Transport, and Scaffolding

COB IS EXTREMELY HEAVY. Dry cob has a density of around 100 pcf (give or take about 20% depending on the mix recipe and ingredients), and it is even heavier when freshly mixed, before the water evaporates. To build a cob structure of any size requires mixing and moving many tons of material. For example, a small circular cob room 10′ across with walls 18″ thick and 8′ high (3 m × 46 cm × 2.4 m) requires around 40,000 lbs (20 tons) of materials for the cob walls alone, not counting water.

Good planning and organization are essential to avoid unnecessary work moving heavy materials. Think about the best places to mix your cob and how to stage clay soil and sand as close by as possible. If you are mixing by foot on tarps, you will need clear flat space inside or immediately outside the building—between 50 and 100 ft² (5–9 m²) of space for each mixer. If you mix further away, you will have to transport the finished cob to the walls. This can be done by loading the material into a truck or wheelbarrow, by dragging or lifting the tarp (which requires many people), or by tossing loaves of cob from one person to the next, bucket-brigade style. While this may be an enjoyable team-building exercise, it is much less time-efficient than mixing close to the wall under construction.

If you are mixing with a skidsteer, tractor, or excavator, you will need a much larger mixing area. But it isn't as critical that the mixing pad be immediately adjacent to the building. Ideally, the machine will be able to pick up the finished mix and transport it to the walls.

A common practice in the UK is to mix cob with a large excavator, pick the finished cob up with the machine's bucket, and place it straight onto the wall. Note that the walls of these buildings are typically quite wide—24″ (60 cm) or more—and lifts are tall (12–24″/30–60 cm) and not manually integrated with a cobber's thumb. Another option is to dump cob from the tractor bucket onto platforms placed at a convenient height for builders to grab. Remember: cob mix is very heavy. Build strong platforms!

Especially when mixing by machine, you can end up with a large pile of cob waiting to be placed on the wall. Some builders choose to mix huge amounts at one time to take advantage of machines being present on site. If the cob mix is on the wet side, it can be left uncovered to dry out for a few hours or overnight. But don't let it get dry and crumbly, or you

A worker scoops cob from the bucket of a skidsteer directly onto the wall under construction. This saves a lot of heavy lifting!
CREDIT:
CONRAD ROGUE

will no longer be able to build a strong wall. By covering the mix tightly with a tarp or a sheet of plastic, you can keep it in good condition for at least a couple of days in warm, dry weather. In cool, moist weather conditions cob can remain usable for two weeks or more. If left too long, eventually either the pile will dry out or the straw in the mix will begin to decompose, feeding anaerobic bacteria, which will make the mix smelly as well as weaker. Make sure that stored cob is still plastic and sticky when you get around to using it; if not, mix in some more water either by foot or by machine. Once cob gets completely dry, the easiest way to reuse it is to soak it in water and then add it to a fresh mix.

If you are able to mix close to the cob walls, you can simply transfer the material directly onto the foundation by hand or with a strong digging fork. Once the walls rise higher than your shoulders, this becomes increasingly difficult. If you don't have a machine such as a backhoe or forklift to lift heavy loads, it is usually most practical to toss loaf-sized lumps of cob to a worker on scaffolding or pass buckets up to them. A rope and pulley suspended from a sturdy beam can help raise buckets.

Some kind of scaffolding is essential for most cob buildings. (Cobbing from a ladder is also possible, but it is less safe and efficient.) There are many options. The goal is to create a safe, level platform on which workers can stand. You can buy or rent portable steel scaffolding or make your own. Remember that there are health and safety standards in every jurisdiction— make sure your platforms meet these standards! Ideally, stand so that the top of the wall is about level with your waist while adding new cob. This allows you to use your upper body weight to push down with your fingers or cobber's thumb to integrate each new lift. If the wall is much higher than that, you will work harder for inferior results. This means that the level of your scaffolding should be easily adjustable to keep workers at the proper height as the walls rise higher.

Stephen Hren uses a concrete-mixing implement on a skidsteer to mix cob (see "Other Mechanical Mixing Options" in Chapter 13). In this photo, a finished batch is being dumped onto a scaffolding close to the wall under construction. The mixing auger is visible inside the bucket.
CREDIT: STEPHEN HREN

Building Strong Walls

Cob walls are always built in *lifts*, or horizontal layers. Depending on drying conditions, the size of the building crew, and other factors, each lift can be anywhere from 6–24" high. Most commonly, a single lift is added to the wall in a day. If the weather is warm and dry, the cob may dry out enough for another lift to be added the following day. However, if the weather is cool and damp—or if the lift is tall—it may be necessary to wait several days or even weeks before adding the next lift. You will soon learn to tell by feel whether the previous lift is stiff enough to support more weight.

As straw is added to a cob mix (especially a large quantity of long, strong straw), it effectively ties the entire mix together into a single integrated unit. When you mix cob on a tarp, you will notice that at some point the entire mix will begin to roll around on the tarp like a big squishy ball. The straw is distributed and knitted together in such a way that it

resists tearing the cob mix into smaller chunks. Unfortunately, it is usually necessary to break the mix apart in order to transport it to the wall. (Mixing cob with an excavator and placing it onto the wall with the bucket minimizes this problem.) The goal when building a cob wall is to reintegrate the cob into a single internally reinforced unit without seams. This is why cob is also called *monolithic adobe.*

Integrating each lump of cob with adjacent lumps makes the wall stronger. A fair amount of this can be achieved just by the way the cob is initially placed; each lump should be overlapped, pinched, and smeared together with its neighbors. The builder can use either their fingers or a tool called a *cobber's thumb*: simply a round-ended stick about the thickness of a thumb (see Figure 5.1). A cobber's thumb penetrates deeper into the wall than fingers can. Its rounded end catches straws in the newly placed cob and *stitches* them into the still-plastic lift below, erasing the seams between lifts

The walls are about halfway completed on this Clay Sand Straw project in Texas. Note that the bottom portion of the wall is a lighter color, indicating that it is dry on the outside, while the upper part is still moist. An integral spine and ribs has been formed into the top course to improve both the drying speed and the bond between lifts. The wall surfaces have been meticulously trimmed. Window bucks are installed and held plumb with temporary braces.

CREDIT: KINDRA WELCH

as well as between adjacent cob lumps. When you are done with it, each lump should be quite difficult to pull off the wall.

One of the advantages of high-fiber cob is that it slumps less than high-density cob, which makes it easy to build taller lifts. Because high-fiber cob is so sticky and fibrous, it seems to bond readily within each lift. If the top of each lift is left rough, the next layer seems to integrate well without much need to stitch the layers together. (High-fiber cob has so far mainly been used in smaller, non-structural scenarios, and the strength of this type of wall has not been formally tested.)

Jess Tong placing cob around vertical rebar reinforcing in a permitted cob building not far from the Hayward seismic fault in Berkeley, CA.
CREDIT: MASSEY BURKE

It's usually most efficient to build cob walls in teams. Two builders working on opposite sides of the wall can most easily make sure that the cob is well integrated everywhere and is going up plumb on both sides. A third worker is often helpful to keep the others supplied with cob. If you are working alone, be sure to walk around to the other side of the wall before adding new cob on that side.

Don't waste time flattening the top of each lift. If it is lumpy and uneven, you will only create a better connection with the subsequent lift. Also, the more surface area the top of the wall has, the faster it will dry. Under poor drying conditions, some people poke holes into the top of the wall with a cobber's thumb to increase the surface area. Another way to increase the surface area is by building a raised *spine* or *key* down the center of the wall, with *ribs* radiating from the spine to each surface. (See photo page 149.)

Reinforcement

In some cases, more reinforcement is required to prevent wall collapse and potential injury than any amount of straw can provide. This is especially true in seismically active regions and in cob shear walls. There are many possible ways to further reinforce a cob wall, but to date only a few of these possibilities have been appropriately tested. The current version of Appendix AU allows several different reinforcement strategies. These include continuous ⅝" threaded steel rod (spaced on center at either 1' or 4' [30 or 120 cm]) as well as 6" x 6" x 6"-gauge steel mesh running vertically either in the center of the wall or inset 2" (5 cm) from both faces of the wall. In each of these cases, the steel reinforcing must be in place before cobbing begins and secured to hold it vertical while the cob wall is built. One way to do this is to install the wooden bond beam in position at the top of the wall before cobbing starts and attach the bars or mesh to that. The bond beam will need to be braced and supported temporarily until the cob wall reaches it. See Chapter 8 for more details on reinforcement and on bond beams.

Plumb or Tapered?

Cob walls can either be *plumb* (straight up and down) or *tapered* (the wall thickness decreases with height). There are several reasons for tapering. The base of the wall carries much more load than the top (all the considerable weight of the cob itself), so it needs to be stronger. Reducing the thickness of the wall as you go up means less cob to mix and less weight to lift. Less weight means less force on the wall if there's an earthquake, though keep in mind that a thinner wall is also not as strong. Another disadvantage is that a thinner wall has less insulation value. This could be compensated for by switching to a lower-density, high-fiber mix at the top of the wall. (But you will want to be confident of the compressive strength of this mix; on permitted projects, you will likely have to test both mixes for compressive strength.) Also, it can be awkward to place furniture such as bookshelves and cabinets against a tapered wall. For that reason, we often taper the outside of the wall and keep the inside plumb.

How much tapering is ideal? There are no set rules. A common slope is about 5%, meaning that for every 20" (50 cm) of height, the wall loses 1" (2.5 cm) of width. The most important consideration is that the top of the wall still needs to be wide enough to support the weight of the roof. We recommend a minimum width of 10" (25 cm) at the top of any load-bearing wall. Short walls that carry no weight can be much thinner.

Beginning cob builders often have a hard time keeping their walls plumb. Once the first

The design of this project (the welcome hut at Shasta Hot Springs shown on the cover of this book) called for ⅝" threaded steel rod reinforcing on 12" (30 cm) centers. To hold the rods plumb during construction, the wooden bond beam was installed before cob building began. The beam is supported on nuts underneath as well as on top. Note the wet straw being used to prevent the top of the cob wall from drying out. CREDIT: ROB POLLACEK

foot or so of cob is in place, plumb surfaces should be established on both sides of the wall either by adding more material where necessary or by trimming off excess. From then on, keep a level within reach and check the wall frequently. If your design calls for a tapered wall, cut a wedge of wood to the desired angle and tape it to one side of a builder's level. To check if the wall is at the correct taper, simply hold the wedge-side of the level against the wall.

At the end of the day, or before adding new cob the next day, trim off any protruding cob. (See more on trimming, below.) As the wall goes up, add lumps of new cob along both faces

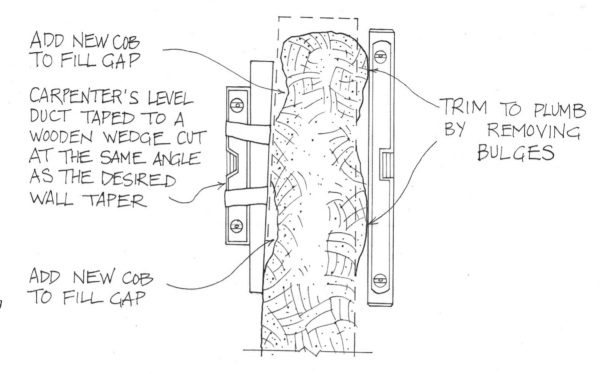

ADD NEW COB TO FILL GAP

CARPENTER'S LEVEL DUCT TAPED TO A WOODEN WEDGE CUT AT THE SAME ANGLE AS THE DESIRED WALL TAPER

ADD NEW COB TO FILL GAP

TRIM TO PLUMB BY REMOVING BULGES

Fig. 14.1: *Keeping walls plumb and tapered.* CREDIT: DALE BROWNSON

of the wall first, using one hand as a form to keep the surface of the newly added cob plumb as you integrate it with the cobber's thumb in your other hand. Then fill in the cavity remaining in the middle of the wall.

Drying

A cob wall needs to dry as it is being built. If a new lift is added before the previous one has dried enough to bear its weight, the cob underneath will compress and bulge. This creates extra work in several ways: 1) extra cob has to be mixed to make up for height lost to settling; 2) the bulging cob will need to be trimmed off to return the wall to its desired surface planes; and 3) the wall may also go out of plumb as it compresses (especially on curves) or dries unevenly, requiring extra work to return it to plumb. If a cob wall is built too quickly it can actually collapse! In poor drying conditions, mold may grow on the surface of the wall. If the cob stays wet for long enough (probably a

matter of months), the straw inside will begin to decompose, weakening the wall.

On the other hand, if the top of the wall is allowed to dry out completely, it becomes much more difficult to integrate the next lift. Extra time and effort must then be spent rehydrating the top of the wall, and the connection between lifts will still be inferior. Ideally, the builder maintains a happy medium where the lower sections of the wall are allowed to dry enough to support the weight above, but the top of the wall remains moist. This can be challenging to achieve.

Generally, cob building is fastest and easiest during warm and dry weather. In many parts of the world, this means that there is a preferred season for cobbing. In most of California, for example, one can expect good cobbing weather between April or May and October. As you go further north or closer to the coast, the length of this good-weather window shrinks. In climates with dependable dry seasons, it is sometimes possible to build the walls first, then add the roof before winter comes.

A season of reliably warm, dry weather makes cob building simpler, faster, and more pleasant, but it isn't absolutely necessary. Cob can also be built in places with higher humidity or summer rainfall, or during mild winter weather, but the drying time will be considerably longer. Both the CobBauge project in the UK and Eco-Sense in British Columbia, Canada, found that it can take a full year for cob walls to dry in those damp climates.

Cob walls under construction can withstand light showers without damage, but if heavy rain is expected, cover the walls to prevent erosion and soaking. Some builders in rainy climates erect a temporary roof, which protects the walls from rain and allows work to continue in inclement weather. A temporary roof erected above wall height promotes drying better than tarps or plastic sheeting, which prevent air flow around the wall. Sometimes the permanent roof of a building can be constructed before the cob walls, supported either on permanent posts or on temporary posts that will be removed after the walls are complete. A roof also provides welcome shade on hot building sites.

Avoid cobbing during freezing weather. If temperatures fall below freezing for long enough while the cob is still wet, the expansive force of freezing water inside the wall can cause structural damage. (This will only happen during a sustained deep freeze, as cob's thermal mass prevents it from changing temperature rapidly.)

When cob is installed inside an already enclosed structure, drying can be very slow—regardless of the weather. Not only that, but large quantities of water evaporating from the cob may condense elsewhere in the building and cause serious problems. Drying can be speeded up by large fans or floor blowers and by either opening windows and doors (if the

outside air is dry) or by using a heavy-duty dehumidifier (though this is only practical when the building is completely enclosed and heated). Another trick for slow drying conditions is to insert dry materials such as stones into the wall to stabilize the cob and reduce the amount of moisture that needs to evaporate out. Be aware that these insertions interrupt the straw reinforcement in the wall and will therefore reduce the wall's flexural strength (see "Niches, Bottles, and Stones" in Chapter 8). Therefore, this is not a good strategy for load-bearing walls in seismic regions, or for shear walls.

Excessively fast drying can also be a problem. In hot, dry climates it is common for the top surface of the wall to crust over rapidly, making integration of the next lift more difficult. The top of the wall can be kept moist by covering it with tarps or plastic sheeting between lifts. Even more effective is placing absorbent material, such as old towels or straw, on top of the wall, soaking it with water, and then covering with a tarp. By this means, the top of the wall can be kept moist and plastic for several days even during the hottest, driest weather.

Expect cob walls to settle and shrink *vertically* as they dry. When the wall is built slowly and/or drying conditions are excellent, most of this shrinkage will happen as the wall is under construction. But if the wall is built quickly relative to drying conditions, there can still be a lot of moisture inside the cob when the wall reaches its full height. Under these conditions it is best to wait as long as possible (say a month of dry weather) for the walls to dry thoroughly before loading them with the weight of the roof. It is also ideal to wait until the wall is dry before plastering; as the cob dries, it continues to shrink and settle, which can crack plaster. Also, plasters (even clay plasters) reduce the rate of evaporation from the wall and therefore slow down drying.

Trimming

Unless a cob wall is built unusually slowly and carefully, some amount of trimming and filling will be necessary to make the wall surfaces straight and even. If the wall is built quickly or carelessly, it can end up deviating from plumb (or the desired taper, if any) either by being too wide or too narrow at the top. These kinds of problems are best caught and corrected as soon as possible. After adding each new lift, check your work with a level to make sure the wall is still plumb. The best solution to bulging walls is to trim them before they get hard with a modified hand saw (see photo in Chapter 5 under "Tools for Cob Wall Building"). If, on the other hand, the wall has become too narrow or has unwanted voids, you can apply more cob to the surface of the wall to thicken it. Use moist, sticky cob for this and work it into the wall horizontally with a cobber's thumb.

As cob dries, it becomes harder to trim. After a day or two of drying, it will become too firm to cut with a hand saw, but it can still be trimmed with a machete, hatchet, or gardener's adze. Once the wall is fully dry, these tools may still work, but they will be slow. A quick way to remove large amounts of dry cob is to use an electric hammer drill with a chisel bit made for cutting through concrete or tile. The claw of a hammer also works well for chipping out channels in the wall for wiring and plumbing.

Colin Ritchie and Féile Butler used plywood forms when building their home. Tamping into the forms allowed them to use a stiffer cob mix to reduce drying time in their cool, damp Irish climate, and the need for trimming was almost eliminated.
Credit: John Butler

Formwork

Cob walls are most often built without forms. However, some cob builders in France and the UK began using wooden forms as early as the 1820s,[1] and this option is becoming increasingly popular today. The biggest advantage is that forms make it easier to keep walls straight and plumb and their thickness consistent. They can save a lot of time by reducing or even eliminating trimming. This needs to be balanced against the extra time and materials required to build and install strong forms.

Large forms like those used for rammed earth walls are not practical because they don't allow the cob to dry inside. (Rammed earth uses a much drier mix than cob.) When the forms are removed, all of that wet cob is likely to slump or collapse. If formwork is used, it should be only the height of a single lift—no more than 24" (60 cm), maximum. This reduces the weight of wet cob and allows builders to reach their hands to the bottom to integrate each new layer. Solid forms should be removed as soon as possible so that the wall can dry. Leaving them in place longer than a few hours can lead to mold growth on the wall surface.

Rob Pollacek of California Cob has used forms to build all of his cob walls since 2001. His standard form system, shown in the picture, uses fabricated pipe clamp brackets attached to strips of plywood 18" (45 cm) tall. They can be adjusted to any wall width from 10" to 24" (25–60 cm). They also taper; the width at the top of the form can be adjusted to between ½" and 1" (1–3 cm) narrower than the bottom. If the form is used five times to build a wall 7'6" (2.3 m) high, the wall will be between 2.5" and 5" (6–13 cm) narrower at the top than it is on the bottom, which is a taper between 2.7% and 5.5%.

Before setting up his forms for a new lift, Rob works in an inch or so of fresh cob onto the previous lift to ensure a good bond. The forms are then clamped onto the wall with the pipe clamps, overlapping the previous lift by about 2" (5 cm). He uses a wooden tamper to fill the forms to the top, then punches some holes in the top of the wet cob with a form stake to improve bonding with the next layer. Then he immediately removes the forms by releasing the clamps. Each lift is allowed to dry for 2–3 days before the next one is placed on top.

As part of their efforts to develop a thermal cob wall for northern climates, the CobBauge project (see "CobBauge: A Structural and Insulating Cob Wall" in Chapter 3) developed a unique forming system. Their forms needed to contain both cob (on the inside of the wall) and an insulating matrix of clay and fiber on the outside. They elected to use metal mesh— with a 2" (5 cm) space between wires—as the form material, which permits the wall to dry while the forms are left in place. Each form is 24" (60 cm) tall, and enough were constructed to allow each tier of forms to remain in place while another layer is added and filled on top.

Rob Pollacek's standard forming system. The pipe clamps can be tightened to give the wall a slight taper. Forms are removed as soon as they are full.
CREDIT: ROB POLLACEK

These small forms are longer on the outside than the inside to match the curvature of the wall. Each form is lined with a loose sheet of aluminum flashing so that the cob doesn't stick to the form. Threaded steel rods spaced 12" (30 cm) apart run from the foundation to the top plate to strengthen the seismic resistance of the wall. CREDIT: ROB POLLACEK

Stephen Hren's formwork for an octagonal cob building. The forms are made of ½" galvanized mesh attached to an 18"-tall (45 cm) stud frame. The vertical 2×6s are temporary. They support both the forms and a temporary rooflet that protects the wall from rain during construction.
CREDIT: STEPHEN HREN

This protects and contains the friable light-earth layer while the next lift of material is added on top. See the color section for a photo of this system.

Building Arches

There are two ways to span across the top of wall openings such as doorways, windows, and niches: *lintels* and *arches*. Lintels, which are structural beams generally made of wood or concrete, are described later in this chapter and also in Chapter 8. Arches made of earth, stone, and brick have been in use in many cultures for millennia. There are still-standing mud-brick vaults at the Ramesseum in Egypt and arched adobe gateways in Iran and Israel—all well over 3,000 years old. These and many other examples attest that in pure compression, earthen arches are extremely strong and durable. Like other masonry building systems, they tend to perform more poorly during earthquakes. We are not aware of any engineering research that

has been done on cob arches, and there is no provision for them in Appendix AU.

Because of the lack of testing data, we can't offer clear guidelines about how to build cob arches safely. We would advise caution when building arches more than three feet across, especially in Seismic Design Categories C and D (see Figure 8.3). Keep in mind that the catenary shape (made by hanging a chain with both ends held level) is stronger than a Roman, or semicircular, arch. It is probably also advisable to reinforce the cob above archways with strips of steel or plastic mesh, rebar, bamboo, or flexible saplings.

Cob arches can be built in two ways: with or without formwork. Especially if the opening is wide, it is faster to build an arch over a form, which can be as simple as a stack of bricks, blocks, or firewood with a sheet of flashing or cardboard over the top to keep the cob separated. When the cob is dry enough to maintain its shape, the form materials are removed and

Using a form to build an arch at a Mud and Wood workshop in Ireland. Note that bent flexible sticks are encased in the cob as reinforcement over the arch.
CREDIT: FÉILE BUTLER

the cob around the opening can be trimmed to refine its shape. Or a form can be built to the exact shape and size of the desired opening. To create a small round opening, you could set a plastic bucket sideways on your wall, cob around it, wait until the cob is mostly dry, and then pull the bucket out, leaving a round opening which can later be filled with glass. Wrapping the bucket with cardboard or plastic sheeting makes it easier to remove.

In the early days of the Cob Cottage Company, around 1993, Ianto, Linda, and Michael developed a technique for building arches without formwork. In those days (taking a cue from the etymological root of the word *cob*, which means a *loaf or lump* in Old English), we kneaded every batch of cob we made into the size and shape of a small loaf of bread before placing it on the wall. We later realized that this was usually unnecessary (although it can be helpful where extra precision is required, and for tossing cob up to a builder high on a wall). We found that if we made our cob loaves quite a bit smaller (perhaps 10" long, 2" thick and 3" wide ($25 \times 5 \times 8$ cm) and worked extra straw into them running lengthwise through the loaf, these modified *corbel cobs* could be used almost like ceramic bricks to construct corbelled arches, vaults, and domes. It's important to go slowly, adding only a couple layers of corbel cobs at a time and integrating them carefully with a cobber's thumb. See *The Hand-Sculpted House* for detailed instructions.

Sculpting

Like ceramics, cob is an extremely sculptural medium. It is easy to sculpt curved walls, round or arched (or practically any shape) windows and niches, and decorative reliefs. There are many ways to build these sculptural details. Sometimes it is helpful to create an armature out of sticks, rebar, bamboo, twine,

Left: *The Yanaguaña-Meek residence in Texas Hill Country, designed and built by Kindra Welch of Clay Sand Straw. The cob walls are complete, and the arched forms have been removed from above the doorway and windows.* CREDIT: KINDRA WELCH

Below: *A finished view of the same house, with window, entrance door, and clay plaster.* CREDIT: JOHN CURRY

This bottle wall was designed and built by Massey Burke. The bottles were first wired onto a piece of vertical wire mesh inserted into the wall, then cob was built between the bottles and the exposed wires snipped away. Note the curved laminated wood lintel which distributes the weight of the roof onto the walls on either side.
CREDIT: MASSEY BURKE

wire fencing, or other materials to serve as a structural skeleton for sculptural elements. For example, if one wished to shape a cob gargoyle with its head projecting 12" (30 cm) from the wall surface, a good way to do that would be to construct a rough shape out of rebar and fencing and bury this partially in the cob wall. After the wall is mostly or completely dry, cob can be sculpted around the armature.

High-fiber cob is particularly useful for sculptural work, since it is stickier, more flexible, and much less prone to slump than normal cob. Rather than *stitching* with a cobber's thumb as described above, we tend to add high-fiber cob to a wall or other cob element with a wrapping and smearing action. This approach offers particular flexibility when you want to create cantilevers such as seats or shelves: the ends of the fibers can be integrated into the top and bottom of the shelf without adding a lot of extra height, allowing a shape to extend horizontally with less vertical thickness than when

corbelling. The wrapping technique can also be used to build vertical walls quickly (this works best for walls that are non-structural and not more than a foot thick, such as landscape walls and interior partitions).

If the sculptural element is low-relief (say a flowering vine that appears to be growing up the wall, projecting just an inch or two from the surface of the wall), it can be sculpted directly into the cob or the base coat of straw-clay plaster without additional reinforcement. If the cob wall is already dry when you come back to add a low-relief sculpture, first drive nails or dowels halfway into the cob where you wish to add the relief, then wet the wall and cob around the protruding nail heads. If you need more control for any kind of detailed sculptural work, you may wish to make a finer cob mix by sifting your clay soil and using chopped straw instead of full-length straw. You can also create relief sculpture by cutting into dry cob with a hatchet, a chisel, the claw of a hammer, a hammer drill, or an angle grinder with a masonry blade.

Bottle Walls

Many cob builders have achieved gorgeous decorative effects by inserting bottles, jars, or small pieces of colored glass into their walls. The design potential is practically unlimited; it creates spots of colored light that crawl across the floor and interior walls as the sun moves through the sky. However, be aware that bottles weaken the wall in at least two ways. First, the wall thickness is often dramatically reduced in order to allow light through the bottles. Second, the matrix of reinforcing straw is broken wherever bottles are inserted (see "Niches, Bottles, and Stones" in Chapter 8). As a result, bottles should not be used in shear walls. To be safe, the bottle-and-cob section should be considered structurally as if it were a window opening, with a lintel to distribute the load of cob wall

and roof above into the full-thickness cob walls on either side.

It is also advisable to reinforce bottle walls, especially in seismic zones. One way to do this is to insert vertical wire mesh into the center of the wall, like those shown in Wall 2 in Figure 8.6. Some of the wires can be snipped away as necessary to create space for the bottle design, but all remaining wires would ideally be surrounded by 2" of cob to prevent them from rusting away over time. It is also possible to use strips of mesh, like those shown in Wall A in Figure 8.6. These flexible strips could run horizontally or diagonally through the cob, but should ideally be continuous from one side of the bottle wall to the other.

Electrical and Plumbing Installation

Electrical and plumbing in permitted cob buildings must comply with electrical and plumbing codes as in any other building. Cob walls present unique challenges to satisfying these codes, especially in regard to the location and attachment of electrical wiring, boxes and panels, and rough plumbing pipes. The simplest solution is to minimize electrical and plumbing in cob walls. This is especially important for plumbing supply pipes; if not caught and repaired quickly, a leaky pipe will damage a cob wall or reduce its structural capacity. However, electrical switches and outlets must be located on some cob walls in order to satisfy the electrical code's maximum outlet spacing and light switching requirements.

Appendix AU contains no requirements for electrical and plumbing in or on cob walls; however, its commentary contains five paragraphs regarding common, successful practices for both. (See under Section 101.1 in the Appendix). Codified or accepted practices for concrete block or concrete walls could provide guidelines for electrical and plumbing in cob walls, especially if a building official needs an example from these somewhat analogous and more familiar wall systems. Many builders have had good success simply burying romex (insulated wire) inside cob walls. Others prefer to use cable rated for direct burial in the ground, which has a thick protective coating. Or conduit can be built into the wall and the wires pulled through later. This last option is the most expensive but also the most flexible because it allows wires to be added or resized in the future without cutting into the wall and damaging the finishes.

When planning for wiring and plumbing inside a cob wall, one critical decision is what to build first. Either the pipes and wires can be assembled first and the cob wall built around them, or the cob wall can be built first and the utilities added later. A combination of both options is common. For example, where pipes and wires come up through the foundation and into a cob wall, they generally have to be installed first. But, depending on your circuit design, it might make sense to install the wiring to the first outlet in the circuit before cobbing begins, then run wires or conduit through the wall horizontally when the cob wall reaches the appropriate height. The primary difficulty in installing utilities first is that they must somehow be suspended securely in the air where the wall is going to be. One good way to do this is to embed rebar or threaded rods into the foundation or the lower lifts of cob and secure the wires and pipes to those. Some building designs require rebar, threaded rod, or wire mesh inside cob walls for structural reinforcement. Those elements provide convenient anchors for wires and pipes.

It is often more convenient to build the cob wall first and install utilities later. Even after the cob is dry, it is not difficult to drill holes through the wall or carve channels into the surface using a heavy-duty hammer drill with concrete bits. Another way to make a channel in dry cob is to use an angle grinder or circular saw with a masonry blade to cut two kerfs into the wall, then chip out the cob between the cuts with a chisel, the claw of a hammer, or a hammer drill. Make the channel deep enough that there is no chance of a future occupant damaging the buried wire when driving a nail into the wall to hang a picture—check local codes for depth. Wires, conduit, or pipes can be secured inside the channel with staples or nails. After the circuit or pipe has been tested, the channel can be filled with fresh cob. Even when using this option, plumbing and wiring should be planned ahead of time so that you don't accidentally create obstacles. It is much harder to drill a hole through a cob wall or carve out a channel if rocks or steel reinforcing are in the way.

Whenever possible, we prefer to keep plumbing out of cob walls and run it through interior walls instead. (Plumbing codes in cold climates often prohibit plumbing in exterior walls anyway.) These walls can be conventionally framed, making the plumbing and wiring simpler and the consequences of leaks less dire. If you choose to run plumbing through a cob wall, the safest option is to avoid placing any pipe joints inside the wall. Even a minor leak inside a cob wall can be extremely detrimental to the structure over time. All plumbing should be pressure tested for leaks before being buried in cob. If you need to run plumbing from one side of a cob wall to the other, you can install a larger pipe during wall construction (sloped slightly toward the exterior in case of leakage) and then run your pipes later through this protective sleeve. Another option is to build a chase into the interior base of a cob wall, place pipes and wires inside of this cavity, and then cover it with a removable wooden trim board for easy access. This technique will structurally weaken the wall, so the depth of the chase should be kept to a minimum and the overall thickness of the wall may need to be increased to compensate. In other words, the structural design of the wall should not include the thickness of the chase.

Connecting Cob to Wood

Few, if any, buildings have ever been constructed entirely out of cob; nearly all include other materials such as lumber, especially for framing the roof, windows, and doors. In many cases, cob walls are load-bearing, meaning that they support the weight of the roof directly. Other designs include structural posts and beams to support the roof or a second story.

Here we offer a few options and considerations for creating secure connections between cob walls and wooden framing elements. Note that due to the *hydrophilic* (water-loving) properties of clay, dry cob acts as a long-term preservative of any wood it touches. This can

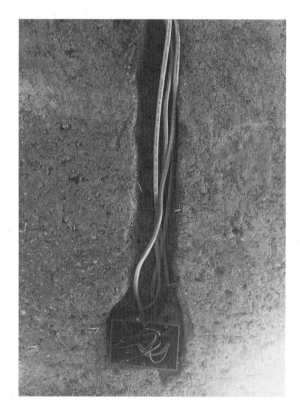

After the wall is dry, channels can be cut to install wiring and plumbing. Make sure that all wires are covered by several inches of cob. CREDIT: KINDRA WELCH

be seen in centuries-old cob homes in the UK, where wooden members such as roof rafters and beams are much more likely to have developed dry rot in places where two wooden members touch than where one of them touches a cob wall. It is therefore not necessary to use rot-resistant wood in contact with cob (except if it also comes in direct contact with the foundation, which may wick water up from the ground); any sound lumber will work.

Structural Posts and Beams

Many builders prefer to keep structural posts out of cob walls. If a post buried inside a cob wall moves at all (during an earthquake, for example), it could initiate cracking in the surrounding cob. Sometimes structural posts are kept completely separate from cob walls, and sometimes they are partially buried and partly exposed on the interior of the building. In order to keep the cob walls strong and continuous, make sure that buried posts don't interrupt more than ⅓ the thickness of the wall. In other words, a 4×4 (10 × 10 cm) is the maximum size of post that can be fully buried in a 12"-thick (30 cm) cob wall.

If you do embed posts inside a cob wall, keep in mind that the cob will move, shrink, and settle as it dries. Especially when the cob wall is built quickly, it commonly pulls away from rigid vertical element such as posts. One way to prevent this is to stud with nails all surfaces of the post that will be surrounded with cob. Large, strong nails such as 16-penny construction sinkers (they needn't be galvanized) are driven into the post at about a 6" (15 cm) spacing, deeply enough to anchor them firmly. The protruding heads of these nails will effectively grab onto the wet cob, preventing it from pulling away from the post.

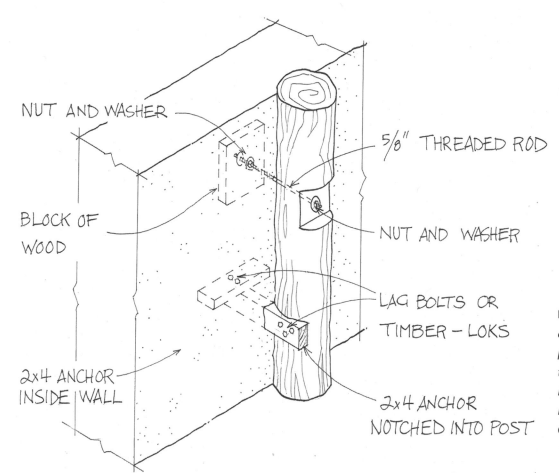

NUT AND WASHER

BLOCK OF WOOD

2×4 ANCHOR INSIDE WALL

⅝" THREADED ROD

NUT AND WASHER

LAG BOLTS OR TIMBER — LOKS

2×4 ANCHOR NOTCHED INTO POST

Fig. 14.2: *Options for attaching structural posts to cob walls.* Note: *the threaded rod option is preferred where this material is available.* Credit: Dale Brownson

There are other ways to anchor a wooden structural element to a cob wall. One is to insert a large bolt or threaded rod into the beam or post, leaving several inches sticking out to be surrounded by cob. Another is to construct a wooden anchor by firmly attaching a piece of structural lumber such as a 2×4 onto the post or beam, running transversely into the cob. At the other end of this piece of wood, secure another piece at right angles to the first, to be fully buried inside the cob (see Figure 14.2).

Wooden bond beams at the tops of cob walls are a special situation. To achieve their purpose of securely connecting roof framing to cob walls, these beams must be strongly built and securely anchored. (See "Bond Beams" in Chapter 8 for detail drawings.) Appendix AU specifies in most cases using ⅝" threaded anchors on 12" centers, with each anchor embedded either 12" or 16" (30 or 40 cm) deep into the cob wall. In the case of wooden bond beams, the easiest way to achieve this is to install the beam before the top 12" of cob have been built. Thread the anchors into holes drilled in the beam and install washers and nuts on each anchor both above and below the beam. The beam will be temporarily supported on the anchors; brace it as necessary so that it doesn't move as you install cob under and around it. Build up with cob until a gap about 3" (8 cm) high remains under the bond beam, then pause to let the cob solidify and shrink. Then back off the nuts underneath the beams as far as possible. When you fill the gap under the beam, use a stiff, sandy mix and ram it in place with a wooden tamper to completely fill the gap. After the cob has dried and settled, tighten the exposed upper nuts to cinch the beam down onto the cob wall. Figure AU106.9.5 in Appendix AU shows a diagram of how roof framing should be connected to bond beams.

Windows, Doors, and Cabinetry

Lightweight items such as pictures and wall hangings are easy to attach to a cob wall after the wall is dry. To support objects weighing up to a few pounds, you can simply drive screws into a finished cob wall. Long, thin lag bolts and timber screws work well as long as there are not many rocks in the cob. But the heavier the item is, the less likely this strategy is to work. A heavy mirror, cabinet, or TV monitor should be mounted to a piece of lumber that is securely attached to the cob wall. This requires planning ahead.

There are many ways to build an anchor to serve this function. One good approach is to set a vertical piece of lumber such as a 2×4 or 4×4 on top of the foundation, holding it in place with temporary braces so that it will remain plumb. One face of the lumber will end up flush with the inside surface of the cob wall, with the remainder buried in the cob. Stud the surfaces of the wood that will end up buried in cob with nails, as described in the previous section. Once the cob is dry and the braces have been removed, the lumber will be securely attached to the cob wall, with one face exposed. This surface can then be used to securely anchor an object such as a cabinet, counter, or window frame. In the case of a door or window frame, it is better to place the wooden anchor inside the wall so that it ends up completely surrounded by cob except where the frame will be attached. Keep in mind that these kinds of anchors may weaken the wall, especially if the wall is very thin to begin with or under high seismic loading. To be safe, they should be no more than ⅓ the thickness of the cob wall.

Another way to accomplish the same goal, borrowed from the adobe tradition in the Southwestern US, is called a *gringo block*. This is a box with no bottom or top, easily made out of 4 pieces of 2×4 or 2×6 and some screws. The block is buried in a cob wall with one face exposed, which can later be used to anchor door or window frames or cabinets. In the case of a door or window frame, use 2 or 3 gringo blocks aligned vertically on each side of the frame to create a strong connection. A similar effect can be achieved by burying almost any piece of wood (called a *deadman*) in the cob wall with one end or face exposed.

These anchoring systems work well when the cob walls are to be built first, then the wooden members attached later. Sometimes the opposite sequence is preferred. For example, it is convenient to place a rough door frame (or *buck*) in place, secured to the foundation, before cob building begins. Commentary Figure AU.105.4.5 in the Appendix shows our preferred way of building a door buck

in this situation. Securing another piece of framing lumber such as a 2×4, called a *stiffener*, perpendicular to the main buck framing accomplishes two important goals. First, the stiffener strengthens the buck framing. If there is no stiffener, the considerable horizontal force exerted by wet cob will likely deform a door buck constructed of 2-by boards until it is no longer straight or plumb. Secondly, the stiffener serves as an anchor to firmly attach the frame to the cob. This is especially important in the case of door frames, since the force of opening and closing doors over time may cause the frame to separate from the cob wall if it is not securely anchored. Always brace any window or door frame securely, keeping in mind the enormous force that wet cob can exert horizontally.

Fig. 14.3:

Anchoring wood to cob using deadmen, gringo blocks, and vertical anchors.

CREDIT:

DALE BROWNSON

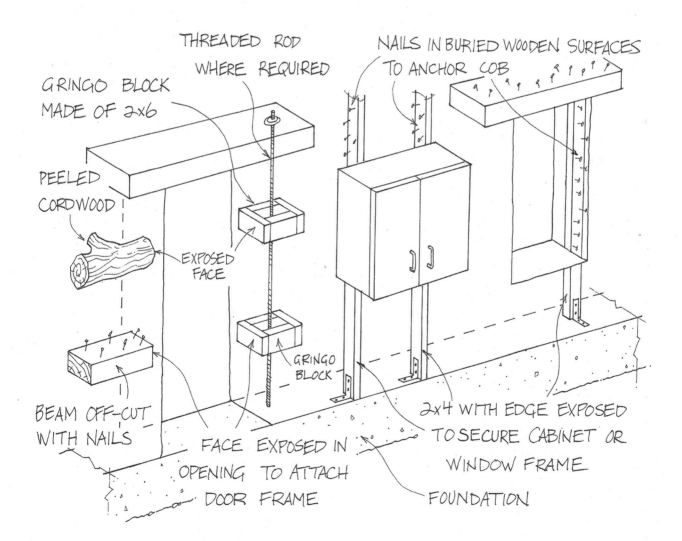

THREADED ROD WHERE REQUIRED

NAILS IN BURIED WOODEN SURFACES TO ANCHOR COB

GRINGO BLOCK MADE OF 2×6

PEELED CORDWOOD

EXPOSED FACE

GRINGO BLOCK

BEAM OFF-CUT WITH NAILS

FACE EXPOSED IN OPENING TO ATTACH DOOR FRAME

2×4 WITH EDGE EXPOSED TO SECURE CABINET OR WINDOW FRAME

FOUNDATION

The usual way to support the weight of cob above a window, door, or other opening in a cob wall is to insert a lintel over the opening. This is a short beam often made of wood (though concrete and stone also work well) and supported by the cob wall on either side. Appendix AU requires that threaded rod be embedded in the foundation on either side of window and door openings and run through holes in the lintel to the bond beam. See "Lintels/Headers" in Chapter 8 for more recommendations on sizing and placement. The embedded sections of wooden lintels should be studded with nails as already described. In the case of concrete lintels, the rebars in the lintel can be left a couple of inches longer than the concrete, serving as anchors between cob and lintel. See the photos on the cover of this book for examples of how lintels can look in finished buildings.

Because cob shrinks and settles as it dries, it is best to wait as long as possible before installing lintels. If a lintel is installed on a thick layer of wet cob, it may lose its level as the cob dries. In good drying conditions, waiting a few days before installing a lintel is probably enough (while keeping the top of the cob moist to ensure good bonding later). Even so, expect the lintel to drop slightly as the cob continues to dry. It is best to insert the lintel at least ½" higher than the desired final height of the opening. Any remaining gap can later be closed with a

Build Your Own Dual-Pane Windows

by Art Ludwig

Regular dual-pane glass comes as sealed units, with silica gel or another moisture-absorbing material inside. When the seal eventually fails due to wall movement or material degradation, more moisture will leak inside than the finite amount of silica gel can absorb. At this point, water begins to condense between the panes of glass. This moisture is acidic from CO_2 in the air, so it can permanently etch the glass, making it more and more cloudy until eventually the whole unit must be replaced. Thus, standard dual pane glass is a single use, throw-away product—albeit usually with a reasonably long life.

How about making your own dual-pane windows instead, taking advantage of clay's natural hygroscopic properties to extend their lifetime indefinitely? My approach is to rip ½" × ½" (13 × 13 mm) redwood spacers on a table saw, place them between the edges of two same-sized sheets of glass, and tie the whole ensemble together with duct tape. The duct tape seems to have just the right permeability to prevent too much water from getting inside the unit when the cob is wet, while allowing the water vapor to escape once the wall is dry. This unit may be built directly into the cob wall without a frame. The literal tons of water-absorbing clay surrounding the window should keep it condensation-free pretty much forever. This way of making dual or triple pane glass represents a huge cost savings. It is probably not suitable for permitted buildings where windows must be laboratory-tested and rated for air-tightness and other thermal properties.

I have extensive experience with this system in the Santa Barbara, CA, climate, which is very dry in summer, but sometimes damp and cold for weeks in winter. I've seen a bit of fog right after the glass was cast into the wet, freshly built walls—and *none* in the decades since, even on walls facing every direction and with direct rain impingement. Would it work as well in a wetter climate? There may be a humidity threshold inside a cob wall where water would begin to condense between the panes. (But see the Bairds' extremely reassuring findings in "Moisture Management and Control" in Chapter 3.) A cool feature of homemade dual glass: it will give a visual indication if your walls are dangerously wet. Please let us know your experience in your climate.

piece of wooden trim. Some builders set their lintels 2 or 3 inches higher and later fill the gap with cob.

Similarly, it is important to wait before placing cob across the top of any large rigid object that has been placed in the wall, such as a door or window buck. Otherwise, the cob on either side of the object will settle vertically, while the cob on top will be held in place; this can lead to severe structural cracks that run diagonally upwards from each upper corner of the frame.

Retrofits and Renovations

Once a cob building is complete, it can be intimidating to contemplate making major alterations. While it's true that renovations can be messy and complicated (think: mud in the house, water on the floor, plaster repairs), adjusting the cob walls or adding new ones is not out of the question. The biggest issues have to do with design and structural integrity. Before adding new windows or doors or otherwise weakening any wall, it's important to remember that the wall may be serving as a shear wall, helping to resist wind and earthquake forces—and not only as a gravity-bearing wall. These are the same considerations involved with any major renovation regardless of building system, and you may want to find an architect or engineer to help you. One of the differences from other building materials is cob's great weight (approximately 100 pcf). You will need a sturdy foundation for any additional walls, and new foundations should be securely connected to the old. Roofline alterations also require careful planning to make sure that you aren't creating any new pathways for water to get into the structure.

Planning a retrofit is much easier and safer when you know the locations of internal steel reinforcing, plumbing, wiring, and any other elements that are no longer visible. This is one of several important reasons to make complete building plans and to make sure those documents stay with the house. Detailed photographs of each wall under construction are similarly invaluable.

One of the most effective ways to remove dry cob is to use a hammer drill with a chisel bit made for cutting through concrete. This is a dusty and noisy process, but it need not be arduous. The material you remove can be soaked and added to new cob mixes; it's one of the beauties of cob!

Before adding fresh cob to dry, be sure to wet the surface thoroughly. It takes a while for water to soak in, so this process is best started a couple of days before the new cobbing will begin. Clay slip works even better than water to reactivate dry clay. When possible, cut back the old cob into a stair step pattern, leaving horizontal planes that water can sit on to soak in more deeply.

When building a new cob wall adjacent to an old one, some kind of pinning is recommended to prevent the walls from separating over time. You can drill into the existing wall with a masonry bit and drive lengths of rebar into the hole, leaving them protruding a foot or more into the new cob. Or you can use helical steel ties, made for connecting and repairing masonry walls. These are simply driven into the old wall with a drill.

In England and other parts of the UK, where thousands of centuries-old cob homes remain in use today, there is an active community of restorationists dedicated to keeping these lovely old buildings alive. During the second half of the 20th century, it was common to repair cob buildings using non-vapor-permeable materials such as concrete patches, tar, cement stucco, and commercial paint. These materials often had the unintended consequence of worsening the problems they were meant to fix, and in some cases led to the collapse of entire buildings. Nowadays, it is well understood that earthen buildings must be repaired and maintained with compatible materials such as earth and lime. If you own a historical cob building in need of repair, please reach out for local expertise. Look in the Resources section for some organizations dedicated to preservation of earthen buildings. The Devon Earth Building Association has a useful pamphlet available on their website called *The Cob Buildings of Devon 2: Repair and Maintenance. Building with Cob: A Step-by-Step Guide* also contains an excellent chapter on cob restoration.

Chapter 15

Finishes

THERE ARE MANY POSSIBLE FINISHES for cob walls. The most common are: no finish (raw cob) and vapor-permeable plasters, such as clay, lime, and/or gypsum (with or without natural paints). Wooden siding is also a good choice in some circumstances.

In theory, the ideal finish material for your cob building would combine the following characteristics: wear-resistant, water-resistant, vapor permeable, reflective, strongly adhesive to cob, safe and non-toxic, inexpensive, low environmental impact, easy to mix and apply, easy to repair, and … beautiful! This is a tall order to fill. Many types of natural finishes combine most of these characteristics, but none has the entire list of desirable qualities. The best choice in any particular circumstances will depend on many factors including climate, building design, budget, and aesthetic preferences.

Unlike some other natural wall systems such as strawbale, which absolutely need to be

Table 15.1: Common Finishes for Cob Walls

Type of finish	Embodied carbon	Permeability	Water resistance	Advantages	Disadvantages	Best applications
No finish (raw cob)	none	high	low-medium	much less work than any other option, reveals the wall material	tends to be dark, crumbly and look "unfinished"	dry or protected exteriors, non-residential interiors
Clay plaster	very low	high	low-medium	very safe, excellent adhesion to cob, easy to apply/repair, excellent air quality	relatively soft, some shrinkage, DIY recipes need testing	interiors, dry or protected exteriors
Gypsum plaster	medium	high	medium	white, reflective, sets quickly, fairly hard, no shrinkage	short working time	interiors
Lime plaster	high	medium	high	mold-resistant, can be maintained/repaired with lime-wash, used for fresco	delamination possible, best to apply in cool weather	weather-exposed exteriors, wet interiors (bathroom, sauna)
Cement stucco	high	low	high	easy to source materials and find experienced workers	tends to crack and delaminate	not recommended due to low permeability
Wooden siding	low-medium	medium	high	excellent protection from extreme weather and abrasion	somewhat complicated to install, especially on curved walls	weather-exposed exteriors, interior wainscoting

plastered to ensure their durability and resistance to fire, rodents, insects, and moisture, plastering cob walls is somewhat optional. The rough texture of unplastered cob makes it moderately weather-resistant, since water running down the surface will be slowed down by lumps and divots, protruding straws, and bits of gravel. The slower the runoff, the less its erosive force. Surprisingly, researchers have found that the average rate of erosion of unplastered cob walls in Devon is only about an inch of wall thickness lost per century,[1] except where roof runoff is blown against the walls or splashes onto the base of the wall. (This figure assumes an intact roof that sheds most of the weather away from the walls.) Emily Reynolds' research on cob storage buildings in Japan found a much greater erosion rate in many cases.[2] The 14 buildings she measured averaged 80–100 years old, and most of them had minimal roof overhangs and no foundations at all. In nearly all of these structures, the degradation due to erosion was most severe in the bottom third of the exterior wall surface. As one would expect, the most weather-exposed walls were the most degraded, while those protected by other buildings or landscape features tended to be in good condition. There was also a correlation to annual rainfall—on average, the more rainy the location, the more severe the erosion damage.

While leaving the exterior surfaces of a cob building unfinished may be a reasonable choice under some circumstances (low rainfall and/or well-protected walls), it is almost always desirable to plaster the interior. One reason is that unfinished cob has a rough, friable texture. Every time you, your cat, or your curtains brush against the wall, bits of grit and straw will fall down into your bed, your soup, etc. Any decent finish plaster is smoother and harder than raw cob, so it resists wear much better. Also, raw cob tends to be dark in hue and swallows up the light. By adding a smooth, light-colored finish plaster, you vastly increase the amount of reflected light inside the space, which is better for your eyes, your mood, and your electricity bills.

Types of Plaster

Many kinds of plaster are more water-resistant than cob, which can be important in circumstances where the wall regularly gets wet. However, as discussed in "Moisture Management and Control" in Chapter 3, cob's relationship with moisture is a bit complicated. It isn't necessarily the case that the most durable finish for a cob wall is the most waterproof. Rather than water*proof*, what is wanted in wet conditions is a water-*resistant* finish. This means that if the material gets wet, it will not get soft and erode away (or it will erode very slowly, allowing plenty of opportunity for maintenance or replastering). At the same time, the finish material should also be *vapor permeable*, meaning that water vapor inside the wall can escape through the finish layer and evaporate away. Appendix AU specifies that all finishes used on cob walls must have a vapor permeability rating of 5 perms or greater. Many water-resistant or waterproof materials do not have sufficient permeability to meet this criterion. For example, a typical cement-based stucco has a rating of less than 1 perm, as do oil-based paints. Latex paints range from 2 to 10 perms depending on the product. Clay plasters, lime plasters, gypsum plasters, and many natural paints are highly vapor permeable, making them the safest finishing options for earthen walls.

Permeability is not a purely hypothetical concern. There are too many sad stories of old cob buildings in the UK (and other kinds of earthen buildings in the American Southwest and elsewhere around the world) that have met an untimely demise as a result of inappropriate surfacing. For example, tar-based materials are sometimes painted onto the base of cob walls in the mistaken belief that this will protect them from moisture damage. Similarly, old lime and earthen plasters have been replaced with more modern "maintenance-free" cement stucco. In both cases, the results are sometimes catastrophic; trapped moisture builds up inside the cob walls to the point where the wall, or the whole building, may collapse.[3,4] Nor is this problem limited to natural buildings. As conventional building envelopes have become more sealed and less permeable, mold growth and other problems related to trapped moisture have proliferated, leading to severe health concerns, especially in damp climates.

It is common to apply all plasters in at least two coats, a *base coat* and a *finish coat*. In the case of cob walls, the primary purpose of the base coat is to level out the surface so that the finish coat can go on in a thin (usually between ⅛" and ¼" thick [3–6 mm]) and even layer. If your goal is a very smooth and fine finish, such as the Japanese *shikkui* or *otsu* (see below), you will need an intermediate flattening layer called a *brown coat*. If you plan to use a coarser finish plaster and are comfortable with some undulation to your finished wall, you can go straight from the base coat to the finish coat.

What follows is a discussion of the most common plaster systems for cob buildings, focused on the advantages and disadvantages of each. We are not providing full instructions on how to use any of these finishes; for that you will have to read a different book. We highly recommend *Using Natural Finishes: Lime- and Earth-Based Plasters, Renders, and Paints,* and *Essential Natural Plasters: A Guide to Materials, Recipes, and Use.* Conventional plastering crews are also a fabulous source of technical plastering know-how, often with overlapping skills in lime and gypsum, though it is unusual for them to be familiar with clay.

Clay or Earthen Plasters

Clay-based plasters have a number of desirable characteristics that make them the most obvious choice for cob walls, at least in dry areas. Compared to other kinds of plaster, clay plasters are safe and non-toxic, inexpensive, environmentally friendly, relatively easy to use and apply, and easy to repair. Because of the similarity of the ingredients and clay's natural stickiness, these plasters adhere easily to cob. They are extremely vapor permeable and also vapor absorbent. Because clay is hygroscopic, earthen plasters absorb moisture out of the air, reducing the likelihood of condensation and mold growth even on other less-permeable surfaces in the house. For this reason, our colleague Athena Steen advises clients and students to always plaster at least one wall of their bathrooms with clay.

The main drawbacks of clay plasters are that they are relatively soft and not very water-resistant. Like cob, earthen plasters are made of unfired clay that will turn back into mud if it gets sufficiently hydrated. In practice, this is not usually a problem. Well-formulated

The cement stucco on this old Irish cob building has cracked and begun to peel away, leaving the cob beneath vulnerable to erosion. Cement has been used to "protect" many historic earthen buildings all over the world—with the unfortunate result that much of our built vernacular is failing or has been irretrievably lost.
CREDIT: FÉILE BUTLER

Straw-clay plaster can be mixed by hand in a wheelbarrow or tub. Simply make clay slip and add chopped straw to the desired consistency. See Chapter 5 for a photo of mixing in a mortar mixer.
CREDIT: ELKE COLE

Opinions are mixed among production plasterers on the topic of starch paste in plaster: some really enjoy the added stickiness and decreased dustiness, whereas others note that starch paste can cause visual mottling and change the texture of the troweling in objectionable ways. Small amounts of linseed oil (less than 1% of the total volume) added to a plaster will also improve its water-resistance and hardness.

Because of the many advantages of earthen plasters, we typically use a clay base plaster no matter what we have chosen for the finish plaster (except sometimes with lime plasters; see below). Following the lead of Bill and Athena Steen, Carole Crews, and other teachers, we usually make our clay base coats out of just three ingredients: clay soil, water, and chopped straw. Since the mixture typically contains no added sand, the key to prevent cracking is to add lots and lots of chopped straw. This mix, called *straw-clay plaster*, is ideal not only for smoothing and leveling the wall surface, but also for adding sculptural details. One of its most surprising properties is that it is amazingly weather-resistant. We have used straw-clay plaster as the finish material on exterior surfaces that get blasted repeatedly by rainstorms, then dry out and look much the same as before. This cycle can go on for many years with little degradation; how much depends on the properties of the clay from which the plaster was made.

Another common way to make an earthen base plaster is by recycling the cob trimmings that have been shaved off the wall during construction. These can be collected and saved until the wall is ready to plaster, then soaked and stomped on a tarp or mixed in a mortar mixer, usually with the addition of more chopped straw. However, if your cob mix contains lots of rocks or gravel, the resulting material can be difficult to work with.

and carefully applied clay plasters are surprisingly hard and durable; enough so that you can brush, wipe, and even sponge the surface without damaging it. Hardness and water-resistance can be improved by adding water-soluble paste to the plaster mix. It is easy to make your own starch paste by cooking white flour in water. You can also use methyl cellulose wallpaper paste or even diluted white glue. Added in relatively small quantities (commonly between ⅛ and ¼ the volume of the clay in the mixture), paste will improve the durability of any clay plaster, even if you start with a silty clay soil.

Make sure to thoroughly wet the surface of the cob wall before applying plaster to improve adhesion. Earthen base coats are usually applied by hand and then smoothed out with a wooden float while still wet. Do not use a steel trowel on your base coat because it will close the pores and polish the surface. This not only slows down drying, but also makes it more difficult for the finish coat to bond, since the suction of moisture and fines into the open pores of the base coat is one of the things that makes the finish plaster adhere. A wood float leaves the surface flat but not polished, ready to receive a clay finish. However, if you plan to apply a lime finish plaster over an earthen base coat, we recommend scratching the surface with deep horizontal grooves before the plaster dries. This provides a better mechanical bond for a dissimilar material like lime, which otherwise has the potential to delaminate.

Clay finish plasters are strongest when made from clay soil with a relatively high clay content and little silt, or from bagged clay, available in many kinds and colors from a ceramic supply. Depending on the purity and expansiveness of the clay source, it is common to add between two and three parts of sand to each one part of clay powder or slip. The sand should be fine, clean, and angular. You can use your cob sand sifted through a $\frac{1}{16}$" (1.5 mm) screen or

Straw-clay plaster over a cob wall (foreground) and a strawbale building (background) at the Canelo Project in Arizona. When exposed to weather, the clay gradually washes away from the surface, leaving a mini-thatch of exposed straw fiber which resists further erosion.

Credit: Athena Steen

Marisol Lopez smoothes clay plaster on the exterior of the Earth Chapel in Davis, CA, using the plastic lid of a yogurt tub. This finish plaster is made of wild clay soil, sand, finely chopped straw, and starch paste.
CREDIT: LAURA SANDAGE

purchase plaster sand or fine silica sand—60 mesh and 90 mesh are commonly used sizes. Other standard ingredients in clay finish plasters include fibers, such as very finely chopped and sifted straw or hemp, horse or cow manure, paper pulp, or cattail fluff, which make the plaster tougher and less likely to chip or crack. You can also add starch paste, pigments for color, and mica flakes for sparkle.

Pre-mixed clay finish plasters are available for purchase in the US and elsewhere. Some of the available products provide excellent results, but they are expensive and not available everywhere. Most are intended to go on in very thin coats over prepared sheetrock, rather than over a rougher earthen wall; they will crack if applied in a thicker layer than they are designed for. With practice you can learn to formulate your own plaster mixes that are as good or better than any you can buy. The key is to make lots of test samples. When working with a new

Clean, well-defined edges wherever the plaster encounters other materials such as wood and stone go a long way toward giving the building a professionally finished look.
CREDIT: ROB POLLACEK

clay soil, we will often try a dozen or more recipes before coming up with one we like.

There is a typical difference between clay finish plaster recipes for interior and exterior applications. Interior plasters are usually smooth and hard, with high proportions of fine sand and very fine fiber, if any. Exterior plasters have a rougher texture and a higher volume of longer fiber—both of these properties make the plaster more weather-resistant. As the clay on the surface erodes away, more and more fiber is revealed, creating a *mini-thatch* effect that slows

water down. A thick layer of straw-rich plaster is much more durable on exterior surfaces than multiple thin coats, which may delaminate when they get wet. It's also interesting to note that many "wild" clay soils (especially those with high clay content) produce more water-resistant finishes than bagged ceramic clays. This is, in part, thanks to the wide range of sand particle sizes naturally occurring in clay soils, which is difficult to match when mixing purchased ingredients.

Recipe

Recipe 2: Earthen Plaster Recipes

	Recipe (measured by volume)	Best applications/notes
Straw-clay plaster	1 part clay slip[1] sifted through ¼" screen 1½ –2 parts coarse chopped straw (1"–4" long)	Interior or exterior base coat, exterior finish, sculpting
Recycled cob filler	1 part cob wall trimmings, soaked (dump out excess water) ¼ part clay slip[1] ½ part coarse chopped straw	Great for filling holes in cob walls; often too rough for a nice finish
"Wild" clay finish plaster	1 part clay slip[1] sifted through ⅛" screen 1½ –2 parts plaster sand, sifted through ¹⁄₁₆" screen 0–¼ part starch paste ¼ part fine chopped straw, sifted through ¼" screen, or horse/cow manure	Interior or exterior finish; will go on ³⁄₁₆" to ¼" thick; good for somewhat irregular walls; starch paste greatly improves water-resistance
Bagged clay finish plaster	1 part dry clay powder 1 part 60 mesh silica sand 1½ parts 90 mesh silica sand ⅛ part starch paste 0–¼ parts fine chopped straw, sifted through ⅛" screen Pigment as desired	Interior finish; will go on about ⅛" thick over a flat brown coat; pigment should be dispersed in water before adding to the plaster
Starch paste	1 cup white flour 2 cups cold water 4 cups boiling water Dissolve flour into cold water with a wire whisk, then pour into boiling water. Continue to whisk over heat until bubbles form and the paste turns translucent. Let cool.	Makes earthen plasters stickier, harder, and more water-resistant; a frequent ingredient in clay paints and adhesion coats

[1] Clay slip is the consistency of a thick milkshake or pancake batter.

Note: These are just starting ratios. Be sure to make test samples and let them dry before you determine your final recipe.

Chopping Straw for Plasters

When building cob walls, we prefer to use straw full length as it comes out of the bale. There can be a wide range of straw lengths, depending both on the crop (some grain varieties have longer stems than others) and the type of equipment used for harvesting and baling. Occasionally you will find bales where the average straw length is less than 6" (15 cm). We often set those aside to use in base plasters. In some regions of North America, it is possible to buy "bales" (actually, compressed bags) of chopped straw from a feed store, where it is stocked as bedding for small animals. It is also sometimes possible to source *chaff* (the leftover residue of winnowing grain) from grain processors. This is a mixture of fine straw, grain hulls, and some seeds. Where none of these options is available, you will find yourself needing to chop your straw for use in earthen plasters and floors.

We have seen people chop impressive amounts of straw with hand tools: machetes in Africa and Latin America, and a paper-cutter-like contraption in Japan. But when large quantities are needed, we typically use machines to do this job.

One efficient way to chop straw is to use a chipper-shredder. These range in size from small gas-powered models made to chip small branches up to PTO-driven tractor implements capable of shredding small trees. Any of these machines will work well on straw. With the smaller models, you may need to feed the straw into the hopper slowly in order not to overwhelm the machine. Chipper-shredders typically chop straw to 2–4" (5–10 cm) long. This is an excellent size range for base plasters. If you

want the straw finer or coarser, the length can sometimes be adjusted by changing screens on the machine.

A flail mower on a tractor will chop a great deal of straw quickly. A home-scale lawn mower can also be used to do this job. To prevent the chopped straw from flying everywhere, make a confined space by stacking straw bales or assembling sheets of plywood into an open-topped box large enough to maneuver the mower inside of.

Another useful tool is an electric leaf-mulcher. These consist of a hopper on a stand, with a rotary head like that of a weed-whacker inside. After being chopped, the straw sifts into a trash can or bag underneath. These machines are quiet and safe, and are able to chop straw to lengths between ½" and 3" (1–8 cm). If you want super-fine straw for finish plasters, you may need to run the straw through the machine a second time and then sift the results through a ¼" hardware cloth screen. These are not industrial machines—we have burned out the motors with overuse and by allowing straw to pile up around the motor, causing it to overheat. It's *possible* to use a weed whacker inside a plastic garbage can the same way, but it is much harder to control the process. You can't see what you're cutting through the clouds of straw dust, and you tend to end up pulverizing some of the straw while other strands remain much longer than you want.

The best tool we know of for producing very fine straw is a *hammermill* made for processing livestock feed. This uncommon and expensive tool rapidly converts an entire bale into tiny fibers between ¼" and 1" (0.5–2.5 cm) long, perfect for finish plasters and floors.

Lime Plasters

Lime plasters have the enormous advantage of being water-resistant *and* highly vapor permeable. This combination has made lime the finish of choice for earthen buildings over the centuries from Wales to Morocco to Iran to Japan to Mexico. When properly mixed and applied, lime plaster is very durable. It is light-colored, reflective, and mold-resistant. (In damp places like bathrooms, a polished lime plaster will be more hygienic and easier to clean than a rough one, since dirt can build up inside pores and become a substrate for mold.) For many centuries, throughout towns and villages all over Europe, the annual spring cleaning ritual included painting a new coat of limewash onto both the inside and outside walls of the home. Lime is also the base material for a wide range of beautiful finishes and effects, including *fresco, venetian plaster, marmorino,* and *tadelakt.*

Unfortunately, the production of lime is highly energy intensive. To make building lime, limestone or seashells (both forms of calcium carbonate) must first be heated in a kiln to about 1650°F (900°C). This process requires huge amounts of energy, generally from the burning of wood or fossil fuels. As a result, the embodied carbon of lime approaches that of cement. Per unit of volume, the embodied CO_2 in lime plaster is about 75% that of cement stucco— and over ten times higher than clay plaster![5] We therefore reserve lime plasters for places where their water-resistance is most needed, such as on weather-exposed exterior walls, near a shower or bathtub, or in a greenhouse or wet sauna.

Burned lime is called *quicklime* or *calcium oxide.* When it is added to water, heat is released and the chunks of quicklime gradually melt into a sticky putty—*calcium hydroxide, hydrated lime,* or *building lime.* This process is called *slaking;* depending on the quality of the quicklime, it can take up to several months. Lime putty is highly caustic, with a pH between 12 and 13. This makes it somewhat hazardous to work with; gloves, protective clothing, and especially eye protection are strongly recommended.

In Europe, it is still possible to source lime putty made in the traditional method described above. But good lime putty is not readily available in North America. Instead, it is common to find dry hydrated lime, usually called *Type S lime.* This product is sold in a bag and can be turned into putty by soaking it in water—or it can be mixed with sand and water to make lime plaster. Powdered lime has a limited shelf life. When exposed to the air (even inside the bag),

it will absorb carbon dioxide and begin to carbonate, eventually converting back to calcium carbonate. The best way to store dry lime is to turn it into putty, which will stay good indefinitely as long as it is covered with water. Lime putty made from dry hydrate is not as reactive as putty made directly by slaking quicklime, and the resulting finishes are less durable.

Some types of limestone naturally contain clay. When burned and then slaked, these produce *natural hydraulic lime* (NHL), so-called because of its ability to set under water. Hydraulic lime sets faster and harder than other types of lime, with properties more similar to Portland cement. Its vapor permeability is intermediate between non-hydraulic lime and cement. NHL has recently become popular for finishing natural wall systems such as strawbale and cob. It produces very durable finishes, and the quick set time is convenient for modern building schedules. However, the cost is high in North America because the material is imported from overseas. It's possible to mimic the properties of NHL by

Sukita Crimmel frescoing a cob sculpture at a Natural Building Colloquium. Fresco is the process of painting mineral pigments onto fresh lime plaster. As the lime sets, it binds chemically with the pigments, creating a highly durable and color-fast surface.
CREDIT:
MICHAEL G. SMITH

mixing a *pozzolan* such as fired clay, brick dust, or volcanic ash with regular lime.

To make lime plaster, sand is mixed with either lime putty or dry lime powder. The standard proportion for a base coat is one lime putty to three sand by volume. This produces a gritty mix that is not very sticky and can be hard to work with. In successive coats, both the proportion and the size of the sand are reduced, resulting in a more workable mixture.

Fiber is sometimes added to lime. The Steens make a lime base coat from Type S lime, medium plaster sand, and chopped straw. As with clay plasters, you can get away with a higher lime content because of the added fiber, which makes it easier to work with. In Europe and North America, the most common fiber used in the past was hair from horses or goats. In Japan, seaweed glue and fine hemp fiber (or sometimes straw) are traditionally mixed with lime to produce a finish called *shikkui*. The mix contains no sand and is generally applied credit-card thin over an earth plaster brown coat. *Shikkui* is designed to last two to ten decades, depending on exposure to wind-driven rain. It is basically a sacrificial layer intended to protect the earthen

wall beneath from erosion. When it eventually begins to separate from the earthen plaster below, it can be easily removed and replaced.

This points to a potential problem with lime plaster over cob: delamination. Even though lime plaster is water-*resistant* (meaning it doesn't get soft when it gets wet), it is also water-*permeable*. If enough water soaks through the lime plaster, the cob or clay base plaster behind it can absorb that water and expand, weakening the connection between the two materials. Even worse occurs when a heavy soaking rain is followed by a hard freeze. Freezing water expands inside the wall with enormous force, which can push the entire layer of lime plaster off of the wall. We saw this happen when we covered an unroofed cob bench with a "protective" lime plaster, beautifully and painstakingly frescoed. After one hard freeze, the plaster was all on the ground! Since that time, we use lime plasters over cob only where the roof overhang is sufficient to keep rain from soaking the plaster.

There are several other potential solutions to the delamination problem. One is to improve the connection between clay and lime by scoring the clay layer with deep horizontal

In Japan, a thin lime plaster called shikkui *has been used for centuries to protect earthen buildings like these storehouses. The same plaster also safeguards the exposed wooden roof structure from fire. Note that the bottom parts of some walls, which are vulnerable to damage from water pouring off of the roofs, are shielded with wooden siding, which has been charred to make it more resistant to both water and fire.* CREDIT: MICHAEL G. SMITH

scratches, as described above. In Mexico, small stones are often embedded into the surface of the earthen base plaster for the lime plaster to adhere to. Lime and lime-clay base coats also reduce the likelihood of delamination. In the UK, the traditional approach to applying lime plaster over cob walls was called *roughcast*. In this technique, a stiff lime plaster was thrown onto the wall using a specialized tool called a *harling trowel*, something like a tiny shovel. The force of collision resulted in good adhesion and also caused the lime plaster to be densest in the back, where it came in contact with the cob wall. The surface was not troweled but only roughly *knocked down* with a wooden float. This left the plaster softest and most permeable at its surface and hardest and most compact deep inside. Each layer of roughcast was a maximum of ⅜" (1 cm) thick, but decades—or centuries—of replastering could result in a very thick protective layer, reducing the likelihood of water soaking through. This is almost opposite to the Japanese approach. *Shikkui* contains a very high proportion of lime and is applied with metal trowels, both of which tend to reduce water absorption, even though the lime plaster layer is so thin.

Recipe

Recipe 3: Lime Plaster Recipes

	Ingredients	Directions
Lime-sand scratch coat	1 part Type S lime or lime putty 2–3 parts plaster sand Water as needed 0–½ part fine chopped straw, horse or cow manure, hemp fiber, or chopped hair Optional: When using Type S lime, 5–10% metakaolin (a natural pozzolan) will increase hardness and weather resistance	Mix in a mortar mixer, a bucket with a hand-held mixer, or in a mixing boat with a hoe. If fiber is to be added, a mortar mixer works best. Start with water, then mix in half the sand, then lime, then the rest of the sand. Add optional fiber slowly and mix well. Add more water as necessary to achieve a stiff but spreadable consistency. It works best to mix the plaster a day or more before you plan to use it, store it in a sealed container, then mix again just before using.
Lime-earth base coat[1]	3 parts clay soil Water as needed 1–2 parts Type S lime or lime putty 2 parts sand 6–8 parts coarsely chopped straw	Sift clay soil through a ¼" screen. With a mortar mixer or hand-held mixer, slowly add lime powder and water to achieve a thin pancake batter consistency. Add sand, then mix in straw slowly, adding more water as necessary.
Lime brown (middle) coat	1 part Type S lime or lime putty 2–3 parts plaster sand screened to 1/16" Water as needed 0–¼ part fine chopped straw, manure, or paper pulp Optional metakaolin (see scratch coat recipe)	Mix as per lime-sand scratch coat. Paper pulp can be either cellulose insulation or toilet paper, which can be removed from the roll, dunked in water and added whole to the mixer. Float with a wooden or poly float.
Lime finish coat	1 part Type S lime or lime putty 1–2 parts fine aggregate (60 mesh silica sand, 90 mesh silica sand, marble dust, marking chalk, and/or dolomite) Water as needed 0–¼ part fine chopped straw, hemp fiber, or paper pulp Pigment if desired	Mix as per lime brown coat. Use a combination of different aggregates for greater strength and smoothness. Pigment if added should not total more than 10% of the volume of the lime, and should be dispersed in water before being added to the mix.

[1] Recipe from Athena Steen

Lime-Clay Plasters

The right combination of lime and clay can provide the best qualities of both. This mixture serves as the binder for sand or fiber or both, depending on the situation. Lime-clay plasters adhere better to earthen walls than lime plasters without clay. Their relative softness allows them to move and flex more easily than harder, stiffer lime plasters, which reduces the likelihood of delamination when the wall gets wet and swells. The lime in these mixes also helps reduce the extreme shrinkage typical of thick straw-clay base coats. Since lime sets chemically, it speeds up the drying process. Without added lime, thick interior earthen base coats tend to dry very slowly and can sometimes grow mold, especially in wetter climates. Lime-clay finishes have enough lime in them that they can be frescoed.

Among the lexicon of traditional Japanese finishes is a beautiful, polished plaster called *otsu*, made from a mixture of clay soil, lime, and very fine fibers, such as hemp or paper pulp. Like a polished lime-sand plaster, otsu is extremely smooth, reflective, and abrasion-resistant. It is also more water-resistant than a typical clay plaster. The colored clay in the mixture lends a rich hue difficult to achieve with lime alone. Lime plasters and clay plasters can both be polished, but lime-clay mixtures seem to polish more easily than either one alone.

A warning: an *otsu* finish is difficult to formulate and apply without specialized training. If the proportions are wrong, lime-clay mixes will turn out soft, crumbly, and powdery. Adding small amounts of lime to a clay plaster can actually *reduce* the binding quality of the clay. The best ratio varies depending on the type of clay soil, including whether it is acidic or alkaline. The goal is to reach a pH of 11.5 or higher. At a lower pH, the mixture will be weak and crumbly. The Steens have found that a ratio of one lime to three clay by volume is suitable for most Southwestern alkaline clay soils. Where the clay soil is acidic, the ratio should be increased to 1:2 or even 1:1.

When working with either lime or lime-clay mixes, it's important not to let the plaster dry too quickly—otherwise, it will not set strong. Ideally, the weather should be cool and damp when these types of plaster are applied. If it isn't, keep the wall surface shaded and humidified during and after application. Because combining lime with clay dramatically increases the number of variables, it is difficult to offer simple guidelines for how to do it right; but it also remains an important and relatively underexplored area for creative effort to expand and improve modern natural plaster methods.

Gypsum Plasters

Gypsum, also known as *plaster of Paris*, has been used as a wall-finishing material at least since the times of the ancient Egyptians. It is made by heating a naturally occurring mineral called *bassanite* to approximately 300°F (150°C) in a kiln. Since this temperature is much lower than the temperatures required to produce lime and Portland cement, gypsum's embodied energy is correspondingly less. Gypsum plaster is white, smooth, reflective, and fairly hard. Gypsum has the unique property (unlike clay, lime, or cement) of not expanding when mixed with water nor shrinking as it dries; therefore, it resists cracking—even without the addition of sand or other aggregate. True gypsum plasters have excellent vapor permeability, similar to earthen plasters.

Gypsum plasters are more water-resistant than clay, but much less so than lime. (This is the same base material from which drywall is made—the near-ubiquitous plaster substitute that must be coated with paint to protect it from moisture.) For the most part, it is used only on interior walls. If gypsum gets soaked with water, it will soften and fall apart. That is why you aren't supposed to get your plaster of Paris medical casts wet.

When water is added to this white powder, a chemical reaction causes the gypsum to harden in about half an hour. The rapid set time has both advantages and disadvantages. An experienced plastering crew can apply several coats of gypsum plaster one after the other and walk away with the job finished much faster than is possible with earthen or lime plasters. Unfortunately, the short working window also makes developing the necessary skills a challenge.

There are still extant traditions of gypsum plastering in Europe and in some areas of the Northeastern US. In these regions, a wide range of commercial gypsum plaster products are available. Elsewhere in the US, the list of available products is typically very short. A commonly available base coat is called Structo-Lite, consisting of gypsum mixed with finely ground perlite aggregate, which allows the plaster to go on more thickly with added insulation and less weight. Pure gypsum finish plasters are also available. Note that true gypsum products are always sold as a dry, bagged powder. Joint compound, spackle, and drywall texturing products commonly sold as premixed white putty in buckets are made of gypsum mixed with synthetic chemicals that retard setting but also severely reduce the vapor permeability of the material. Don't use these products over natural wall materials such as cob.

Gypsum was the base material for the craft of decorative plasterwork that flourished in Europe starting in the 14th century and spread around the world from there. Various mixtures of gypsum, lime, sand, and hair have been used to make ornate decorative ceilings, moldings, and friezes, employing complex techniques including casting, sculpting, and carving. Even today, fine finishes are sometimes made with a mixture of gypsum and lime. The lime slows down the set time to give the craftsman more working time. Conversely, gypsum can be added to clay- and lime-based plaster to speed up the hardening process.

Preparing Cob Walls for Plaster

We recommend waiting for your cob walls to dry before beginning the plastering process. One reason is that cob shrinks as it dries, so plaster applied prematurely is more likely to crack. Also, each successive coat of plaster smoothens the wall surface, reducing surface area and porosity. By plastering too soon, you are actually slowing down the rate at which moisture can escape from the wall. How long to wait? At the minimum, until all surfaces of the walls have reached the lighter color of dry cob. The timing for this varies enormously depending on drying conditions, as explained in Chapter 14.

Trimming and filling are appropriate at any time; you don't need to wait for the wall to dry. In fact, the sooner you do these the easier it will be. See "Trimming" in Chapter 14 for techniques. Depending on the recipe, a base coat plaster will be able to fill voids up to 2" or 3" thick at one time, so only deeper holes than that really need to be filled before plastering begins. Large depressions or voids in cob walls can be filled in with cob (add less sand than usual to improve adhesion, and substitute chopped straw if you have it available, for better workability), or you can use an earthen base coat plaster (see above for recipes). In general, the larger the voids in the wall that need to be filled, the more appropriate it is to use regular cob and the sooner the patching material should be applied, since all patches and base coats need to be fully dry before you can apply your finish coat.

The base coat stage is also an ideal time to add or refine sculptural elements. If you get too sculptural while the wall is still under construction, it is much harder to trim the wall and keep it straight and plumb. Base plasters made with sifted soil and chopped straw are easier to

A strip of split bamboo trim helps create clean lines between two types of plaster in the Emerald Earth Sanctuary meeting room in Northern California. The wall plaster is clay-lime otsu, *and the ceiling is plastered with* tosa shikkui, *a Japanese finish consisting of lime and finely chopped straw. Plaster work was led by Kyle Hozheuter.* CREDIT: TOM SHAVER

A strip of dark-colored plaster at the base of the wall is a good solution where dark floors meet light plasters. When the trim plaster is sealed with the same oil as the floor, it becomes very hard and water-resistant, helping to protect the wall plaster from damage and keep it clean. CREDIT: MICHAEL G. SMITH

sculpt than regular cob. Because the base plaster will not show once the building is complete, it is OK to base plaster different walls, or different parts of the same wall, at different times. It is also fine to apply a second coat of base plaster (the *brown coat*) when further wall shaping or a very flat surface is desired. Remember that a well-detailed base coat makes it much easier to install your finish plaster in a thin, even coat. Take a detailed tour of every square foot of wall area (especially around wall edges, windows and doors, and other details) before you consider the base plaster step finished.

It is very helpful to figure out *early* in the construction process where the finish plane of your plaster will end up. This is especially important with regard to electrical outlets. Electrical boxes should be installed so they will be flush with the finish plaster; this way, the face plates can seamlessly cover both. Luckily, electrical box extenders are available at many hardware stores, in case you end up recessing your boxes too far. Understanding the finish level of your plaster is also helpful when planning out foundations, bond beams, windows and doors, counters, cabinets, and trim details.

Be mindful that plastering can be messy, especially for less-experienced workers. It's also hard to keep other finish materials from getting stained during the plastering process. Mask any finished wood adjacent to the wall you intend to plaster, or, better yet, wait to install finished wood until after the plaster is complete. It also makes sense to work *with* gravity—finish your ceilings first, then your walls, then your floor—knowing that materials are likely to drop at every stage. Where finish plasters abut ceilings and floors, it can be challenging to make clean boundary lines. One good solution is to use some kind of trim to cover the junction. Wooden trim works well for straight walls. For curved walls, consider a more flexible material such as split bamboo, tile, metal, or even rope. Or you can make a kind of trim with a different colored plaster—for example, a strip of dark plaster along the base of a wall where it comes in contact with an earthen floor. If this plaster is the same color as the floor, it will not show mud stains when the floor mix inevitably splashes onto it. And the plaster trim can be oiled along with the floor, making it hard and water-resistant.

Natural Paints

There are many traditional recipes for natural paints. Each uses some kind of a binder to glue the pigment and other ingredients securely to the wall surface. As with plasters, paints are generally referred to by their binder: *clay paints, lime paints, milk paints,* and so on.

One common reason to paint a plastered wall is to change its color; this allows you to use dark-colored wild clay in the plaster and just a small fraction as much light- or bright-colored clay—along with pigment if desired—to determine the final surface color. Paints also improve the durability and water-resistance of the wall to a surprising degree. We once worked on a strawbale building at a demonstration center in California that was unfortunately sited in a flood plain. Soon after the building was finished, a 100-year flood caused the nearby river to overflow its banks. This particular building ended up about 3' deep in water for 24 hours. The interior walls had been finished with clay base coat and finish plasters. For decorative effect, two of those walls had then been painted with homemade clay paint containing starch paste. When the flood waters receded, the finish plaster on both unpainted walls had washed completely away. Much to our surprise, the plaster was intact on the two painted walls. Evidently, the starch paste in the paint provided enough protection to keep the plaster intact!

The clay plaster on the outside of the same building also made it through the flood in good condition; it had been sealed with linseed oil for weather resistance.

A few common types of paints are described briefly below. For more information and recipes, see: *The Natural Paint Book,* or *Using Natural Finishes.* Another fascinating read is *Clay Culture: Plasters, Paints and Preservation* by Carole Crews; this is the best resource on clay paints and alis, and it also covers casein and egg paints.

Clay Paint (*Alis*)

Alis is a kind of clay paint that comes to us from the adobe tradition of the American Southwest. In Taos, New Mexico, a special mica-bearing clay soil called *tierra blanca* was traditionally made into slip, then painted on the interior walls of adobe houses and polished with a piece of sheep's fleece. The resulting finish was smoother and more durable than raw adobes, with a pale silvery color to which the mica lent a reflective luster. Carole Crews adapted this tradition to make it accessible to those of us who don't have access to the micaceous clay beds in Taos. Read Carole's book for recipes and application techniques. Because of its whiteness, it is common to start with *kaolin,* an easy-to-source clay used for making porcelain ceramics. Mineral pigments will tint the paint to a wide range of pastel colors. But any color of clay can be used for a richer, more intense color.

In our experience, clay paints are a natural fit over clay plasters. They are extremely simple to make and apply (although to get a sheen like Carole does takes a lot of practice—see the color section for an example of her work), durable, safe, and nontoxic. They are so easy to use that Michael once repainted the interior of his cob home three times over a period of weeks—until he was completely satisfied with the color.

Unlike other kinds of homemade paint, we've never experienced any problems with clay paints dusting or flaking off the wall.

Clay paints adhere well to porous substrates such as clay plaster, unpainted drywall, or rough wood. If you plan to put *alis* over a slick surface such as latex paint, you may want to apply an *adhesion coat* first. The simplest adhesion coat is made out of paste and fine sand, although Carole advises using acrylic cement fortifier instead of starch paste because it won't dissolve when the *alis* is applied.

Clay paints are most commonly used on interior surfaces. They are also appropriate on weather-protected exterior walls. They will erode away over time if subjected to rain, but in our experience they are at least as durable over earthen plasters as limewash; on an exposed exterior wall they will need to be reapplied every couple of years. Even on interiors, this relatively soft finish can get scratched, revealing the color and texture of the plaster below. But it's easy to dry your leftover alis and rehydrate it when you need to make repairs.

Danielle Ackley of the Mud Dauber School of Natural Building in North Carolina applies clay paint over a clay plaster. Clay paints are beautiful, durable, nontoxic, and easy to make yourself. CREDIT: DANIELLE ACKLEY

Recipe

Recipe 4: Basic Paint Recipes

	Ingredients	Directions	Application
Simple clay paint	1 part clay slip ¼ part starch paste[1] Pigment, if desired 0–½ part fine sand, mica powder, or whiting powder	Start with a high-clay soil or powdered clay. Soak in water and whip with a hand-held mixer to make a paint-like slip, then sift through a 1/16" screen or paint filter. Using a wire whisk or hand-held mixer, add paste and pigment dispersed in water, then sand or mica powder. Should be thicker than regular paint.	Apply with a brush or roller over clay plaster. Stir frequently. Usually takes 2 coats. When leather hard, can be polished with a stainless steel finish trowel or buffed with a sponge to remove brush marks and increase hardness. Use quickly or store in refrigerator.
Carole's favorite *alis*[2] (using commercial ingredients)	1 gallon water 5 quarts kaolin clay powder 2 quarts aggregate: whiting powder, 90 mesh silica sand, fine mica flakes and/or mica powder Binder: 1 quart starch paste[1] or 1 cup buttermilk or 1 cup prepared casein Optional: pigment, fine chopped straw, large mica flakes	Start with water in a clean bucket. Using a wire whisk or hand-held mixer, mix clay into water. Then mix in aggregates, binder, and measured, hydrated pigment. Use sand and whiting in the first coat; whiting and mica for the second coat. Add water if needed to the consistency of cream.	Use small brushes to apply edges first, then a larger brush for the bulk of the wall. Paint smoothly and quickly, not overworking it. Allow to dry, then do a second coat. When leather-hard, buff with a clean, damp tile sponge. Then shortly after, use a dry flannel cloth to polish mica chips and/or straw.
Limewash	1 gallon lime putty[3] 3 gallons water ¼–½ cup casein powder soaked in 1 additional gallon hot water 0–1 cup pigment	Dissolve casein powder and pigment in hot water and let sit for at least 2 hours. Slowly add water to lime putty, mixing thoroughly. Add casein/pigment mixture slowly and mix well. Should be the consistency of whole milk.	Use for color and protection over lime or clay plaster. Apply at least 2 coats with a brush, stirring every couple of minutes. Mix thoroughly if limewash sits overnight. Use quickly or store in refrigerator.
Casein glaze[4]	1 cup casein powder 1 quart cold water ¼–1 cup pigment 2 tablespoons borax 2 quarts hot water	Stir casein and pigment into cold water and let sit overnight. Dissolve borax in hot water and let cool. Thoroughly mix the two solutions together.	Use as a transparent glaze over clay plaster for decoration and water-resistance. Stir frequently. Use quickly or store in refrigerator.
Adhesion coat	1 part starch paste[1] or other binder ½ part water 1 part sand (no coarser than the sand in the plaster or paint that is going over it) Optional: ½ part fresh manure	Combine ingredients with a wire whisk or hand-held mixer. Other suitable binders include buttermilk, casein, and acrylic cement fortifier.	Paint over a slick surface such as painted drywall to get a clay finish plaster to adhere. Stir constantly. Wait for it to dry before plastering.

[1] See recipe for starch paste in Recipe 2: Earthen Plaster Recipes
[2] Recipe from Carole Crews
[3] Use high-quality lime putty made from quicklime if available
[4] Adapted from *Using Natural Finishes* by Adam Weisman and Katy Bryce
Note: Fine silica sand, mica powder, whiting powder, pigments, and powdered clays, including kaolin, are all available from ceramics supply stores.

Limewash

Limewash or whitewash has been used in combination with lime plasters for centuries all over Europe. The most basic limewash is simply lime putty watered down to a very thin consistency, like skim milk. Painting this mixture periodically over a lime plaster has multiple benefits. It whitens the surface, kills mold, and deters insects. Perhaps most importantly, the limewash penetrates cracks, where it crystallizes to repair the damage. In this way, properly maintained lime plasters have been known to last for centuries.

The water-resistance of a limewash can be improved by the addition of a fat, such as tallow or raw linseed oil, without compromising vapor permeability. There are tricks to getting these materials to mix and stay mixed (oil and water …), and care must be used to add the proper amount of fat. *Using Natural Finishes* has a simple recipe. You can also tint your limewash by adding mineral pigments. However, only a small amount of pigment—no more than 7% of the weight of the lime—can be added before it weakens the limewash, which is why limewashes typically have pastel hues.

It is possible to paint a limewash over nearly any porous surface including clay plasters, stone, and rough wood. However, limewash adheres best to lime plasters. In our experience, limewash over clay plaster usually turns out disappointingly soft and dusty. This may, in part, be the result of using inferior bagged hydrated lime, rather than putty made by slaking quicklime. Greg Allen in North Carolina reports good results with limewash over clay plasters on both interior and exterior surfaces. He starts with high-quality lime putty from Lancaster Limeworks, and adds powdered casein as an additional binder to prevent dusting. In the UK, there are many historical examples of cob walls protected only with layer upon layer of limewash, which can build up to considerable thickness over the decades and centuries. Keep in mind that in weather-exposed areas, this kind of finish needs to be renewed every year or two or it will erode away. A single heavy rain or hail storm can break through the thin layer of lime and expose the clay plaster beneath.

Casein (Milk Paint)

Casein, or milk protein, has been used for millennia as a binder for paint and glue. Like other natural paints, milk paints serve to seal and protect the surface, while still allowing vapor to escape. They can be a good choice to seal a soft or dusty earthen plaster. You can either purchase ready-made casein paints, buy casein powder and make your paint from that, or start from scratch by curdling skim milk with vinegar or lemon juice, then blending the rinsed curds with water and borax or lime. This mixture makes a glossy, translucent glaze. By adding dispersed mineral pigments, you can create beautiful shiny color washes that look fantastic on top of a clay plaster or paint. You can also make a matte casein paint by adding an opaque filler such as kaolin clay or whiting to the mix.

In our experience, casein washes and paints, while beautiful and fun to experiment with, are less predictable than clay paints. We have seen them peel off of a wall months after being applied. We've also had slugs come inside a cob building and eat the homemade casein paint off the wall! Very likely, all of these problems could be solved with the right guidance or experimentation.

Oils

Boiled linseed oil is a drying oil made from pressed flaxseed. When exposed to air, the oil polymerizes into a water-resistant substance similar to plastic. Other drying oils include tung oil, hemp seed oil, walnut oil, and poppy seed oil. Other kinds of vegetable oils are not suitable for paints because they don't polymerize. This chemical reaction is the basis for oil paints, which have a long history both as an artistic medium and as a protective finish for buildings, boats, and bridges.

The simplest oil glaze can be made by thinning boiled linseed oil with a solvent, such as turpentine or citrus thinner. With or without added pigments, this mixture can be painted onto a clay plaster to harden it and reduce water penetration. This can be a good option where the plaster is exposed to a lot of water and/or severe wear, such as a bench, windowsill, or the bottom of a wall. Linseed oil will darken the appearance of the plaster substantially. If applied too thickly, or in multiple coats, linseed oil will eventually reduce vapor permeability of the plaster to an unacceptable level. We

have also had problems with mildew growing on linseed-treated earthen plasters, especially on exterior surfaces that get little direct sun in moist climates. According to Athena Steen, mold can be avoided by applying a single coat of sun-thickened raw linseed oil instead of boiled linseed oil.

Waterglass and Silicate Dispersion Paints

Waterglass is the common name for sodium silicate and/or potassium silicate. These minerals are water-soluble under special conditions, creating a solution that is akin to glass dissolved in water. Waterglass is used as a masonry sealant; it reduces water penetration without compromising vapor permeability. It can be purchased either in powder form or in solution. You can use diluted waterglass alone as a clear protective

Kiko Denzer did this beautiful cob and plaster work at Intaba's Kitchen restaurant in Corvallis, Oregon. The clay-plastered bench was treated with linseed oil to improve durability and make it easier to clean. Note the dark and somewhat uneven color from the oil. CREDIT: MICHAEL G. SMITH

coating or add pigment to make a colored glaze. The bond will be much better over a plaster containing silica sand.

Several companies are now manufacturing and selling waterglass-based paints called *silicate dispersion paints* or *silicate mineral paints*. These paints are reputed to provide long-lasting protection to masonry surfaces including lime and earthen plasters. They are increasingly used in the UK to seal and protect exterior lime plasters, as a more durable substitute for limewash. They bond primarily to the silica sand in a plaster mix, so they are not effective over the high-fiber, low-sand earthen plaster mixes that are commonly used on exteriors.

Waterproof Finishes

There are certain limited circumstances where *water-resistant* finishes don't seem good enough; what is really needed is *waterproof*. These situations include a shower enclosure, a backsplash behind a sink, and an exterior windowsill—places where large amounts of water are likely to accumulate and could soak through an absorptive surface. There are several ways to create truly waterproof finishes for cob walls; our advice is to use them sparingly. If a relatively small area is sealed so thoroughly that moisture inside the wall cannot escape through the finish material, that moisture should be able to migrate through the cob and escape somewhere else. When using an impermeable finish on one side of a wall (such as the inside of a shower stall), make sure to choose a highly vapor permeable material on the other side.

As discussed above, stucco made of Portland cement and similar cementitious products can be both water-resistant and impermeable. Unfortunately, cement stucco is also hard and brittle and adheres poorly to cob. If you plan to use it, you will need to first secure some kind of reinforcing mesh, such as stucco wire, to the wall.

Another good choice of finish material for very wet areas is ceramic tile. We have installed ceramic tile mosaics in cob showers, using conventional thin-set tile adhesive, cementitious grout, and commercial grout sealer. In one case, the shower is over 20 years old and there have been no signs of moisture problems. The opposite side of the cob wall is covered in vapor-permeable clay and lime plasters. We've also heard reports of people successfully tiling cob showers using lime-based adhesives and grout, which are both more absorbent and more vapor-permeable. Tile, brick, and stone all make great waterproof surfaces for exterior window sills.

Another possible finish in these situations is *tadelakt*. Originating in Morocco centuries ago, *tadelakt* is a highly polished lime plaster, sealed with olive oil soap before the lime sets. The soap undergoes a chemical reaction with the lime to create, in combination with intensive polishing, a surface that is truly waterproof, water-repellant, wear-resistant, and mold-resistant. *Tadelakt* has been used to seal earthen and masonry cisterns and baths, as well as walls. It is undergoing a revival today due to both its beauty and durability. But be warned: because of the polishing required, this is an extremely labor-intensive process. And like other waterproof finishes, *tadelakt* has very low vapor permeability, so it should be limited to fairly small areas.

Wooden Siding

Although some sort of plaster is by far the most common finish for cob walls, other options are available. One worth considering is wooden siding. If properly installed, wood can be durable and make a nice aesthetic contrast with plaster. Wooden siding can be used both on interior and exterior surfaces. For centuries in Europe and North America, it was common practice to install *wainscoting* or wooden paneling to protect the lower portions of plaster walls, which are subject to wear and tear from furniture, children, pets, and cleaning. The paneling should be attached to wooden sleepers built into the cob wall. Obviously the installation will be simpler if the wall is straight and flat.

Any kind of water-resistant siding (commonly wood or metal) can also be used to keep the weather off exterior walls. A solution that is gaining popularity for natural buildings in very wet climates is called *rainscreen siding*. Although we know of few, if any, examples of rainscreen siding over cob walls, the methods that have been developed for other wall systems could easily be adapted. See *Essential Building Science* for instructions.

When employing this strategy, there are a number of details to get right. You will need to attach the siding to the wall by setting wooden framing members into the cob. The most important thing is to leave a ventilated airspace between the cob wall and the siding, allowing moisture to evaporate out of the wall and be vented harmlessly away. Make sure to leave a wide enough gap that air can flow quickly between the cob and the siding—a minimum of ¼″ (6 mm). Vertical air spaces ventilate and drain more readily than horizontal spaces. Assuming that the siding is attached to vertical wooden supports fastened to the cob, each airspace must be vented at the top and the bottom of the wall in a way that prevents rain from getting in at the top, allows water to drain out the bottom with properly installed and detailed flashing, and also excludes insects and other creatures. Rainscreen siding can also be integrated with external insulation such as light straw-clay, hempcrete, or any other vapor-permeable material.

Wooden wainscoting protects the lower part of this cob wall from wear and tear. The work is by Adam Weismann and Katy Bryce of Clayworks Ltd. on a new cob house in Cornwall, UK. CREDIT: RAY MAIN

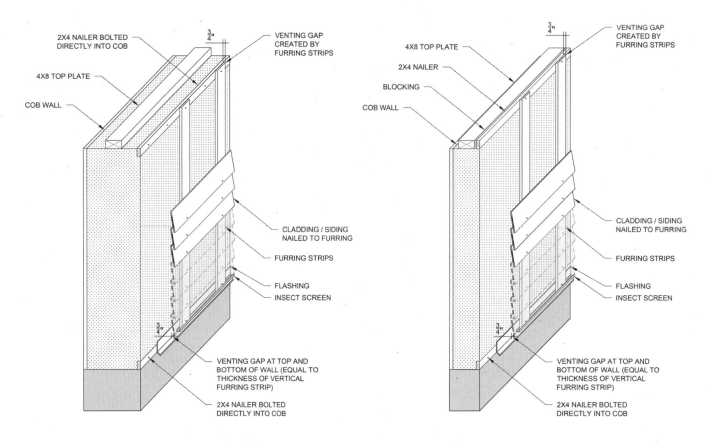

2X4 NAILER BOLTED
DIRECTLY INTO COB

4X8 TOP PLATE

COB WALL

3/4"

VENTING GAP
CREATED BY
FURRING STRIPS

CLADDING / SIDING
NAILED TO FURRING

FURRING STRIPS

FLASHING
INSECT SCREEN

3/4"

VENTING GAP AT TOP AND
BOTTOM OF WALL (EQUAL TO
THICKNESS OF VERTICAL
FURRING STRIP)

2X4 NAILER BOLTED
DIRECTLY INTO COB

EXTERIOR RAIN SCREEN FOR 24" COB WALLS

4X8 TOP PLATE

2X4 NAILER

BLOCKING

COB WALL

3/4"

VENTING GAP
CREATED BY FURRING STRIPS

CLADDING / SIDING
NAILED TO FURRING

FURRING STRIPS

FLASHING
INSECT SCREEN

3/4"

VENTING GAP AT TOP AND
BOTTOM OF WALL (EQUAL TO
THICKNESS OF VERTICAL
FURRING STRIP)

2X4 NAILER BOLTED
DIRECTLY INTO COB

EXTERIOR RAIN SCREEN FOR 12" COB WALLS

Fig. 15.1: *Rainscreen siding detail.*

CREDIT: RACHEL TOVE-WHITE, VERDANT STRUCTURAL ENGINEERS

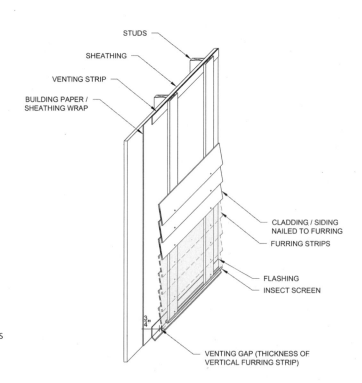

STUDS

SHEATHING

VENTING STRIP

BUILDING PAPER /
SHEATHING WRAP

CLADDING / SIDING
NAILED TO FURRING

FURRING STRIPS

FLASHING
INSECT SCREEN

3/4"

VENTING GAP (THICKNESS OF
VERTICAL FURRING STRIP)

EXTERIOR RAIN SCREEN FOR FRAMED WALLS
(FOR COMPARISON)

Appendix

Appendix AU:
Cob Construction (Monolithic Adobe)

The provisions contained in this appendix are not mandatory unless specifically referenced in the adopting ordinance.

General Comments

Cob is an earthen material made of clay subsoil, straw and added sand when needed. These are mixed together with water and placed onto a foundation in horizontal layers to produce a monolithic wall. Because cob's constituent materials and density are very similar to those of adobe bricks, this building technique is sometimes known as "monolithic adobe." Under many names, cob has been used for thousands of years around the world, notably in England and Northern Europe, the Middle East, West Africa, China and the southwestern United States. An estimated 20,000 cob homes are still inhabited in the English county of Devon alone, some dating from the fifteenth century.

Historically, the materials were mixed in shallow pits by either draft animals or human labor. The stiff, plastic mixture was extracted manually from the pit and applied to the wall in lumps. The term "cob" derives from an Old English word for "lump" and has evolved to be the name of the material. Today, cob is mixed either manually by human feet or mechanically using tractors, excavators or rotary mixers. Wall construction is still typically manual and without formwork. However, mechanical placement and the use of formwork are recent trends that speed construction, and cob has potential for robotic 3D printing.

The provisions in this appendix are based on a combination of heritage methods of cob building successfully used for centuries, recent decades of innovation and practice, relevant existing US and international codes and standards, recent materials and system testing, and modern building science.

Cob buildings feature raised foundations and extended roof eaves to protect the walls from moisture and weather. Cob walls are often plastered with clay or lime plasters, which protect and beautify the cob without leading to the moisture problems associated with less vapor-permeable finishes such as conventional cement stucco on historic adobe structures.

Since the 1990's, there has been a resurgence of interest in cob construction in the United States and much of the world. Compared to manufactured building materials, the use of cob and other earthen materials can greatly reduce the embodied energy and life-cycle CO_2 emissions of buildings. With good design, construction and maintenance, and their natural resistance to rot and fire, cob buildings have proven to last many centuries. The oldest continuously inhabited building in the United States, the Taos Pueblo, is over 700 years old and made of a combination of adobe brick and cob.

The constituent materials are inexpensive and typically require little transportation compared with lumber, steel, concrete and other remotely produced building materials. Cob is highly fire resistant, easily recyclable and nontoxic in all stages of construction and use. Cob's thermal mass and moisture management properties modulate interior temperature and humidity, helping to create thermally comfortable and healthy buildings.

Many credible sources such as books, websites and associations provide information, training, building methods and best practices for cob construction for those who need them. Such guidance is beyond the scope of this appendix and its commentary, which address the minimum requirements for safe and durable cob structures.

Purpose

While adobe masonry is included in the masonry chapter of the *International Building Code®* (IBC®), and cob building codes or guidelines exist in England and New Zealand, prior to the approval of this appendix there has been no building code for cob construction in North America. As a result, the permitting of cob buildings has been left to building officials on a case-by-case basis. Designers, builders and building officials may be unaware of the design and construction practices that make cob buildings safe and durable. The desire to utilize cob construction is growing and promises to accelerate in response to economic and environmental pressures. There is also a particular need for fire-resistant construction systems that can withstand the increased frequency and intensity of wildfires in the western United States. The lack of a cob building code has been an impediment to the safe and widespread adoption of this building system.

The purpose of this appendix is to enable permitting of cob buildings while aligning the use of cob with safe practices derived from both historical evidence and recent testing. The goal is to provide builders with the flexibility to make use of time-tested regional variations as well as viable new methods as cob construction continues to evolve.

APPENDIX AU—COB CONSTRUCTION (MONOLITHIC ADOBE)

SECTION AU101
GENERAL

AU101.1 Scope. This appendix provides prescriptive and performance-based requirements for the use of *natural cob* as a building material. Buildings using *cob* walls shall comply with this code except as otherwise stated in this appendix.

❖ This appendix covers the use of natural cob as an earthen building material and system. It does not cover cob with added stabilizers such as cement or asphalt emulsion that are often used in other earthen building systems. (See the commentary to definitions of "Natural cob," "Stabilized" and "Unstabilized.") It is not intended to preclude the use of stabilized cob; however, such use should be evaluated as an alternative material.

All components and aspects of cob buildings other than their cob walls—including foundations; non-cob walls; roof structure; energy efficiency; and mechanical, plumbing and electrical systems—must comply with the code unless otherwise stated in this appendix. Integrating cob construction with provisions designed for use with wood-frame and masonry construction will likely require flexibility in meeting the intent of some aspects of the code.

Historically, many variations of cob construction have been practiced, influenced by climate, high-wind and seismic risk, available materials, local building practices and regional architecture. This appendix, through prescriptive and performance-based requirements, is intended to be inclusive of as many safe and durable methods of cob construction as possible. See Commentary Figure AU101.1 for typical cob building components, Figure AU101.4 for a typical cob wall, and their corresponding commentary. See also Sections AU101.4 and AU105.1 and their commentary.

Electrical wiring systems in cob buildings must meet the same criteria for other wall systems in the code such as protection of wiring from damage in service, secure attachment of wiring, conduit systems and electrical boxes, and air sealing of penetrations in exterior walls.

Decades of electrical installations in cob buildings have yielded common practices. Wiring in conduit or armored cable has been surface mounted to cob walls with code-compliant attachments or embedded in the cob wall. Protective chase spaces are sometimes created on the surface of the wall or in the floor to run nonmetallic sheathed cable. Where wiring is directly embedded in a cob wall, UF cable rated for direct burial in earth has been commonly used and approved.

Electrical boxes in cob walls are typically fastened to the wooden frames surrounding doors or windows (see Section AU105.4.5) or to a nail-studded wood block embedded in the cob. Surface-mounted electrical boxes are also common.

Wiring and electrical boxes must not be embedded in cob structural walls except in their "surface voids" as allowed in Section AU106.5, Item 3, or in a location and dimension that does not reduce the wall's required thickness. See Section AU106.2, Item 2. Wiring and electrical boxes can be embedded anywhere in nonstructural cob walls, including below windows. See the commentary to the definition of "Nonstructural wall."

Plumbing supply, waste and vent pipes have historically been kept out of cob walls where possible, and instead are run in wood-framed walls or secured to the surface of cob walls. Where unavoidable, pipes in cob walls can be run in shallow chases and wrapped or sleeved as needed. The same restrictions that apply to electrical wiring and boxes in structural walls apply to plumbing pipes and fixtures.

Commentary Figure AU101.1 shows typical components of a cob building. Like other building systems, many design configurations and variations are possi-

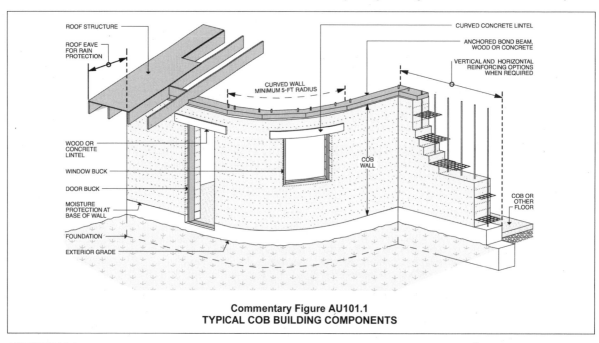

Commentary Figure AU101.1
TYPICAL COB BUILDING COMPONENTS

ble within the limits of the code. Not all components shown are required or present in all cob buildings, for example the wall reinforcing shown is not required for all wall types or situations. See also Figure AU101.4.

AU101.2 Intent. In addition to the intent described in Section R101.3, the purpose of this appendix is to establish minimum requirements for cob structures that provide flexibility in the application of certain provisions of the code, to permit the use of site-sourced and local materials, and to permit combinations of historical and modern techniques.

❖ The purpose of this appendix is to establish the minimum requirements for cob construction (described broadly in Section R101.3) and to provide needed flexibility in the application of certain provisions of the code. Because many cob construction materials and methods are not covered in the code, or may not align with conventional materials and systems, there is greater reliance on the intent of certain code provisions. This includes the use of site-sourced subsoil and other local materials, as well as innovative combinations of proven historical and modern techniques that reduce life-cycle impacts and/or increase affordability where shown to be equivalent in performance to materials and systems in the code. The flexibility is intended for both the approval and inspection processes.

AU101.3 Tests and empirical evidence. Tests for an alternative material, design or method of construction shall be in accordance with Section R104.11.1, and the *building official*

shall have the authority to consider evidence of a history of successful use in lieu of testing.

❖ Where cob construction varies substantially from this appendix, or other components of the building vary substantially from the requirements of the code, the building official has the authority to approve the design as an alternative material, design or method of construction as described in Section R104.11. When the building official determines a test is necessary to approve the design, Section R104.11.1 specifies the way the tests shall be conducted. However, for cob construction and its associated construction, the building official also has the authority to consider a history of successful use, as presented by the applicant, in lieu of such testing. This is similar to Section 2109 on adobe construction in the IBC, which uses empirical evidence as the basis for its provisions.

AU101.4 Cob wall systems. *Cob* wall systems include those shown in Figure AU101.4 and *approved* variations.

❖ "Cob wall systems" is a general term encompassing the systems of cob construction shown in Figure AU101.4 as well as approved variations. Each element in the figure references a section in this appendix or the code. Historically, many variations of cob construction have been practiced, and the illustrated systems are not meant to preclude viable variations. However, systems that vary substantially from what is shown in Figure AU101.4 must be approved by the building official.

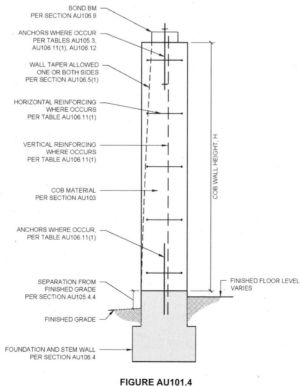

BOND BM
PER SECTION AU106.9

ANCHORS WHERE OCCUR
PER TABLES AU105.3,
AU106.11(1), AU106.12

WALL TAPER ALLOWED
ONE OR BOTH SIDES
PER SECTION AU106.5(1)

HORIZONTAL REINFORCING
WHERE OCCURS
PER TABLE AU106.11(1)

VERTICAL REINFORCING
WHERE OCCURS
PER TABLE AU106.11(1)

COB MATERIAL
PER SECTION AU103

ANCHORS WHERE OCCUR,
PER TABLE AU106.11(1)

SEPARATION FROM
FINISHED GRADE
PER SECTION AU105.4.4

FINISHED GRADE

FOUNDATION AND STEM WALL
PER SECTION AU106.4

COB WALL HEIGHT, H

FINISHED FLOOR LEVEL
VARIES

FIGURE AU101.4
TYPICAL COB WALL

APPENDIX AU—COB CONSTRUCTION (MONOLITHIC ADOBE)

Figure AU101.4 illustrates the elements of a typical cob wall. Each element references a section in this appendix or the code. Where "where occur(s)" appears in the reference note, the element is not necessarily part of the cob wall system but is sometimes required, depending on the structural demands of the building's location and design.

AU101.5 Definitions. The words and terms in Section AU102 shall, for the purposes of this appendix, have the meanings shown herein. Refer to Chapter 2 for general definitions.

❖ Section AU102 clarifies the terminology used in this appendix. The terms take on unique and specific meanings, with many of the terms used solely in the context of cob construction.

SECTION AU102
DEFINITIONS

BRACED WALL PANEL. A *cob* wall designed and constructed to resist in-plane shear loads through the interaction of the cob material, its reinforcing and its connections to its bond beam and foundation. The panel's length meets the requirements for the particular wall type and contributes toward the total amount of bracing required along its braced wall line in accordance with Sections AU106.11 and R602.10.1.

❖ The term "Braced wall panel" in this appendix refers to cob walls constructed in accordance with Section AU106.11 for prescriptive use as braced wall panels and is synonymous with "shear wall." Cob walls with various means of reinforcement and connections at the top and bottom of the wall have been tested in laboratory settings to determine their ultimate strength and allowable capacity to safely resist in-plane lateral loads. This testing is the basis for the prescribed use of cob wall types included in Table AU106.11(1) as braced wall panels.

BUTTRESS. A mass set at an angle to or bonded to a wall that it strengthens or supports.

❖ A buttress provides support perpendicular to the plane of the wall. See the commentary to Section AU106.14.

CLAY. Inorganic soil with particle sizes less than 0.00008 inch (0.002 mm) and having the characteristics of high to very high dry strength and medium to high plasticity, used as the binder of other component materials in a mix of *cob* or of clay plaster.

❖ Clay has been used for thousands of years as a building material. This includes unfired clay in adobe bricks; rammed earth; cob; light straw-clay; earthen plasters and earthen floors; and fired clay in bricks, roofing tiles, and floor and wall tiles. In all of these materials, clay is the binder, sometimes along with another binder such as lime or cement, that holds together other materials such as sand or straw. For this appendix, clay is the binding material in cob and is typically found in the clay subsoil used in the mix. Clay can also be obtained as a commercially quarried and bagged material. See also the commentary to the definitions of "Clay subsoil" and Section AU103.4.

CLAY SUBSOIL. Subsoil sourced directly from the earth, containing clay, sand and silt, and containing not more than trace amounts of organic matter.

❖ The word "soil" is commonly associated with topsoil, which contains organic matter. Clay subsoil (below the layer of topsoil) used to make cob is an inorganic mineral soil containing clay, sand and silt in varying proportions. It is often obtained from the building site. Experience shows that trace amounts of organic matter have no discernible effect on performance and are therefore acceptable. The term "trace amounts" is subjective and should be judged based on reason and the experience of the cob practitioner and building official as to what amount will not affect performance; for example, fine roots. See the commentary to Section AU103.1 regarding clay subsoil suitability.

COB. A composite building material consisting of refined *clay* or *clay subsoil* wet-mixed with loose straw and sometimes sand. Also known as "*Monolithic adobe*."

❖ Cob is a composite earthen building material that has been used for centuries in many parts of the world . It consists of clay subsoil or refined clay, straw and sand (added and/or in the subsoil) that are wet-mixed and used to create walls and other architectural elements such as benches and fireplaces. Cob is also known as "monolithic adobe" because its materials are similar or identical to those in adobe blocks, but instead are used to create monolithic walls. See also the General Comments at the beginning of this appendix, the commentary to Sections AU103.1 through AU103.6 and the commentary to the definition of "Monolithic adobe."

COB CONSTRUCTION. A wall system of layers or lifts of moist *cob* placed to create monolithic walls, typically without formwork.

❖ Cob construction is a building system using cob, placed in layers, to create monolithic earthen walls. During construction, the wet cob mix is self-supporting and does not require mechanical compaction, so cob walls are typically built without formwork. The lack of formwork is conducive to creating curved walls and other free-form building elements. Formwork is allowed and is advantageous for some applications and designs. See also the General Comments at the beginning of this appendix and the commentary to Section AU101.4.

DRY JOINT. The boundary between a layer of moist *cob* and a previously laid and significantly drier, nonmalleable layer of *cob* that requires wetting to achieve bonding between the layers.

❖ Dry joints between layers of cob should be minimized or avoided, when possible, to allow proper integration of layers for a monolithic wall. See the commentary to Section AU103.7.

FINISH. Completed combination of materials on the face of a *cob* wall.

❖ See Section AU104 for acceptable finishes on cob walls.

LIFT. A layer of installed *cob*.

❖ The term "lift" is commonly used for a layer of cob material in a cob wall. It is similar to a course in masonry construction, but without mortar. Layers of cob are instead integrated with the layer below when both layers are wet and pliable. See the commentary to Section AU103.6.

LOAD-BEARING WALL. A cob wall that supports more than 100 pounds per linear foot (1459 N/m) of vertical load in addition to its own weight.

❖ The definition of this term is consistent with the definition of "Wall, load-bearing" for stud walls in the IBC. Vertical loads include superimposed dead and live loads from the roof and/or ceiling.

MONOLITHIC ADOBE. See "*Cob*."

❖ In recent decades, modern cob is increasingly referred to as "monolithic adobe" because its materials are similar or identical to those in adobe blocks, but instead are used to create monolithic walls. Adobe construction is a modular masonry system that uses dry adobe blocks stacked in courses with earthen mortar. Cob construction uses a wet cob mix installed in layers (lifts) where each layer bonds and is integrated with adjacent layers, effectively becoming monolithic (see commentary, Section AU103.6). The term helps people understand and accept the material and building system by relating them to the more familiar material and system of adobe, which is included in the IBC.

NATURAL COB. *Cob* not containing admixtures such as Portland cement, lime, asphalt emulsion or oil. Synonymous with "*Unstabilized cob*."

❖ This appendix covers only natural cob. Though synonymous with unstabilized cob, the term "natural cob" is preferred because "unstabilized" incorrectly implies the material is not stable. Though stabilizers are sometimes added to earthen materials for greater durability and increased compressive strength, natural cob is sufficiently stable, strong and durable for applications meeting the requirements of this appendix. Cob's comparative high straw content increases tensile strength and durability, its monolithic nature increases wall integrity, and like all thick-walled systems, it is inherently stable.

In many parts of the world, unstabilized or natural cob buildings remain in service after more than 600 years. Though appropriate in some circumstances, stabilizers can add cost and have negative environmental and human health effects. See also the definition and commentary to "Unstabilized."

NONSTRUCTURAL WALL. Walls other than *load-bearing walls* or *shear walls*.

❖ This definition is consistent with the definition in ASCE 7, a standard used as the basis for defining structural loads imposed on buildings in the IRC and IBC. Nonstructural cob walls do not carry superimposed loads (from the roof and/or ceiling) and are not designed or constructed as shear walls to resist in-plane lateral loads (from wind and/or earthquakes). The cob material in nonstructural walls serves only as enclosure and as a substrate for any finishes. Nonstructural cob walls

must be capable of withstanding out-of-plane lateral loads (see Section AU105.3). Nonstructural cob walls include portions of walls below window openings, which are neither load-bearing nor shear walls, and can include portions of walls above window and door openings if they are not load bearing. See the definitions of "Load-bearing wall" and "Shear wall."

PLASTER. Clay, soil-cement, gypsum, lime, clay-lime, cement-lime or cement plaster as described in Section AU104.

❖ In the code, the word "plaster" is used only for cement and gypsum plasters. The word "plaster" in this appendix is used with all plaster types described in Section AU104.

SHEAR WALL. A *cob* wall designed and constructed to resist in-plane lateral seismic and wind forces in accordance with Section AU106.11. Synonymous with "*Braced wall panel*."

❖ The term "shear wall" is used interchangeably with the term "braced wall panel" in this appendix. See the definition of "Braced wall panel."

STABILIZED. *Cob* or other earthen material containing admixtures, such as Portland cement, lime, asphalt emulsion or oil, that are intended to help limit water absorption, stabilize volume, increase strength and increase durability.

❖ The use of stabilizers in earthen construction such as adobe, rammed earth and compressed earth block has become common in recent decades and has potential for cob construction. However, stabilizers are outside the scope of this appendix (see commentary, Section AU101.1). Stabilizers can be appropriate to achieve the intended qualities stated in the definition; however, they are unnecessary for many applications in all types of earthen construction. In cob construction, some stabilizers may interfere with bonding between layers. Stabilizers also add cost and can have negative environmental and human health effects. The term "stabilized" in this appendix refers to stabilization achieved by chemical rather than mechanical means. See the definition of "Unstabilized."

STRAW. The dry stems of cereal grains after the seed heads have been removed.

❖ Straw is an agricultural byproduct of grain plants after the nutrient grains have been harvested. Straw is not hay. See Section AU103.3 regarding the types of straw used in cob buildings.

STRUCTURAL WALL. A wall that meets the definition for a "*Load-bearing wall*" or "*Shear wall*."

❖ This definition is consistent with the definition in ASCE 7, a standard used as the basis for structural loads on buildings in the IRC and IBC. See the definitions of "Load-bearing wall," "Shear wall" and "Nonstructural wall."

UNSTABILIZED. A *cob* or other earthen material that does not contain admixtures such as Portland cement, lime, asphalt emulsion or oil.

❖ This appendix covers only unstabilized, or natural, cob. (See also the definitions and commentary to "Natural cob" and "Stabilized.") The term "unstabilized" is

APPENDIX AU—COB CONSTRUCTION (MONOLITHIC ADOBE)

commonly used for earthen materials such as adobe where stabilizers are not used, but it has an undeserved negative connotation because it incorrectly implies the material is not stable.

UNSTABILIZED COB. See "*Natural cob.*"

❖ See the commentary to the definitions for "Unstabilized" and "Natural cob."

SECTION AU103
MATERIALS, MIXING AND INSTALLATION

AU103.1 Clay subsoil. *Clay subsoil* for a *cob* mix shall be acceptable if the mix it produces meets the requirements of Section AU103.4.

❖ Clay subsoil that is appropriate for cob construction typically contains sand and silt in addition to clay. However, only the clay binds the straw and sand or other aggregate. Cob practitioners use various methods to determine the suitability of a clay subsoil for cob, including whether the subsoil contains sufficient clay to bind the mix. The "Ribbon Test" and "Ball Test" (of the subsoil alone) described in ASTM E2392, *Standard Guide for Earthen Wall Building Systems*, and/or test bricks of different cob mixes are often used for this purpose. However, ultimately the cob mix with all of its constituent materials, including the clay subsoil, must meet the requirements of Section AU103.4. See also Section AU103.4 and the definition of "Clay subsoil."

AU103.2 Sand. Sand or other aggregates such as, but not limited to, gravel, pumice and lava rock, when added to *cob* mixes, shall yield a mix that meets the requirements of Section AU103.4.

❖ Most cob mixes require added sand or other aggregate so that the material will not exceed the shrinkage limits in Section AU103.4.1. This reduces or eliminates cracking in the dried material and helps ensure the wall performs adequately. Some clay subsoils contain sufficient aggregate to meet the requirements of Section AU103.4.1 with only the addition of water and straw. Where additional sand is needed, it is ideally sharp (not rounded river sand) and should not contain salt (not ocean beach sand) and is best well-graded (of varying sizes) with minimal fines. Unlike sand used in concrete, it can contain silt or clay, materials already present in clay subsoil. See Section AU103.4.1.

Other aggregates can be used in cob mixes in addition to or in lieu of sand, including but not limited to gravel, pumice and lava rock. They can be of varying size and quantity but still need to meet the requirements of Section AU103.4.1 along with the other constituent materials in the cob mix. Pumice and lava rock are less dense than sand and gravel and are used to create lighter, more insulating cob.

AU103.3 Straw. *Straw* for *cob* mixes shall be from wheat, rice, rye, barley or oat, or similar reinforcing fibers with similar performance. Before mixing, the straw or other reinforcing fibers shall be dry to the touch and free of visible decay.

❖ Straw for cob is dry, intact stems of wheat, rice, rye, barley or oat plants that is free of visible decay or discoloration and should be free of contaminants such as

insects, topsoil and green plant material because mold and mildew can grow in the presence of these microbial food sources. Straw is best harvested dry and kept dry until mixed with the other constituent materials of cob.

Straw is not hay, which includes grasses and other plants cultivated as livestock feed. Hay is baled green, contains nutrients and supports active decomposition. Hay is unacceptable for building. See the commentary to the definition of "Straw." Alternative reinforcing fibers can be used if similar performance is demonstrated.

AU103.4 Mix proportions. *Cob* mixes shall be of any proportions of refined *clay* or *clay subsoil*, added sand (if any) and straw that produce a dried mix that passes the shrinkage test in accordance with Section AU103.4.1, complies with the compressive strength requirements of Section AU106.6 and complies with the modulus of rupture requirements of Section AU106.7.

❖ Historically, cob was made by simply mixing site soil with water and adding plant fiber—usually straw and/or animal manure. The best cob soils had a low proportion of clay (about 5 percent in English cob) and a high proportion of sand and gravel; however, a wide range of soil types and mixtures have been used successfully. Cob builders today commonly add sand to the subsoil to increase compressive strength and reduce shrinkage, though some soils are well suited to making cob with only added straw. Modern cob typically contains a much higher ratio of straw than was common historically, which improves tensile and shear strength, decreases weight and increases thermal resistance. If the subsoil has insufficient clay to hold the mix together, clay can be added.

Because the makeup of clay subsoil and the properties of its clay vary widely, acceptable cob mix proportions also vary widely. Experimentation is typically necessary to find a mix that yields desired qualities and meets requirements. Practitioners commonly create samples with clay subsoil to sand ratios ranging from 1:0 to 1:3, both with and without added straw. Practitioners have a variety of ways to assess these samples for suitability. Added sand or refined clay might not be needed for an acceptable mix, but clay subsoil and straw are required.

Ultimately, the mix can be any proportion of the materials stated in this section, and as described in Sections AU103.1, AU103.2 and AU103.3, as long as the mix passes the shrinkage test required in Section AU103.4.1. Furthermore, mixes for all cob walls must be tested for and comply with the compressive strength requirements of Section AU106.6, and mixes for cob braced wall panels must be tested for and comply with the modulus of rupture requirements of Section AU106.7. The materials in the tests should be representative of those that will be used in construction of the walls, within reasonable limits of the inherent variability of the materials. See also Sections AU103.1, AU103.2 and AU103.3 and their commentary regarding each constituent material.

AU103.4.1 Shrinkage test for cob mixes. Each proposed cob mix of different mix proportions shall be placed moist to completely fill a 24-inch by $3^1/_2$-inch by $3^1/_2$-inch (610 mm by 89 mm by 89 mm) wooden form on a plastic or paper slip sheet and dried to ambient moisture conditions, or oven dried. The total shrinkage of the length shall not exceed 1 inch (25 mm), as measured from the dried edges of the material to the insides of the form. Cracks in the sample greater than $^1/_{16}$ inch (1.5 mm) shall first be closed manually. The shrinkage test shall be shown to the building official for approval before placement of the *cob* mix onto walls

❖ The test method described in this section must be used to determine the acceptability of a proposed cob mix and to demonstrate its acceptability to the building official. If mixes of significantly different materials and/ or proportions will be used in construction, they should each be tested and approved. A slip sheet is necessary to allow movement of the material as it dries. Oil instead of a sheet material has also been used and is acceptable.

The sample must dry to ambient moisture conditions to ensure the mix has finished shrinking before shrinkage is measured. The specimen may dry in any location and during any season because relative humidity has negligible effect on the amount of shrinkage. It does, however, affect drying time. Oven drying and sun drying are acceptable, but typically lead to greater shrinkage than air drying and are therefore conservative.

The 1-inch measurement limit refers to the total shrinkage at both ends of the sample. Cracks in the length of the material not greater than $^1/_{16}$ inch are not counted in the shrinkage length.

AU103.5 Mixing. The *clay subsoil*, sand and straw for *cob* shall be thoroughly mixed by manual or mechanical means with water sufficient to produce a mix of a plastic consistency capable of bonding of successively placed layers or *lifts*.

❖ The goal of mixing cob is to achieve a stiff, homogenous, plastic mixture that is easy to form and will bond to itself to the previous layer while holding its shape when placed on the wall. The clay subsoil and sand (where part of the mix) are first blended with water. If the subsoil is dry and powdery, it can be premixed with any sand before water is added. If the subsoil is clumpy, it is helpful to first soak it with water until soft. After the subsoil, water and any sand are uniformly blended, the straw is mixed in. Both manual mixing (typically by human feet) and mechanical mixing (using equipment such as an excavator, tractor, front loader or rotary mortar mixer) are acceptable.

AU103.6 Installation. *Cob* shall be installed on the wall in lifts of a height that supports itself with minimal slumping.

❖ Cob must be installed in lifts (layers) of a height limited by the mix's ability to support itself. The term "minimal slumping" is subjective, based on experience and the practical need for each lift to add maximally to the height of the wall while maintaining a consistent wall thickness. Some slumping is common and acceptable, and the excess material can be trimmed after partial drying. Formwork can be used to eliminate slumping, increase lift heights and speed construction, or to achieve highly controlled wall shapes or surfaces.

APPENDIX AU—COB CONSTRUCTION (MONOLITHIC ADOBE)

A stiff plastic mixture is typically installed in lumps of varying size, depending on the manual or mechanical application limits, on top of the previous moist and plastic lift. A stiff mix reduces slumping, allows lifts of greater height and reduces drying time between lifts. Too dry a mix can adversely affect cohesion between layers and leave voids that are detrimental to the cob wall's strength.

In Britain, cob is placed on the wall manually or with an excavator, where it is trodden by foot in the traditional manner. With a stiff cob mix and a wide wall (3 to 5 feet thick), lifts up to 18 inches high are possible. In the United States, cob walls are typically narrower (1 to 2 feet thick) and manually installed. Narrower walls, and seismic risk in some regions, have led to more careful installation methods, with lesser lift heights calibrated to the conditions so that slumping is not excessive.

Bonding and integration of successive lifts is important for the strength and integrity of the wall. The clay in the wet cob enables successive layers to bond as the cob dries. Straw from one layer pushed into the previous layer integrates them mechanically. One technique for integrating lifts is by pushing one's thumbs or a rounded dowel called a "cobber's thumb" down through the new layer into the top of the previous layer, which must be plastic for both bonding and integration. Indentations in the top of a layer can also allow keying of the next layer.

Where vertical reinforcement is part of the wall assembly, the cob must be worked tightly around the reinforcement. Any horizontal reinforcement must be integrated between lifts, and where required, must be installed in the proper locations and with the proper spacing. See Table AU106.11(1).

Commentary Figure AU103.6(1) shows a cob wall being installed with only integral straw reinforcing. Commentary Figure AU103.6(2) shows a cob wall being installed with added vertical reinforcing. See the commentary to Section AU103.6. Added horizontal reinforcing is sometimes used or required but is not shown in either figure.

AU103.7 Dry joints. Each layer of *cob* shall be prevented from drying until the next layer is installed, to ensure bonding of successive layers. The top of each layer shall be kept moist and malleable with one or more of the following methods:

1. Covering with a material that prevents loss of or holds moisture.

2. Covering with a material that shades it from direct sun.

3. Wetting.

Where *dry joints* are unavoidable, the previous layer shall be wetted prior to application of the next layer.

❖ While each installed layer must dry to a sufficient rigidity to support the installation and weight of the next layer, the top 1 to 2 inches of cob should remain in a moist and plastic state to allow bonding and integration with the next layer. In cool or damp weather conditions, extra measures may not be needed to ensure this. In hot and dry weather, or if construction is interrupted for hours or days, employing one or more of the

APPENDIX AU—COB CONSTRUCTION (MONOLITHIC ADOBE)

methods specified in Section AU103.7 may be needed. If the top of the most recent lift has hardened, it must be wetted (ideally, repeatedly over several hours) and worked to soften the top few inches before adding the next layer. Because this is less effective than preventing the top surface from drying, dry joints should be avoided when possible.

AU103.8 Drying holes. Where holes to facilitate drying are used, such holes shall be of any depth and not exceeding 3/4 inch (19 mm) in diameter on the face of *cob* walls. Drying holes shall not be spaced closer than 10 hole-diameters. Drying holes shall not be placed in *braced wall panels*. The design load on *load-bearing walls* with drying holes shall not exceed 90 percent of the allowable bearing capacity as determined in accordance with Section AU106.8. Drying holes shall be filled with *cob* before final inspection.

❖ Drying holes are used where drying must be accelerated due to damp climate conditions or construction schedule. After their purpose is fulfilled, the holes must be filled with a moist and similar mix of cob.

AU103.9 Adding roof loads to walls. Roof and ceiling loads shall not be added until walls are sufficiently dry to support them without compressing.

❖ Wet cob is pliable and can compress under load. Dry cob is stiff and can carry significant loads without compressing. As the cob material dries, its load-carrying capacity increases. A cob building's roof and ceiling must not be installed until its supporting cob walls are dry enough to support their dead and live loads without compressing. It can be difficult to determine at what point a cob wall is dry enough to carry a particular load and is largely based on experience.

It is the builder's responsibility to provide the drying time and assessment needed to ensure walls are sufficiently dry prior to the application of roof and ceiling loads. If compression occurs from the application of

roof and ceiling loads to an inadequately dried wall, the load should be removed and the compressed portion of wall repaired/restored to its previous condition.

SECTION AU104
FINISHES

AU104.1 General. *Cob* walls shall not require a *finish*, except as required by Section AU104.2. *Finishes* applied to *cob* walls shall comply with this section and Chapters 3 and 7 unless stated otherwise in this section.

❖ Where cob walls are not substantially rain exposed (see Section AU104.2), they do not require a finish.

Commentary Figure AU103.6(2)
COB WALL INSTALLATION
WITH VERTICAL REINFORCING

Commentary Figure AU103.6(1)
COB WALL INSTALLATION

Minor erosion has proven acceptable on cob walls and is a matter of maintenance, similar to the need to periodically repaint the exterior of buildings of conventional construction. However, where cob walls are susceptible to excessive erosion from rain, finishes are necessary to protect the wall while ensuring that any moisture that might enter the wall can escape without causing harm.

A range of plaster types are allowed as described in Section AU104.4. Nonplaster exterior wall coverings must comply with the applicable sections in Section AU104 and Chapter 7. Other exterior finish systems are allowed with the specifications in Section AU104.1.2. Interior finishes must comply with Chapter 3 as described in Section AU104.1.1.

AU104.1.1 Interior wall finishes. Where installed, interior wall *finishes* and interior fire protection shall comply with the applicable provisions of Section R302, and shall be *plasters* in accordance with Section AU104.4 or nonplaster wall coverings in accordance with Section R702.

❖ Although no interior finish is required, it is common practice to install an interior plaster on cob walls for functional or aesthetic reasons. Plasters must comply with Section AU104.4. Nonplaster interior finishes can also be used and must comply with Section R702. All interior finishes are subject to the applicable interior fire protection provisions of Section R302. Cob has been used for centuries to build ovens, fireplaces and kilns, demonstrating that cob walls do not burn or support combustion. Similarly, the plasters in this section are not flammable and therefore pose no fire hazard (see Section AU108). See Section AU104.3 regarding restrictions on vapor retarders on cob walls.

AU104.1.2 Exterior wall finishes. Where installed, exterior wall *finishes* shall be *plasters* in accordance with Section AU104.4, nonplaster exterior wall coverings in accordance with Section R703, or other *finish* systems in accordance with the following:

1. Specifications and details of the *finish* system's means of attachment to the wall or its independent support and means of draining or evaporating water that penetrates the exterior finish shall be provided.

2. The vapor permeance of the combination of *finish* materials shall be 5 perms or greater to allow the transpiration of water vapor from the wall.

3. *Finish* systems with weights greater than 10 pounds per square foot (48.9 kg/m) and less than or equal to 20 pounds per square foot (97.8 kg/m) of wall area shall require that the minimum total length of *braced wall panels* in Table AU106.11(3) be multiplied by a factor of 1.2.

4. *Finish* systems with weights greater than 20 pounds per square foot (97.8 kg/m) of wall area shall require an engineered design.

❖ Where installed, whether required or not, exterior finishes must meet the requirements of Section AU104.4 for plasters and Section R703 for nonplaster exterior wall coverings. All other exterior finish systems must meet the requirements in Items 1–4.

The purpose of these requirements is to ensure sufficient mechanical support for the finish system and drainage of any water that may penetrate it, ensure adequate permeability for drying and account for the weight of the finish system in the building's structural design. Section AU104 addresses the issues in Items 1–4 for plasters. Nonplaster wall coverings in Section R703 should be evaluated for compliance with Items 1–4. In particular, conventional attachment methods may not be appropriate for cob walls. Some finishes may need an alternative means of attachment or may not have a perm rating of 5 or greater. See also Section AU104.3.

The perm rating of a finish system of more than one material is determined by adding the reciprocal of each material's perm rating (for the given thickness) and then taking the reciprocal of that sum. The perm rating of a finish assembly will always be less than the perm rating of the least permeable element.

For example, if three materials in a finish system have perm ratings of 5, 10 and 3, the perm rating of the entire system is calculated as follows:

$$1/P_1 + 1/P_2 + 1/P_3 = 1/P_{TOTAL}$$

$$1/5 + 1/10 + 1/3 = 0.63$$

$1/0.63 = 1.59$ perms (not vapor permeable per the code threshold of 5 perms)

AU104.2 Where required. *Cob* walls exposed to rain due to local climate, building design and wall orientation shall be *finished* or clad to provide protection from excessive erosion.

❖ Only cob walls substantially exposed to rain are required to be protected by an exterior finish or cladding. Climate, microclimate, building design and wall orientation should all be considered when determining if a wall is exposed to rain. Dry climates, deep roof overhangs, veranda or porch roofs, and walls oriented away from prevailing storm winds can all factor into determining that a wall is not exposed to rain. Where a wall is partially protected by a roof overhang, it can also be determined that the bottom portion of a wall is exposed to rain, whereas the upper portion is not. Where required, the finish can be plaster or a nonplaster cladding, or a rain screen that protects the wall from rain and allows drying.

Other local conditions such as a history of insect damage may be present that necessitate particular finishes to protect the wall.

AU104.3 Vapor retarders. Class I and II vapor retarders shall not be used on *cob* walls, except at *cob* walls surrounding showers or as required or addressed elsewhere in this appendix.

❖ High vapor permeability of both interior and exterior finishes is desirable for cob walls to allow for dispersion of any moisture that enters the walls. Cob's component materials of clay and straw are highly vapor permeable, as are the plasters allowed for cob walls. Other finishes allowed should also be vapor permeable. For these reasons, Class I and Class II vapor retarders are not allowed on cob walls except in extreme situations, such as on walls surrounding showers, where the importance of keeping water vapor from entering the wall exceeds the importance of enabling moisture to exit the wall.

This section does not prohibit the use of Class III vapor retarders (> 1.0 to ≤ 10 perms), but it is recommended that all finishes or finish systems, including any vapor retarder, have a vapor permeance of at least 5 perms, which is the definition of vapor permeable in the code. Section AU104.1.2, Item 1 explicitly requires this for nonplaster exterior wall finish systems not covered in Chapter 7, but it is important for all finishes and finish systems. All plasters as allowed in this appendix achieve a minimum vapor permeance of 5 perms by virtue of their requirements. Care should be taken with nonplaster wall coverings allowed in Section R703 to achieve this as well. See Section AU104.1.2.

AU104.4 Plaster. *Plaster* applied to *cob* walls shall be any type described in this section. *Plaster* thickness shall not exceed 3 inches (76 mm) on each face except where an *approved* engineered design is provided.

❖ The plasters as allowed in this section have a history of successful use on cob and other earthen wall systems. Plaster thickness on each face of the wall is limited to 3 inches (76 mm) because the additional weight of thicker plaster has potential structural consequences, especially in areas of moderate to high seismic risk. Thicker plaster may also unacceptably reduce vapor permeability, depending on the plaster. However, there are potential benefits of thicker plaster, including improved thermal performance in some climates due to the increased mass. The building official may allow thicker plaster with an approved engineered design.

AU104.4.1 Plaster and membranes. *Plaster* shall be applied directly to *cob* walls to facilitate transpiration of moisture from the walls and to secure a mechanical bond between the *plaster* and the *cob*. A membrane shall not be located between the *cob* wall and the *plaster*.

❖ Cob construction has historically been practiced with plaster applied directly to the cob without any membrane, air barrier or water-resistive barrier between the plaster and the cob. This allows the plaster to mechanically bond with the cob. Also, under certain conditions, the presence of a membrane may impede the dispersion of moisture through the plaster to the interior or exterior.

While the code requires a water-resistive barrier for exterior cement plaster over wood-frame construction in accordance with Section R703.7.3, a water-resistive barrier is not required for cement plaster or any exterior plaster allowed over cob walls by this appendix. The moisture management characteristics of cob walls compared with wood-frame construction account for the differing requirements. The clay and straw in cob walls give them considerably more capacity than a wood-frame wall to safely store and disperse moisture.

Where a membrane is allowed or required by this appendix, such as for walls enclosing showers or steam rooms, adequately attached mesh or lath is necessary to ensure adequate support of the finish.

AU104.4.2 Plaster lath. The surface of *cob* walls shall be permitted to function as lath for *plaster*, with no other lath required. Metal, plastic, and natural fiber lath shall be permitted to be used to limit *plaster* cracking, increase the *plaster* bond to the wall, or to bridge dissimilar materials.

❖ The cob wall surface, with its typically irregular texture, can serve as lath for plaster, and thus no other lath is required. However, lath of the listed types is allowed where needed or desired to reduce plaster cracking, increase adhesion to the wall or create a bridge where plaster crosses different materials. See the commentary to Sections AU104.4.8 and AU104.4.9 regarding lath for cement-lime and cement plaster.

AU104.4.3 Clay plaster. *Clay plaster* shall comply with Sections AU104.4.3.1 and AU104.4.3.2.

❖ Clay plasters are the most commonly and historically used plasters for cob walls because they can be easily repaired and maintained, are comparatively inexpensive and bond well to the cob since they both contain clay.

AU104.4.3.1 General. *Clay plaster* shall be any *plaster* having a *clay* or *clay subsoil* binder. Such *plaster* shall contain sufficient clay to fully bind the sand or other aggregate and any reinforcing fibers. Reinforcing fibers shall be chopped straw, sisal, hemp, animal hair or other similar approved fibers.

❖ The relative amounts of clay subsoil and added sand in the plaster mix depend on the amount of clay, sand and silt in the subsoil. Experimentation or experience is necessary to determine mixes that are workable and yield a plaster with minimal or no cracking. In all clay plasters, clay is the binder, sand and other aggregates reduce cracking, and fibers reduce cracking and resist erosion and abrasion.

See Section AU104.4.3.2 regarding the suitability of clay subsoil for clay plasters. Sand and other fine aggregate should be sharp, well graded (of differing sizes) and free of salt.

AU104.4.3.2 Clay subsoil requirements. The suitability of *clay subsoil* shall be determined in accordance with the Figure 2 Ribbon Test and the Figure 3 Ball Test in the appendix of ASTM E2392/E2392M.

❖ This section requires the use of both tests referenced in ASTM E2392 to determine the suitability of clay subsoil for clay plaster. The first tests for plasticity and cohesion of the damp soil, and the second tests for strength and durability of the dried soil. These tests are commonly used by clay plaster and other earthen building practitioners. However, this section is not intended to prohibit the use of other tests that demonstrate clay subsoil suitability, such as a laboratory soils analysis.

AU104.4.4 Soil-cement plaster. Soil-cement *plaster* shall be composed of *clay subsoil*, sand, not more than 7 percent Portland cement by volume and, where provided, reinforcing fibers.

❖ The relative amounts of clay subsoil and added sand in the plaster mix depend on the amount of clay, sand and silt in the subsoil. Site subsoil can sometimes be used, thus requiring less or no imported sand. Experimentation or experience is necessary to determine workable mixes that yield a durable plaster with minimal or no cracking.

AU104.4.5 Gypsum plaster. Gypsum *plaster* shall comply with Section R702.2.1 and shall be limited to interior use.

❖ Gypsum plaster on cob walls must comply with Section R702.2.1. However, it is not subject to lathing requirements because all plasters in this appendix can be applied directly to cob walls in accordance with Section AU104.4.2. It is limited to interior use because it is not weather resistant.

AU104.4.6 Lime plaster. Lime *plaster* is any *plaster* with a binder composed of calcium hydroxide including Type N or S hydrated lime, hydraulic lime, natural hydraulic lime or slaked quicklime. Hydrated lime shall comply with ASTM C206. Hydraulic lime shall comply with ASTM C1707. Natural hydraulic lime shall comply with ASTM C141 and EN 459. Quicklime shall comply with ASTM C5.

❖ Lime plaster provides a durable, vapor permeable finish that can be readily repaired and maintained. There are many types of lime that can be used and this section lists the appropriate standards for each. Though successful applications are common and have a long historical precedent, delamination has occurred in some cases, depending on many variables. Plaster tests on the wall with a 28-day cure time are recommended to evaluate the bond. Applying a lime wash to the cob before plaster can increase the bond.

AU104.4.7 Clay-lime plaster. Clay-lime *plaster* shall be composed of refined *clay* or *clay subsoil*, sand, lime and, where provided, reinforcing fibers.

❖ Clay-lime plaster on cob walls provides a durable, vapor permeable finish that can be readily repaired and maintained. Site subsoil can sometimes be used, thus requiring less or no imported sand. Clay-lime plaster often bonds better to cob than lime plaster because both contain clay. Experimentation or experience is necessary to determine workable mixes that yield a durable plaster with minimal or no cracking.

AU104.4.8 Cement-lime plaster. Cement-lime *plaster* shall be plaster mix types CL, F or FL, as described in ASTM C926.

❖ Cement-lime plasters use Portland cement and lime together as the binder in roughly equal proportions. The mix types listed in this section are shown in Tables 3 and 4 of ASTM C926 and are the only types allowed on cob walls because only they contain sufficient lime to achieve an acceptable vapor permeability. The Portland cement-lime proportions for CL, F and FL plaster range from 1:0.75 to 1:2. While not required, cement-lime plaster can benefit from properly attached lath because it has shown greater tendency to crack or delaminate than other plasters where applied to earthen walls. See Section AU104.4.2.

AU104.4.9 Cement plaster. Cement *plaster* shall have not less than 1 part lime to 4 parts cement and be not thicker than $1^1/_2$ inches (38 mm), to ensure minimum acceptable vapor permeability.

❖ The only subsections of Section R703.7 that pertain to cement plaster used on cob walls are Sections R703.7.4 and R703.7.5. Cement plaster is required to contain lime in the proportion stated in order to achieve a vapor permeability. While not required, cement plaster can benefit from properly attached lath because it has shown greater tendency to crack or delaminate than other plasters where applied to earthen walls. See Section AU104.4.2.

SECTION AU105
COB WALLS—GENERAL

AU105.1 General. *Cob* walls shall be designed and constructed in accordance with this section and Figure AU101.4 or an *approved* alternative design. In addition to the general requirements for *cob* walls in this section, *cob structural walls* shall comply with Section AU106.

❖ The provisions of this section apply to all cob walls, both nonstructural and structural (see the definitions of "Nonstructural wall" and "Structural wall"), except where a subsection states that the provision(s) apply only to nonstructural walls. In addition, structural cob walls must be designed and constructed in accordance with Section AU106.

Figure AU101.4 illustrates an acceptable and typical cob wall system. Each element in the figure references a section in this appendix or the code. Historically, many variations of cob construction have been practiced, and the illustrated system is not meant to preclude viable variations. However, systems that vary substantially from those shown in Figure AU101.4 must be approved by the building official.

AU105.2 Building limitations and requirements for cob wall construction. *Cob* walls shall be subject to the following limitations and requirements:

1. Number of stories: not more than one.

2. Building height: not more than 20 feet (6096 mm).

3. *Seismic design categories*: limited to use in *Seismic Design Categories* A, B and C, except where an approved engineered design is provided.

4. Wall height: in accordance with Table AU105.3, and with Table AU106.11(1) for *braced wall panels*.

5. Wall thickness, excluding *finish*, shall be not less than 10 inches (254 mm), not greater than 24 inches (610 mm) at the top two-thirds, not limited at the bottom third and, for structural walls, shall comply with Section AU106.2, Item 2. Wall taper is permitted in accordance with Section AU106.5, Item 1.

6. Interior *cob* walls shall require an *approved* engineered design that accounts for the seismic load of the interior *cob* walls, except in Seismic Design Category A for walls with a height to thickness ratio less than or equal to 6.

❖ **Items 1 and 2:** The number of stories and the building height for buildings using cob construction are limited in this appendix to help ensure safe use of the building system. This is not meant to exclude multistory or taller cob buildings designed by a registered design professional who adequately demonstrates their safety to the building official.

Item 2: See the definition of "Height, building" in Chapter 2, which is the vertical distance from grade plane to the average height of the highest roof surface.

Item 3: Cob walls are limited to use in Seismic Design Categories A, B and C to help ensure safe use of the building system. However, cob walls are allowed in Seismic Design Categories D and E with an approved engineered design because this provides additional assurance of safe use in these higher seismic risk categories.

Item 4: The height of all cob walls is limited by Table AU105.3, and the height of cob braced wall panels is also limited by Table AU106.11(1), whichever is more restrictive. Wall height means the cob portion of the wall only. See Figure AU101.4.

Item 5: All cob walls must be at least 10 inches thick [except where surface voids reduce wall thickness as allowed in Section AU106.5(3)], and not greater than 24 inches thick in the top two-thirds of the wall height (cob portion only). The wall thickness is not limited for the bottom third of the cob wall because additional thickness in this portion only provides structural benefit (greater load-bearing capacity, greater out-of-plane stability) without increasing seismic load on roof or ceiling diaphragms or shear walls in the building. Tapered walls are allowed per Section AU106.5(1).

Item 6: Where a building includes interior cob walls, an approved engineered design is required that accounts for the seismic load of those walls. These walls can present seismic hazards if not properly designed and detailed, especially their connection at the top of the wall to other building elements. Improper connection can result in out-of-plane instability or over-stressed connections to ceiling or roof elements or to exterior cob shear walls.

AU105.3 Out-of-plane resistance methods and unrestrained wall height limits. *Cob* walls shall employ a method of out-of-plane load resistance in accordance with Table AU105.3, and comply with its associated height limits and requirements.

❖ All cob walls must employ a method of out-of-plane load resistance in accordance with Table AU105.3 because all walls are subject to out-of-plane wind and/or seismic loads. See Table AU105.3.

The maximum unrestrained height of a cob wall, whether nonstructural or structural, is a function of its method of out-of-plane resistance and other parameters listed in Table AU105.3. See Notes c and h for allowable increases in wall height.

One method of improving out-of-plane resistance is to employ curved walls, whose geometry improves their out-of-plane stability. A design benefit resulting from this improved stability is a less restrictive wall height limit compared with straight walls. See Table AU105.3, Note h.

TABLE AU105.3. See page Appendix AU-13.

❖ Table AU105.3 includes Wall Types 1 and 2 and Wall Types A through E, each with their out-of-plane load resistance method. All wall types in Table AU105.3 can be used as nonload-bearing or load-bearing walls, but only Wall Types A through E can be used as braced wall panels.

The out-of-plane resistance methods (anchors and vertical reinforcing) in Table AU105.3 for Wall Types A through E are also part of their in-plane load-resisting system as braced wall panels. However, where used as braced wall panels, all requirements in Table AU106.11(1) must be met, including the more restrictive maximum height and the aspect ratio, and inclusion of horizontal steel reinforcing for Wall Types B, C and D.

The reinforcing materials in this appendix are limited to straw and steel. Other plant fibers, other metals and plastics have also been successfully used to reinforce cob walls but are outside the scope of this appendix.

Note c allows the height of a cob wall to exceed the unrestrained height limits of Table AU105.3 where an approved engineered horizontal restraint is employed at an intermediate height between the wall's foundation and bond beam. This intermediate restraint effectively creates two cob wall heights in one wall. Each of those separate wall heights must not exceed the limits in Table AU105.3.

Note h allows the H/T factor and the absolute wall height limit in Table AU105.3 to be increased for curved walls that meet or exceed an arc length:radius ratio of 1.5:1. The arc length and radius are intended to be ARC_c and R_c as described in Section AU106.11.3 and Figure AU106.11.3. The H/T factor is the number in the "Limit based on wall thickness" column in Table AU105.3.

Note i requires the modulus of rupture bending strength test for wall types not containing full-height vertical reinforcing. These wall types with only straw reinforcing rely on the cob material's bending strength to resist out-of-plane loads, whereas wall types with vertical steel reinforcing rely primarily on the steel. Thus, the test is required for Wall Types 1, A and B.

AU105.3.1 Determination of out-of-plane loading. Out-of-plane loading for the use of Table AU105.3 shall be in accordance with the ultimate design wind speed and seismic design category requirements of Sections R301.2.1 and R301.2.2, respectively. An *approved* engineered design shall be required where the building is located in a special wind region or where wind design is required in accordance with Figure R301.2.1.1.

❖ Out-of-plane loading for cob walls for the use of Table AU105.3 is determined, as for other building systems in the code, by finding the ultimate design wind speed and the seismic design category for the building's location per Sections R301.2.1 and R301.2.2. This information is also used for designing the building's in-plane lateral load-resisting system and its braced wall panels per Section AS106.11.

Cob buildings require an approved engineered design where located in either a Special Wind Region or Wind Design Required location. Approximate locations for both can be found in Figure R301.2.1.1. Local jurisdiction ultimate design wind speeds or other related local requirements take precedence.

AU105.3.2 Bond beams for nonstructural walls. Nonstructural *cob* walls shall be provided with a bond beam at the top of the wall that complies with Section AU106.9, except for requirements relating to roof and/or ceiling loads or *braced wall panels*.

❖ A bond beam is required for all cob walls, including nonstructural walls, because it is necessary for their resistance to out-of-plane loads and collection of in-plane loads and the transfer of both. Other than cob walls in post-and-beam systems, which are not covered by this appendix (see Section AU106.12), it is rare for a cob building to contain a nonstructural exterior wall (see definition of "Nonstructural"). In such cases a bond beam is still required in accordance with Section AU106, and only Section AU106.9.4 does not apply.

Although Section AU105.3.2 states that bond beams for nonstructural walls do not need to meet requirements relating to roof and/or ceiling loads, it is not meant to exempt them from Section AU106.9.3 because this connection is needed to transfer out-of-plane wall loads to the roof diaphragm.

The most common type of nonstructural cob wall in a cob building is an interior cob wall. These walls will likely require a bond beam as part of the approved engineered design required in Section AU105.2, Item 6. See Section AU105.2.

AU105.3.3 Lintels in nonstructural walls. Door, window and other openings in nonstructural cob walls shall require a lintel in accordance with Section AU106.10, except for requirements relating to roof and/or ceiling loads or braced wall panels.

❖ The weight of the cob above openings is still present in nonstructural cob walls and must be supported by a properly sized lintel. Lintel designs in Table AU106.10, as required by this section, are conservative for nonstructural walls because the table includes roof and/or ceiling loads. Using this section's exemption from requirements related to these loads and to braced wall panels would need an approved engineered design.

TABLE AU105.3
OUT-OF-PLANE RESISTANCE METHODS AND UNRESTRAINED WALL HEIGHT LIMITS

WALL TYPE[a, g, h] AND METHOD OF OUT-OF-PLANE LOAD RESISTANCE	FOR ULTIMATE DESIGN WIND SPEEDS (mph)	FOR SEISMIC DESIGN CATEGORIES	UNRESTRAINED COB WALL HEIGHT H[b, c, h]		TOP ANCHOR[e] SPACING (inches)	TENSION TIE[f] SPACING (inches)
			Absolute Limit (feet)	Limit Based on Wall Thickness T[d] (feet)		
Wall 1[i]: no anchors, no steel wall reinforcing	≤ 110	A	$H \le 8$	$H \le 6T$	None	48
Wall 2: top anchors[j], continuous vertical 6″ × 6″ × 6″ gage steel mesh in center of wall embedded in foundation 12 inches	≤ 140	A, B, C	$H \le 8$	$H \le 8T$	12	24
Wall A[i]: top anchors, no vertical steel reinforcing	≤ 120	A, B	$H \le 8$	$H \le 6T$	12	48
Wall B[i]: top and bottom anchors, no vertical steel reinforcing	≤ 130	A, B	$H \le 8$	$H \le 6T$	12	48
Wall C: top and bottom anchors, continuous vertical threaded rod at 4 feet on center embedded in foundation and connected to bond beam	≤ 140	A, B, C	$H \le 8$	$H \le 8T$	12	24
Wall D: continuous vertical threaded rod at 1 foot on center embedded in foundation and connected to bond beam	≤ 140	A, B, C	$H \le 8$	$H \le 8T$	N/A	24
Wall E: top anchors, continuous vertical 6″ × 6″ × 6″ gage steel mesh 2 inches from each face of wall embedded in foundation	≤ 140	A, B, C	$H \le 8$	$H \le 8T$	12	24

For SI: 1 inch = 25.4 mm, 1 foot = 304.8 mm, 1 mile per hour = 0.447 m/s.

N/A = Not Applicable

a. See Table AU106.11(1) for reinforcing and anchorage specifications for wall Types A, B, C, D and E.

b. *H* = height of the cob portion of the wall only. See Figure AU101.4. The maximum *H* is the absolute limit or the limit based on wall thickness, whichever is more restrictive.

c. Bond beams or other horizontal restraints are capable of separating a wall into more than one unrestrained wall height with an approved engineered design.

d. *T* = Cob wall thickness (in feet) at its minimum, without plaster.

e. $\frac{5}{8}$-inch threaded rod anchors at prescribed spacing with 12-inch embedment in cob, full embedment in concrete bond beams or full penetration in wood bond beam with a nut and washer.

f. Attach rafters to bond beam with 4-inch by 3-inch by 3-inch by 18 gage tension tie angles at prescribed spacing. See Figure AU106.9.5. Where rafters are attached to tension ties, roof sheathing shall be edge nailed.

g. All walls shall be tested for compressive strength in accordance with Section AU106.6.

h. For curved walls with an arc length to radius ratio of 1.5:1 or greater, the *H/T* factor shall be increased by 1, and the absolute height limit by 1 foot.

i. Wall type requires a modulus of rupture test in accordance with Section AU106.7.

j. See wall Type A in Table AU106.11(1) for top anchor requirements.

Arches over wall openings are an alternative to lintels but need an approved engineered design. See Section AU106.10.

AU105.3.4 Reinforcing at wall openings. Reinforcing shall be installed at window, door, and similar wall openings and penetrations greater than 2 feet (610 mm) in width in accordance with Sections AU105.3.4.1 through AU105.3.4.3. Surface voids deeper than 25 percent of the wall thickness shall be considered an opening.

❖ Reinforcing is required above and/or below and on each side of wall openings, such as windows and doors, to address increased out-of-plane loading on wall segments next to openings from the portions of wall above and/or below the opening.

Cob walls often contain recessed shelving, niches or built-in seating. If these or other surface voids decrease the wall thickness to less than 75 percent of its full thickness, then the reinforcing requirements of this section apply.

AU105.3.4.1 Opening size limit. Openings shall not exceed 6 feet (1829 mm) in width, and the height of the *cob* wall below openings shall not exceed 6 feet (1829 mm) above the top of the foundation.

❖ Openings can create significant additional out-of-plane stresses on surrounding wall segments (see commentary, Section AU105.3.4). The height of a cob wall below an opening is restricted to 6 feet to limit the out-of-plane forces transferred to the adjacent full-height wall segments. Openings that exceed the limits of this section require an approved engineered design.

AU105.3.4.2 Horizontal reinforcing. Two-inch by 2-inch (51 mm by 51 mm) 14-gage galvanized steel mesh shall be embedded 4 inches (102 mm) in the *cob* above the rough opening and below the rough opening for windows, and shall extend 12 inches (305 mm) beyond the sides of the opening. Walls below rough window openings greater than 4 feet 6 inches (1372 mm) in height shall be provided with additional horizontal reinforcing at midheight.

❖ The required horizontal steel mesh above and below wall openings transfers the out-of-plane loads of those portions of wall to the adjacent full-height wall segments and their vertical reinforcing. Where a wall portion below a window is greater than 4 feet 6 inches in height, reinforcing is required at its midheight in addition to the reinforcing required 4 inches below the opening. Nonferrous horizontal reinforcement can be used as an alternative to the prescribed mesh above and below openings, but there needs to be an approved engineered design or other documentation of equivalent performance.

AU105.3.4.3 Vertical reinforcing. Full-height $^5/_8$-inch (16 mm) threaded rod shall be installed 4 inches (102 mm) from each side of the opening, centered in the thickness of the *cob* wall. The threaded rods shall be embedded 7 inches (178 mm) in the foundation, and 4 inches (102 mm) in concrete bond beams or shall penetrate through wood bond beams and be secured with a nut and washer. The threaded rods shall be embedded in concrete lintels or pass through a drilled hole in wood lintels.

❖ The required vertical steel rod on each side of an opening transfers out-of-plane loads from wall portions above and below the adjacent opening to the foundation and to the bond beam with connections to each, as described in this section. This reinforcing is intended to be within the central two-thirds of the cob wall thickness.

AU105.3.5 Minimum length of cob walls. Sections of *cob* walls between openings shall be not less than 2 feet 6 inches (762 mm) in length. Wall sections less than 4 feet (1219 mm) and not less than 2 feet 6 inches (762 mm) in length shall contain vertical reinforcing in accordance with Section AU105.3.4.3.

❖ Wall segments less than 2 feet 6 inches in length are considered cob columns and are outside the scope of this appendix. The vertical reinforcement required in this section is to ensure adequate out-of-plane load resistance for walls within that length range.

AU105.4 Moisture control. *Cob* walls shall be protected from moisture intrusion and damage in accordance with Sections AU105.4.1 through AU105.4.5.

❖ Preventing intrusion of moisture into cob walls and ensuring that any such moisture can get out is as important for cob walls as it is for other materials and wall systems. A colloquial metaphor commonly used by practitioners of cob and other related construction types summarizes the basic principles for keeping a wall dry: "Good boots, a good hat, and a coat that breathes." This translates into keeping the bottom of the wall protected from ground and weather-related moisture, providing ample roof overhangs (especially in wet climates) to shield the walls and their openings from rain, and providing a protective wall finish that is vapor permeable. This dictum is also considered wise for buildings constructed of wood frame and other materials.

Sections AU105.4.1 through AU105.4.5 contain requirements to minimize the possibility of moisture intrusion from rain or snow and condensation (from uncontrolled flow of relatively warm, moist air into the wall), and from moisture rising into the wall from the ground.

In addition to the importance of preventing moisture from entering a cob wall, moisture must also be allowed to readily exit the wall. Thus, the finish on each side should be as vapor permeable as possible. Exterior wall finish systems are required to have an overall vapor permeance rating of 5 perms or greater in Section AU104.1.2, Item 2. However, this is important for all finishes and finish systems on cob walls. See Sections AU104.1.2 and AU104.3.

AU105.4.1 Water-resistant barriers and vapor permeance. *Cob* walls shall be constructed without a membrane barrier between the *cob* wall and *plaster* to facilitate transpiration of water vapor from the wall, and to secure a mechanical bond between the cob and plaster, except as otherwise required elsewhere in this appendix. Where a water-resistant barrier is placed behind an exterior *finish*, it shall be considered part of the *finish* system and shall comply with Item 2 of Section AU104.1.2 for the combined vapor permeance rating.

❖ Plaster is optional for cob walls, but where it is used, it must be applied directly to the cob wall without a membrane barrier between the plaster and the cob, except where a membrane is required in this appendix. This is to provide a good mechanical bond for the plaster and to facilitate the transpiration of moisture out of the wall. Where a water-resistant barrier is placed behind a nonplaster exterior finish system, it must be sufficiently vapor permeable so that the combined system meets the minimum 5-perm requirement in Section AU104.1.2, Item 2.

AU105.4.2 Horizontal surfaces. *Cob* walls and other *cob* elements shall be provided with a water-resistant barrier at weather-exposed horizontal surfaces. The water-resistant barrier shall be of a material and installation that will prevent erosion and prevent water from entering the wall system. Horizontal surfaces, including exterior window sills, sills at exterior niches and exterior buttresses, shall be sloped not less than 1 unit vertical in 12 units horizontal to drain away from *cob* walls or other *cob* elements.

❖ As with wood-frame walls, horizontal (or nearly horizontal) surfaces in or on cob walls can be vulnerable to weather-related moisture intrusion. Window sills are especially vulnerable and should be carefully detailed and constructed to meet the prescriptive and performance criteria of this section.

Horizontal surfaces that are not weather exposed do not require a water-resistive barrier. Exception 2 in the definition of "Weather-exposed surfaces" in the IBC provides guidance for determining whether a horizontal surface is considered weather exposed, subject to the evaluation of the local building official, with additional consideration of local climate. Ample roof overhangs and wrap-around porch roofs have been employed, especially in wet climates and heavy snowfall areas, to protect walls and their openings from weather-related moisture intrusion.

AU105.4.3 Separation of cob and foundation. A liquid-applied or bituminous Class II vapor retarder shall be installed between *cob* and supporting concrete or masonry.

Exception: Where local climate, site conditions and foundation design limit ground moisture migration into the base of the *cob* wall, including but not limited to the use of a moisture barrier or capillary break between the supporting concrete or masonry and the earth.

❖ A Class II vapor retarder is required to separate the base of the cob wall from supporting concrete or masonry to prevent "rising damp" from entering the wall from below. In certain situations where a combination of conditions, including the climate, site and foundation design, preclude or greatly reduce the ability of ground moisture to enter the base of the wall, the vapor retarder is not required. One example is the use of a moisture barrier or capillary break between the foundation and the earth.

AU105.4.4 Separation of cob and finished grade. *Cob* shall be not less than 8 inches (203 mm) above finished *grade*.

Exception: The minimum separation shall be 4 inches (102 mm) in dry climate zones as defined in Section N1101.7.2, and shall be 2 inches (51mm) on walls that are not weather exposed.

❖ The required 8-inch vertical separation between finished grade and the bottom of the cob wall is intended to minimize exterior wetting at the base of the wall from rain splash back, vegetation, landscape sprinklers and other sources. This requirement is for all cob walls whether plastered or not. The reduced required separations in the exception are due to reduced risk of wetting in those conditions.

AU105.4.5 Installation of windows and doors. Windows and doors shall be installed in accordance with the manufacturer's instructions to a wooden frame of not less than nominal 2-inch by 4-inch (51 mm by 102 mm) wood members anchored into the *cob* wall with 16d galvanized nails half-driven at a maximum 6-inch (152 mm) spacing, with the protruding half embedded in the *cob*. The wood frame shall be embedded not less than $1^1/_2$ inches (38 mm) in the cob and shall be set in from each face of the wall not less than 3 inches (76 mm). Alternative window and door installation methods shall be capable of resisting the wind loads in Table R301.2.1(1). Windows and doors in *cob* walls shall be installed so as to mitigate the passage of air or moisture into or through the wall system. Window sills shall comply with Section AU105.4.2.

❖ The common term for the wooden frame used to create rough openings for windows and doors is a "buck," which is typically made of nominal 2x4, 2x6, or 2x8 lumber. The hydrophilic quality of clay helps preserve wood, and typically cob is dry enough in service not to cause decay in embedded wood. Therefore, it is generally not necessary to separate the wood buck from cob, or use wood that is treated or naturally resistant to decay, unless local experience and conditions necessitate it.

Vertical settling, shrinkage and lateral movement are normal in cob walls as they dry. The settling and movement can exert significant pressure on the buck until the cob has dried. Therefore, window and door bucks must be sufficiently strong and braced, both in the wall plane and perpendicular to it, to keep them square and plumb. It is common to leave space above a buck to allow for settling, which is later filled with cob or plaster (see Commentary Figure AU105.4.5).

Commentary Figure AU105.4.5 shows a typical method of buck construction and installation. The 2x4 "stiffeners" shown at the sides of the buck are not required by this appendix but are a common and stronger anchoring alternative. Both "stiffener" and "nonstiffener" anchoring options are shown.

The cob on both sides of the window or door buck should be dry enough to have fully settled before installing the lintel and cob above. Otherwise, diagonal structural cracks can occur at the upper corners of the window or door buck or lintel. The top surface of the cob on each side of the opening must be kept moist per Section AU103.7 during this time of settling and

APPENDIX AU—COB CONSTRUCTION (MONOLITHIC ADOBE)

drying. For similar reasons, a 1-inch gap should be left between the top of the buck and the lintel.

In order to minimize the area of exposed sill that needs a water-resistant barrier per Section AU105.4.2, windows can be installed at or near the exterior plane of the wall. To achieve this while maintaining the required minimum 3-inch embedment of the buck in the cob wall, a 2x buck extension can be added to the inside of the buck.

Because there is no weather-resistive barrier to integrate with, sealant is recommended between the window or door frame and/or its nail-on fin and the buck to mitigate the passage of air and moisture through the wall system, as required in this section. For the same purpose, all gaps between the buck and (1) the window or door jamb, (2) the lintel and (3) the surrounding cob due to shrinkage should be filled with a flexible sealant capable of bonding with the cob. A plaster finish or wood trim is often used to protect and visually conceal the sealed interface between the window or door and the buck and surrounding cob, flashed as needed to prevent the intrusion of water.

As an alternative to installing a wood buck, the frames of manufactured windows or unframed glass can be embedded directly in the cob wall. Such installations must be capable of resisting the wind loads in Table R301.2.1(1). As with a buck installation, a 1-inch gap between the top of such windows and the bottom of the lintel is necessary.

AU105.5 Inspections. In addition to ensuring compliance with Section R109.1, the building official shall inspect the following aspects of *cob* construction:

1. Anchors and vertical and horizontal reinforcing in *cob* walls, where required in accordance with Tables

AU105.3 and AU106.11(1) and Sections AU105.3.4 through AU105.3.5.

2. Reinforcing in any concrete bond beams or lintels, in accordance with Section AU106.9.2 and Table AU106.10.

❖ The building official shall inspect anchors and reinforcing required by the stated tables and sections to verify proper installation. Where anchors or reinforcing are installed voluntarily, no inspection is required. The "required tests" refer to the shrinkage test in Section AU103.4.1, the compressive strength test in Section AU106.6.1 and the modulus of rupture test in Section AU106.7.1. These inspections and tests are in addition to the normal inspections required in Section R109.1.

SECTION AU106
COB WALLS—STRUCTURAL

AU106.1 General. *Cob* structural walls shall be in accordance with the prescriptive provisions of this section. Designs or portions of designs not complying with this section shall require an *approved* engineered design.

❖ The provisions of Section AU106 apply to structural cob walls (load-bearing and/or shear walls). Structural designs or portions that do not comply with this section require an approved engineered design.

AU106.2 Requirements for cob structural walls. In addition to the requirements of Section AU105.2, *cob* structural walls shall be subject to the following:

1. Wall height: shall be in accordance with Table AU105.3 for load-bearing *cob* walls or Table AU106.11(1) for *cob braced wall panels*, as applicable and most restrictive.

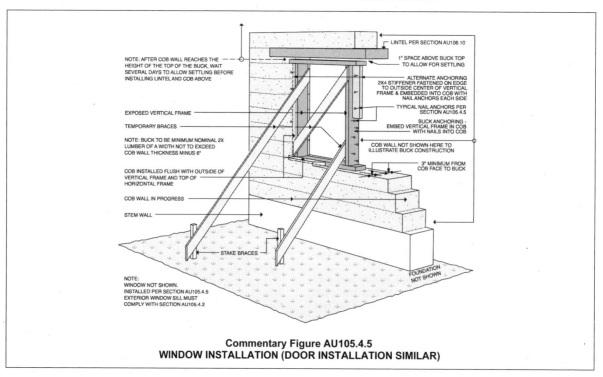

Commentary Figure AU105.4.5
WINDOW INSTALLATION (DOOR INSTALLATION SIMILAR)

2. Wall thickness: shall be in accordance with Sections AU105.2, Item 5 and Section AU106.8.1 for load-bearing *cob* walls or Table AU106.11(1) for *cob braced wall panels*, as applicable and most restrictive.

3. *Braced wall panel* lengths: for buildings using *cob braced wall panels*, the greater of the values determined in accordance with Table AU106.11(2) for wind loads and Table AU106.11(3), AU106.11(4) or AU106.11(5) for seismic loads shall be used.

❖ All cob buildings and all cob walls are subject to the limitations and requirements of Section AU105.2. Structural cob walls (load-bearing and/or shear walls) must also meet the requirements for wall height, thickness and length in the applicable sections as identified in Items 1, 2 and 3. The wall design must always meet the most restrictive applicable requirement(s). Structural cob walls also require a higher minimum compressive strength than nonstructural cob walls per Section AU106.6 and a minimum modulus of rupture for braced wall panels per Section AU107.

AU106.3 Loads and other limitations. Live and dead loads and other limitations shall be in accordance with Section R301, except that the dead load for cob walls shall be determined by Equation AU-1.

$$CW_{DL} = (H \times T_{avg} \times D) \qquad \text{(Equation AU-1)}$$

CW_{DL} = *Cob* wall dead load (in pounds per lineal foot of wall).

H = Height of *cob* portion of wall (in feet).

T_{avg} = Average thickness of wall (in feet).

D = Density of *cob* = 110 (in pounds per cubic foot), unless a lesser value at equilibrium moisture content is demonstrated to the building official.

❖ Equation AU-1 must be used for the dead load design of affected building elements such as foundations supporting cob walls (see commentary, Section AU106.4) and for engineered designs where required or otherwise provided. The weight of the cob walls is already accounted for in cob buildings using cob braced wall panels per Section AU106.11.

AU106.4 Foundations. Foundations for *cob* walls shall be in accordance with Chapter 4. The width of foundations for *cob* walls shall be not less than the width of the *cob* at its base, excluding *finish*.

❖ Foundations that satisfy the requirements of Chapter 4 are acceptable for buildings with cob walls. Three tables in the code are used to determine footing width and thickness, depending on the type of wall construction. Cob walls, with any finish allowed in this appendix, are closest to the weight of cast-in-place concrete or fully grouted masonry walls in Table R403.1(3); therefore, that table should be used. See Section AU106.3.

Figure AU101.4 shows a typical foundation and stem wall supporting a cob wall across its entire width as required in this section. The figure also shows a cob structural wall with its required anchors (except for Wall Type A) to the stem wall. Normal 7-inch (178 mm) embedment for these anchor bolts is required by Section R403.1.6, except that cob braced wall panels B,

C, D and E require 8-inch (203 mm) embedment per Table AU106.11(1).

The use of materials or designs for foundations other than those in Chapter 4 are not covered by this appendix and code and require an approved engineered design.

AU106.5 Wall taper, straightness and surface voids for cob walls. *Cob* walls shall be in accordance with the following:

1. *Cob* structural and *nonstructural walls* shall be vertical or shall taper from bottom to top with the wall thickness in accordance with Section AU105.2, Item 5 and the wall height in accordance with Section AU105.2, Item 4.

2. *Cob* structural and *nonstructural walls* shall be straight or curved. Curved *braced wall panels* shall be in accordance with Sections AU106.11.2 and AU106.11.3.

3. Niches and other surface voids in *load-bearing walls* are limited to 12 inches (305 mm) in width and height and 25 percent of the wall thickness, and shall be located in the top two-thirds of the wall. Surface voids that exceed these limits shall be considered wall openings, and shall receive a lintel in accordance with Section AU106.10 and be reinforced in accordance with Section AU105.3.4. Surface voids are prohibited in *braced wall panels*.

❖ To transmit loads safely to the foundation, irregularities in cob walls, such as taper, curvature and voids must be designed within the limits of this section. Wall taper is acceptable if it conforms with Section AU105.2(5). Both straight and curved cob walls may be used. However, curved walls used as braced wall panels must meet the requirements of Sections AU106.11.12 and AU106.11.13. Curved walls, with their inherent out-of-plane stability, may also take advantage of Table AU105.3, Note h. Finally, niches and other surface voids are restricted by the limits in Item 3 because voids exceeding those limits can create unsafe stress concentrations on adjacent wall segments.

AU106.6 Compressive strength of cob structural and nonstructural walls. All *cob* walls shall have a minimum compressive strength of 60 psi (414 kPa). *Cob* in walls used as *braced wall panels* shall have a minimum compressive strength of 85 psi (586 kPa).

❖ Compressive strength is a critical material property where using cob either structurally or nonstructurally. In nonstructural walls, the weight of the cob itself creates significant compressive forces at the base of the wall, hence the minimum required value in this section. Compressive strength is also important for all walls in their resistance to out-of-plane forces. Where used as a braced wall panel, the higher required value ensures that the wall remains safe under potential combined stresses from static and dynamic loading. See Section AU106.8 regarding compressive strength and a wall's load bearing capacity.

Though the minimum required compressive strength for cob in this appendix is 60 psi or 85 psi, depending on the use, the compressive strength of natural cob (unstabilized cob) can exceed 250 psi when fully dried.

APPENDIX AU—COB CONSTRUCTION (MONOLITHIC ADOBE)

AU106.6.1 Demonstration of compressive strength. The compressive strength of the *cob* mix to be used in structural walls and *nonstructural walls* as required in Section AU106.6 shall be demonstrated to the building official before the placement of *cob* onto walls, with compressive strength tests and an associated report by an *approved* laboratory or with an *approved* on-site test as follows:

1. Five samples of the proposed *cob* mix shall be placed moist to completely fill a 4-inch by 4-inch by 4-inch (102 mm by 102 mm by 102 mm) form and dried to ambient moisture conditions.

2. Samples shall not be oven dried.

3. Any opposite faces shall be faced with plaster of paris if needed to achieve smooth, parallel faces, after which the sample shall reach ambient moisture conditions before testing.

4. The horizontal cross section of the dried sample as tested, and the maximum applied load at failure shall be used to calculate the sample's compressive strength.

5. The fourth-lowest value shall be used to determine the mix's compressive strength.

❖ The required compressive strength test is best done in conjunction with the required shrinkage test in Section AU103.4.1 because both tests must be satisfied for a mix to be approved and used in a cob wall. The minimum required compressive strength depends on the application as described in Section AU106.6.

The testing described in Item 1 must be conducted by a laboratory approved by the building official or with a means of on-site testing approved by the building official. A report of the test results must accompany the laboratory or on-site testing.

Because of variation in cob's constituent materials, especially the clay subsoil, cob test data often exhibits variability, thereby requiring multiple tests. Using the fourth lowest test value to assign the mix's compressive strength is conservative, ensuring the mix will provide the minimum required strength and performance.

AU106.7 Modulus of rupture of cob structural walls. *Cob* in walls used as *braced wall panels* shall have a minimum modulus of rupture of 50 pounds per square inch (345 kPa).

❖ Modulus of rupture—a measure of bending strength—is an important material property for cob in braced wall panels; therefore, the minimum stated value is required in that application. Clay and sand alone have a low modulus of rupture. The straw in the mix increases bending strength through its ability to resist tension forces.

Modulus of rupture is also important to resist out-of-plane loads, especially for cob walls without steel reinforcing. This section applies to three wall types even where they are not braced wall panels. See Table AU105.3, Note i.

AU106.7.1 Demonstration of modulus of rupture. The modulus of rupture of *cob* used in structural walls shall be demonstrated to the building official before the placement of *cob* onto walls, with modulus of rupture tests and an associated report by an *approved* laboratory or with an *approved* on-site test as follows:

1. Five samples of the proposed *cob* mix shall be placed moist to completely fill a 6-inch by 6-inch by 12-inch (152 mm by 152 mm by 305 mm) form and dried to indoor ambient moisture conditions.

2. Samples shall not be oven dried.

3. Each sample shall be tested with the 12-inch (305 mm) dimension horizontal.

4. The fourth-lowest value shall be used to determine if the mix meets the minimum required modulus of rupture.

❖ Where a cob mix is to be used in a braced wall panel or for wall types described in Table AU105.3, Note i, the required modulus of rupture test is best done in conjunction with the required tests for shrinkage in Section AU103.4.1 and for compressive strength in Section AU106.6.1. All three tests must be satisfied for a mix to be approved and used in these applications.

See the commentary to Section AU106.6.1 regarding the subjects of approved testing and use of the fourth lowest value of the results. These requirements and their explanations are identical for both sections.

AU106.8 Bearing capacity. The allowable bearing capacity for *cob load-bearing walls* supporting vertical roof and/or ceiling loads imposed in accordance with Section R301 shall be determined by Equation AU-2.

$$BC = 144\,(C \times T_{min})/3 - (H \times T_{avg} \times D) \qquad \textbf{(Equation AU-2)}$$

BC = Allowable bearing capacity of wall (in pounds per lineal foot of wall).

C = Compressive strength (in psi) as determined in accordance with Section AU106.6.

T_{min} = Thickness of wall (in feet) at its minimum.

H = Height of *cob* portion of wall (in feet).

T_{avg} = Average thickness of wall (in feet).

D = Density of *cob* = 110 (in pounds per cubic foot), unless a lesser value at equilibrium moisture content is demonstrated.

❖ A wall's allowable bearing capacity is determined primarily by the compressive strength of the approved cob mix and the wall's minimum thickness. Allowable bearing capacity is typically expressed in pounds per lineal foot; thus, Equation AU-2 uses the mix's compressive strength in psi times the minimum wall thickness in feet and divides it by the commonly used safety factor of 3. The number 144 at the front of the equation represents 144 in²/ft² to yield units of pounds per lineal foot from the first half of the equation.

The second half of the equation subtracts the weight of the wall because a cob wall must support its own substantial weight. The remaining value is the allowable bearing capacity of the wall to support roof and/or ceiling loads determined in accordance with Section R301 (see Section AU106.8.1). The average wall thickness is used to determine the wall's weight, along with the wall's height and density.

AU106.8.1 Support of uniform loads. Uniform roof and/or ceiling loads shall be supported by *cob load-bearing walls* not exceeding their allowable bearing capacity, as demonstrated in accordance with Equation AU-3.

$$BL \leq BC \qquad \text{(Equation AU-3)}$$

BL = Design load on the wall (in pounds per lineal foot) determined in accordance with Sections R301.4 and R301.6.

BC = Allowable bearing capacity of wall (in pounds per lineal foot of wall) determined in accordance with Section AU106.8.

❖ The design roof and/or ceiling loads on cob walls determined per Sections R301.4 and R301.6 must not exceed the wall's allowable bearing capacity as calculated with Equation AU-2.

AU106.8.2 Support of concentrated loads. Concentrated roof and ceiling loads shall be distributed by structural elements capable of distributing the loads to the *cob load-bearing wall* and within its allowable bearing capacity as determined in accordance with Section AU106.8. Concentrated loads over lintels or over bond beams spanning openings shall require an *approved* engineered design.

❖ The bearing capacity of cob walls (see Section AU106.8) is typically large enough to support uniform roof and ceiling loads with significant remaining capacity (see Section AU106.8.1) to also support moderate concentrated loads. However, in a similar way that Equation AU-3 checks that uniform roof and ceiling loads do not exceed the wall's bearing capacity, concentrated loads from the roof or ceiling (for example, from a ridge beam post or a hip rafter) must be checked to be sure they do not exceed the wall's available bearing capacity.

Concentrated loads are first delivered to the wall's bond beam. Conservatively, the bond beams required on cob bearing walls (see Section AU106.9) are capable of distributing a moderate concentrated load along a 3-foot length of bearing wall. The bond beam suffices as the key horizontal element of the "structural elements" required in this section. Other elements, such as a ridge beam post and its connections, must also be adequate.

Any uniform roof and ceiling load (in pounds per foot) must first be subtracted from the wall's allowable bearing capacity (Equation AU-2) to find its remaining capacity (in pounds per foot) to support a concentrated load. Multiplying the remaining capacity by 3 (for 3 lineal feet of supporting wall, assuming 1.5 feet of uninterrupted wall on each side of the load) yields the maximum concentrated load that that portion of wall can support.

If a concentrated load exceeds a wall's available bearing capacity, structural posts can be employed to carry the load directly to the foundation. This condition is outside the scope of this appendix, as are concentrated loads over lintels or bond beams spanning openings.

AU106.9 Bond beams. *Cob* structural walls shall require a bond beam at the top of the wall in accordance with Section AU106.9.1, AU106.9.2 or AU106.9.3, and shall be anchored to the *cob* below in accordance with Tables AU105.3,

AU106.11(1) and AU106.12 as applicable and most restrictive. Bond beams spanning openings shall be in accordance with Section AU106.9.4.

❖ Bond beams are very important for the structural integrity of cob buildings and are required for all cob walls. Bond beams for structural walls must comply with this section, and nonstructural walls must comply with Section AU105.3.2. The bond beam on load-bearing walls distributes vertical roof and ceiling loads into the cob wall and sometimes across openings. The bond beam on cob braced wall panels (shear walls) distributes lateral wind and seismic loads from the roof (or other walls in the same line) into the cob braced wall panel. Bond beams on nonstructural walls, as well as bond beams on structural walls, distribute out-of-plane wind and seismic loads to the roof diaphragm in order to deliver them to perpendicular braced wall panels elsewhere in the building.

Bond beams can be of wood or concrete with their respective requirements in Section AU106.9.1 and Section AU109.2. Bond beams of other materials are possible but require an approved engineered design per Section AU106.9.3.

Bond beams above openings perform like lintels and must meet the requirements in Section 106.9.4. The connections of roof framing to bond beams are covered in Section AU106.9.5, and the significantly different condition of bond beams at gable and shed roof end walls is covered in Section AU106.9.6.

AU106.9.1 Wood bond beams. Wood bond beams shall be not less than nominal 4 inches high by 8 inches wide and shall comply with Sections AU106.9.1.1 through AU106.9.1.3.

❖ Wood bond beams are acceptable for structural walls. They must be of at least the stated size in this section and must comply with Sections AU106.9.1.1 through AU106.9.1.3 regarding species and grade, discontinuity, corners and curved walls.

AU106.9.1.1 Wood species and grade. Wood bond beams shall be of a species with an extreme fiber in bending (F_b) of not less than 850 psi (5.9 MPa), a modulus of elasticity (E) of not less than 1,300,000 psi (8964 MPa), and No. 2 grade or better. Composite lumber bond beams shall have an F_b of not less than 850 psi (5.9 MPa), and an E of not less than 1,300,000 psi (8964 MPa).

❖ The *National Design Specification (NDS) Supplement: Design Values for Wood Construction* lists the extreme fiber in bending (F_b) and the modulus of elasticity (E) for wood and composite lumber, depending on the species, grade, and size of the lumber. Wood bond beams must be at least No. 2 grade. The wood or composite lumber used for bond beams must have at least the F_b and E values stated in this section.

"Extreme fiber in bending" is the tensile or compressive stress experienced by the outermost material of a beam in bending and is the common measure of the bending strength of lumber. Modulus of elasticity measures the stiffness of lumber. Both values directly affect the ability of a wood bond beam to distribute vertical and lateral loads into and from the wall.

AU106.9.1.2 Discontinuity. Discontinuous wood bond beams shall be spliced on top with a metal strap with not less than the

APPENDIX AU—COB CONSTRUCTION (MONOLITHIC ADOBE)

allowable wind or seismic load tension capacity in accordance with the following, whichever is more restrictive:

1. For *seismic design categories*: A, 2,500 pounds (11 kN); B, 4,500 pounds (20 kN); C, 6,000 pounds (26.7 kN).

2. For braced wall line lengths, when wind governs: 10 feet, 2,500 pounds (11 kN); 20 feet, 3,400 pounds (15.1 kN); 30 feet, 5,000 pounds (22.2 kN).

❖ In certain loading conditions bond beams distribute tension forces along the length of its wall line. If a bond beam is discontinuous, the tension forces must be transmitted to the next section of bond beam with a metal strap with the capacity to transmit the larger of the tension forces of the applicable seismic and wind loading conditions in Items 1 and 2.

AU106.9.1.3 Corners and curved walls. Wood bond beams at corners and discontinuities atop curved walls shall be connected across their exterior faces with a metal strap with a capacity of not less than that determined in accordance with Section AU106.9.1.2.

❖ Straps across wood bond beam discontinuities transmit lateral tension loads and assist with out-of-plane stability. The required straps on the exterior face of bond beams atop curved walls and at corners will remain tight against the bond beam while transmitting tension forces. In contrast, if applied to the interior face of the bond beam, the strap could separate from the bond beam when loaded and its fasteners would be subject to withdrawal.

AU106.9.2 Concrete bond beams. Concrete bond beams shall be not less than 6 inches (152 mm) high by 8 inches (305 mm) wide. Concrete bond beams shall be reinforced with two No. 4 bars, 2 inches (51 mm) clear from the bottom and 2 inches (51 mm) clear from the sides. Lap splices shall comply with Table R608.5.4(1). Reinforcing at corners shall be in accordance with the horizontal reinforcing requirements in Section R608.6.4. The concrete shall have a compressive strength of not less than 2,500 pounds per square inch (17.2 MPa) at 28 days.

❖ Concrete bond beams are acceptable for structural walls and must be at least the size stated in this section to accept and distribute both uniform and concentrated loads. The concrete must be reinforced to ensure the safe transmission of tension forces, hence the reinforcing requirements in this section, Table R608.5.4(1) and Section R608.6.4 for corners. A minimum concrete compressive strength of 2500 psi (17.2 MPa) is required.

AU106.9.3 Other bond beams. Bond beams of other materials, including earthen materials, require an *approved* engineered design.

❖ Many materials and designs may serve safely as a bond beam for cob walls. However, the use of alternative materials and designs requires an approved engineered design.

AU106.9.4 Bond beams spanning openings. Bond beams that support uniform roof and/or ceiling loads and span openings in *cob* walls shall be in accordance with Table AU106.10.

Bond beams shall be continuous across the opening and not less than 1 foot (305 mm) beyond each side of the opening.

❖ Bond beams may also serve as lintels over openings if the opening occurs at the top of a wall. The design of the bond beam over an opening is the same as that for a lintel over a same-sized opening. For safe load transmission, the bond beam must be continuous over the opening, must extend at least 1 foot beyond each side, and must adhere to the other requirements for lintels in Table AU106.10.

AU106.9.5 Connection of roof framing to bond beams. Roof and ceiling framing shall be attached to bond beams in accordance with Table R602.3(1), Items 2 and 6, and Figure AU106.9.5. Roof sheathing shall be attached to roof framing in accordance with Figure AU106.9.5. A minimum nominal 2-inch by 6-inch (51 mm by 152 mm) wood plate shall be installed on concrete bond beams with $^5/_8$-inch (16 mm) diameter anchor bolts with 5-inch (127 mm) embedment at 2 feet (610 mm) on center to allow the required fastening of roof and ceiling framing, including tension ties and straps.

❖ The required connections between bond beams and roof and ceiling framing are as detailed in Table R602.3(1), Items 2 and 6, and in Figure AU106.9.5. Figure AU106.9.5 also shows the required attachment—particular to cob buildings—of roof sheathing to tension tie rafters and additional nailing where applicable [see Tables AU106.11(4) and (5)], in addition to the normal attachment requirements in Table R602.3(1). If a concrete bond beam is used, the connection to roof and ceiling framing is achieved through a wood plate attached to the bond beam as described in this section.

AU106.9.6 Bond beams and connections at gable and shed roof end walls. Bond beams and connections at end walls of buildings with gable or shed roofs shall comply with Figure AU106.9.6 and the following:

1. End walls shall not exceed 20 feet (6096 mm) in length.

2. Bond beams shall be continuous and straight for the entire wall line.

3. Wood bond beams shall comply with the following:

   3.1. Not less than nominal 4 inches by 8 inches (102 mm by 203 mm) where wind design governs in accordance with Table AU106.11(2) and where seismic design governs in accordance with Table AU106.11(3), AU106.11(4) or AU106.11(5) for wall lengths less than or equal to 20 feet (6096 mm) in Seismic Design Category A or wall lengths less than or equal to 10 feet (3048 mm) in Seismic Design Categories B and C.

   3.2. Not less than nominal 4 inches by 10 inches (102 mm by 254 mm) for wall lengths less than or equal to 20 feet (6096 mm) in Seismic Design Category B.

   3.3. Not less than nominal 6 inches by 12 inches (152 mm by 305 mm) or 4 inches by 16 inches (102 mm by 406 mm) for wall lengths less than or equal to 20 feet (6096 mm) in Seismic Design Category C.

APPENDIX AU—COB CONSTRUCTION (MONOLITHIC ADOBE)

3.4. Corners shall be connected in accordance with Section AU106.9.3.

4. Concrete bond beams when used shall be in accordance with Section AU106.9.2 in Seismic Design Categories A, B and C and for ultimate design wind speeds less than or equal to 140 mph (63.6 m/s).

5. Walls between the bond beam and roof shall be of wood-framed construction in accordance with Section R602. The ratio of its greatest height to its length shall not exceed 1:2. The wall shall not contain openings.

❖ Where roof framing is parallel to a wall line, such as at the end walls of buildings with a gable or shed roof, it is not possible to directly connect the roof framing to

those walls to resist and transfer out-of-plane wall loads (as shown in Figure AU106.9.5). Though there are many ways to resist and transfer these loads in this condition, this appendix requires the bond beam to transfer the out-of-plane loads to the end walls and their integral braced wall panels.

Section AU106.9.6 and Figure AU106.9.6 give the detailed requirements and connections between and including the bond beam and roof for this condition.

An important note in Figure AU106.9.6 states that its requirements pertain to short wood-framed walls that support a shed roof between a cob wall's bond beam and the shed roof rafters. This condition results in a similar inability for the roof framing to directly connect

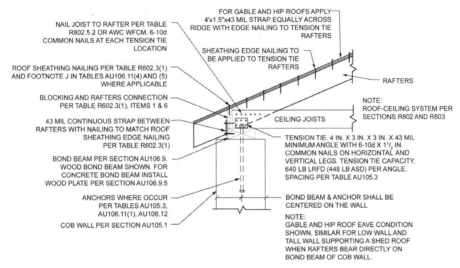

FIGURE AU106.9.5
CONNECTION OF ROOF FRAMING TO BOND BEAMS

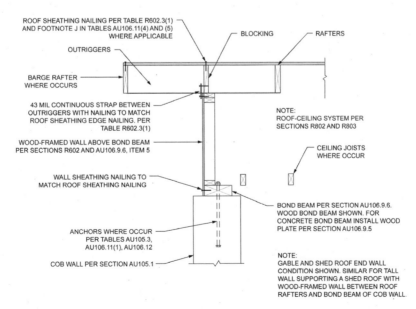

FIGURE AU106.9.6
CONNECTIONS AT GABLE AND SHED ROOF END WALLS

APPENDIX AU—COB CONSTRUCTION (MONOLITHIC ADOBE)

to the bond beam; therefore, the connection requirements in Figure AU106.9.6 apply. This condition occurs where a cob wall is not tall enough to directly support the shed roof due to the height limit of the cob wall or because of design preference.

AU106.10 Lintels. Door, window and other openings in load-bearing *cob* walls shall be provided with a lintel of wood or concrete in accordance with Table AU106.10.

❖ Openings in load-bearing cob walls must be provided with a lintel that supports the weight of the cob above it in addition to the roof and ceiling loads. These loads are transferred to each side of the opening as concentrated loads. In Table AU106.10, the required lintel length of 1 foot beyond each side of the opening provides sufficient bearing capacity in the wall to support the live and dead loads for the conditions in the table, except where indicated as NP (not permitted). Table AU106.10 also provides required specifications and sizes of lintels in either wood or concrete for various building sizes, wall thicknesses and spans. This section is also used for bond beams spanning openings.

TABLE AU106.10. See below.

❖ See the commentary to Section AU106.10.

AU106.11 Cob braced wall panels. *Cob braced wall panels* shall be in accordance with Section R602.10 and Tables AU106.11(1), AU106.11(2), AU106.11(3), AU106.11(4) and AU106.11(5). Wind design criteria shall be in accordance with Section R301.2.1. Seismic design criteria shall be in accordance with Section R301.2.2. An *approved* engineered design shall be required in accordance with Section R301.2.1 where the building is located in a special wind region or where wind design is required in accordance with Figure R301.2.1.1.

❖ Every building must have a lateral force-resisting system to resist wind, and in most regions, seismic forces. The system includes the building's roof structure and diaphragm, floor structure and diaphragm (where raised floors are present), shear walls (braced wall panels in code terminology), foundation and the connections between them. These elements must be of sufficient strength for the building's size, construction system, configuration and location for its design wind

TABLE AU106.10
LINTELS AND BOND BEAMS SPANNING OPENINGS

GROUND SNOW LOAD ≤ 30 PSF			WOOD: • F_b ≥ 850 psi • E ≥ 1,300,000 psi • No. 2 Grade or better • Oriented flat • 1 piece or 2 equal-width pieces • Extend 1 foot beyond opening sides		CONCRETE: • 2500 psi compressive strength • Height = 6 inches • Extend 1 foot beyond opening sides • Reinforcement two No. 4 bars[a] • 2 inches clear from bottom • 2 inches clear from sides[a]	
Building Width (feet)	Cob above Lintel (feet)	Total Cob Wall and Plaster Thickness (inches)	Size of Wood Lintel or Bond Beam—*H × W* (nominal inches)		Width of Concrete Lintel or Bond Beam (inches)	
			For Span ≤ 4 ft	For Span ≤ 6 ft	For Span ≤ 6 ft	For Span ≤ 8 ft
10	0	≤ 27	4 × 8	4 × 8	8	8
10	1	15	4 × 12	4 × 12	12	12
10	1	19	4 × 16	4 × 16	16	16
10	1	27	4 × 24	4 × 24	24	24
10	2	15	4 × 12	6 × 12	12	12
10	2	19	4 × 16	6 × 16	16	16
10	2	27	4 × 24	4 × 24	24	24
20	0	≤ 27	4 × 8	6 × 8	8	8
20	1	15	4 × 12	6 × 12	12	12
20	1	19	4 × 16	6 × 16	16	16
20	1	27	4 × 24	4 × 24	24	24
20	2	15	4 × 12	6 × 12	12	NP
20	2	19	4 × 16	6 × 16	16	**NP**
20	2	27	4 × 24	6 × 24	24	NP
30	0	≤ 27	4 × 8	6 × 8	8	NP
30	1	15	4 × 12	6 × 12	12	NP
30	1	19	4 × 13	6 × 16	16	NP
30	1	27	4 × 24	6 × 24	24	NP
30	2	15	4 × 12	6 × 12	12	NP
30	2	19	4 × 16	6 × 16	16	NP
30	2	27	4 × 24	6 × 24	24	NP

For SI: 1 inch = 25.4 mm, 1 foot = 304.8 mm, 1 pound per square inch = 6.895 kPA.

NP = Not Permitted.

a. Concrete bond beams spanning openings, and lintels greater than 16 inches in width, shall have an additional No. 4 bar in the center of their width.

speed and seismic design category as determined by Sections R301.2.1 and R301.2.2, respectively.

The code contains prescriptive provisions for all elements of the lateral force-resisting system and their connections. Buildings with cob walls cannot use the conventional braced wall panels in the code (unless an approved engineered design is provided) because cob walls are heavier and impart seismic loads much greater than the code braced wall panels with the lengths in the IRC tables are designed to resist. The building's cob walls can be used for its braced wall panels where it is constructed in accordance with Table AU106.11(1) and meets the minimum total lengths in Table AU106.11(2) for wind loads and Table AU106.13(3), (4) or (5) for seismic loads. These wind and seismic tables fit the same format as those for conventional braced wall panels in the code. The cob walls must connect to the roof structure in accordance with Sections AU106.9.5 and AU106.9.6 instead of the connections prescribed in the code.

To use the prescriptive braced wall panel tables in this appendix, it is necessary to select two orthogonal directions. Each orthogonal direction requires at least two braced wall lines. These tables assume a building with two sets of parallel wall lines orthogonal (perpendicular) to each other. (See Section AU106.12 for non-orthogonal braced wall panels.)

Cob braced wall panels (shear walls) are reinforced, compression-based, monolithic walls similar in concept to some forms of reinforced concrete. Like fiber-reinforced concrete, cob walls have ductility by virtue of their straw as a microfiber reinforcement throughout. Like steel bar and mesh-reinforced concrete, cob wall ductility is increased by steel bar and mesh reinforcing. Though reinforced cob has significantly lower strength than reinforced concrete, it has considerable strength sufficient for many applications.

The reinforcing materials in this appendix are limited to straw and steel. Other plant fibers, other metals and plastics have also been successfully used to reinforce cob walls but are outside the scope of this appendix.

Testing of straw- and steel-reinforced cob walls has demonstrated their capacity to safely absorb and dissipate energy under in-plane loading. The resulting test data were used to determine the braced wall panel provisions of this appendix. The braced wall panel tables in this section are limited to Seismic Design Categories A, B and C because the entire appendix is likewise limited, except with an approved engineered design in Seismic Design Categories D and E. See Section AU105.2, Item 3.

The braced wall panel tables based on seismic design category contain two columns indicating the minimum percentage of wall openings in the braced wall lines: 0 percent, 25 percent or 50 percent. The percentage of openings affects the seismic load on the building and the total required braced wall panel length because the weight of cob walls is a significant portion of the building's seismic load. Windows, doors and other openings reduce that load. A column for percent-

age of wall openings does not appear in Table AU106.11(2) for wind design because wind loads are independent of a building's weight.

The minimum length of a braced wall panel is determined by the maximum aspect ratio (*H:L*) in Table AU106.11(1) for the wall type used. For example, if the maximum aspect ratio (*H:L*) is 2:1, and the height of the cob portion of the wall is 7 feet, then the minimum braced wall panel length is 3.5 feet. For a 1:1 maximum aspect ratio, the minimum braced wall panel length would be 7 feet. If a wall meets the minimum required length for that wall type and height, its length counts toward the minimum total braced wall panel length in the applicable table(s) in Section AU106.11.

The total length of braced wall panels along each braced wall line must comply with Table AU106.11(2) for wind loads and Table AU106.11(3), (4) or (5) for seismic loads. The greater value of the minimum total length in each table must be used in the building's design.

Meeting the braced wall panel requirements in this appendix provides a prescriptive lateral load-resisting system for cob buildings. This appendix does not provide the necessary data for an engineered lateral load-resisting design using cob shear walls, such as an *R*-factor and allowable shear for cob braced wall panels (shear walls). Design professionals must rely on testing from credible sources or credible published literature or test reports (for example, those in the commentary bibliography) and use accepted engineering principles and judgment for such a design.

Cob buildings require an approved engineered design where located in either a special wind region or wind design required location. Approximate locations for both can be found in Figure R301.2.1.1. Local jurisdiction ultimate design wind speeds or other related local requirements take precedence.

TABLE AU106.11(1). See page Appendix AU-24.

❖ See the commentary to Sections AU106.11 and AU106.11.3.

TABLE AU106.11(2). See page Appendix AU-25.

❖ See the commentary to Sections AU106.11, AU106.11.1 and AU106.11.2.

TABLE AU106.11(3). See page Appendix AU-25.

❖ See the commentary to Sections AU106.11, AU106.11.1 and AU106.11.2.

TABLE AU106.11(4). See page Appendix AU-26.

❖ See the commentary to Sections AU106.11 and AU106.11.1.

TABLE AU106.11(5). See page Appendix AU-27.

❖ Though Wall Type A has sufficient in-plane load-resistance capacity as shown in this table, Wall Type A is not allowed in Seismic Design Category C because it is not allowed in Table AU105.2 due to its out-of-plane load-resistance limitations. For other aspects of Table AU106.11(5), see the commentary to Sections AU106.11 and AU106.11.1.

APPENDIX AU—COB CONSTRUCTION (MONOLITHIC ADOBE)

TABLE AU106.11(1)
COB BRACED WALL PANEL TYPES

WALL TYPE[a] DESIGNATION	ANCHORS TO FOUNDATION[b]	ANCHORS TO BOND BEAM[c]	VERTICAL STEEL REINFORCING[b, c]	HORIZONTAL STEEL REINFORCING	MAXIMUM HEIGHT H[d] (in feet)	MAXIMUM ASPECT RATIO (H:L)
A	none	$5/_8$″ threaded rod @ 12″; 4″ from wall ends; 12″ embedment in cob	none	none	7[e]	1:1
B	#5 bar @ 12″; 16″ embedment in cob	$5/_8$″ threaded rod @ 12″; 4″ from wall ends; 16″ embedment in cob; 2″ × 2″ × $1/_4$″ washer and nut at cob end	none	2″ × 2″ × 14 gage welded wire mesh[f] @ 18″; 6″ from foundation and bond beam	7[e]	1:1
C	#5 bar @ 12″; 16″ embedment in cob	$5/_8$″ threaded rod @ 12″; 16″ embedment in cob	$5/_8$″ threaded rod; 4″ from each end of braced wall panel; continuous from foundation to bond beam	2″ × 2″ × 14 gage welded wire mesh[f] @ 18″; 6″ from foundation and bond beam	7[e]	2:1
D	(see vertical steel reinforcing)	(see vertical steel reinforcing)	$5/_8$″ threaded rod; 4″ from each end of braced wall panel and @ 12″; continuous from foundation to bond beam	2″ × 2″ × 14 gage welded wire mesh[f] @ 18″; 6″ from foundation and bond beam	7[e]	2:1
E	6″ × 6″ × 6 gage welded wire mesh; 12″ embedment in foundation	$5/_8$″ threaded rod @ 12″; 4″ from wall ends; 12″ embedment in cob	6″ × 6″ × 6 gage welded wire mesh; 2″ from each wall face	none	7.5	1:1

For SI: 1 inch = 25.4 mm, 1 foot = 304.8 mm.

a. Braced wall panel Types A, B, C and D shall be not less than 16 inches thick. Braced wall panel Type E shall be not less than 12 inches thick. All braced wall panels shall be not greater than 24 inches thick.

b. Not less than 8-inch embedment into foundation, unless otherwise stated.

c. Not less than 4-inch embedment into concrete bond beams. Full penetration through wood bond beam, secured with nut and washer.

d. H = height of the cob portion of the wall only. See Figure AU101.4.

e. Maximum height shall be 8 feet when wall thickness is increased to 18 inches.

f. Galvanized mesh.

APPENDIX AU—COB CONSTRUCTION (MONOLITHIC ADOBE)

TABLE AU106.11(2)
BRACING REQUIREMENTS FOR COB BRACED WALL PANELS BASED ON WIND SPEED

• EXPOSURE CATEGORY B[d] • 25-FOOT MEAN ROOF HEIGHT • 10-FOOT EAVE-TO-RIDGE HEIGHT[d] • 10-FOOT WALL HEIGHT[d] • 2 BRACED WALL LINES[d]			MINIMUM TOTAL LENGTH (FEET) OF COB BRACED WALL PANELS REQUIRED ALONG EACH BRACED WALL LINE[a, b, c, d]			
Ultimate Design Wind Speed (mph)	Story Location	Braced Wall Line Spacing (feet)	Cob Braced Wall Panel[e] A; (aspect ratio $H{:}L \leq 1{:}1$)	Cob Braced Wall Panel[e] B; (aspect ratio $H{:}L \leq 1{:}1$)	Cob Braced Wall Panel[e] C, D; (aspect ratio $H{:}L \leq 2{:}1$)	Cob Braced Wall Panel[e] E; (aspect ratio $H{:}L \leq 1{:}1$)
≤ 110	One-story building	10	6.0	6.0	3.7	6.0
≤ 110	One-story building	20	7.9	7.4	7.4	6.0
≤ 110	One-story building	30	11.8	11.0	11.0	6.9
≤ 115	One-story building	10	6.0	6.0	4.1	6.0
≤ 115	One-story building	20	8.7	8.1	8.1	6.0
≤ 115	One-story building	30	13.0	12.1	12.1	7.6
≤ 120	One-story building	10	6.0	6.0	4.4	6.0
≤ 120	One-story building	20	9.4	8.8	8.8	6.0
≤ 120	One-story building	30	14.1	13.1	13.1	8.3
≤ 130	One-story building	10	6.0	6.0	5.1	6.0
≤ 130	One-story building	20	11.0	10.3	10.3	6.5
≤ 130	One-story building	30	16.5	15.4	15.4	9.7
≤ 140	One-story building	10	6.0	6.0	5.9	6.0
≤ 140	One-story building	20	12.7	11.9	11.9	7.5
≤ 140	One-story building	30	19.1	17.8	17.8	11.2

For SI: 1 foot = 304.8 mm, 1 mile per hour = 0.447 m/s.
a. Linear interpolation shall be permitted.
b. Braced wall panels shall be without openings.
c. Braced wall panel Types A, B and E shall have an aspect ratio ($H{:}L$) ≤ 1:1. Braced wall panel Types C and D shall have an aspect ratio ($H{:}L$) ≤ 2:1.
d. Subject to applicable wind adjustment factors associated with Items 1 and 2 of Table R602.10.3(2).
e. Cob braced wall panel types indicated shall comply with Section AU106.11 and Table AU106.11(1).

TABLE AU106.11(3)
BRACING REQUIREMENTS FOR COB BRACED WALL PANELS BASED ON SEISMIC DESIGN CATEGORY A

• SOIL CLASS D[f] • TOTAL WALL HEIGHT = 10 FEET (INCLUDING STEM WALL AND BOND BEAM) • COB WALL HEIGHT PER TABLE AU106.11(1) • 15 PSF ROOF-CEILING DEAD LOAD[d] • STORY LOCATION: ONE-STORY BUILDING • SEISMIC DESIGN CATEGORY A • 1.5″ PLASTER THICKNESS EACH SIDE[g]				MINIMUM TOTAL LENGTH (FEET) OF COB BRACED WALL PANELS REQUIRED ALONG EACH BRACED WALL LINE[a, b, c, d, e]		
Braced Wall Line Spacing (feet)	Braced Wall Line Length (feet)	Min. Braced Wall Line % Openings	Min. Perpendicular Braced Wall Line % Openings	Cob Braced Wall Panel[e] A, B	Cob Braced Wall Panel[e] C, D	Cob Braced Wall Panel[e] E
10	30	0	0	—	3.4	6.0
20	20	0	0	—	3.5	6.0
20	30	0	0	—	4.5	6.0
30	30	0	0	—	5.6	6.0

For SI: 1 inch = 25.4 mm, 1 foot = 304.8 mm, 1 pound per square foot = 0.0479 kPa.
a. Interpolation is not permitted.
b. Braced wall panels shall be without openings.
c. Braced wall panel Types A, B and E shall have an aspect ratio ($H{:}L$) ≤ 1:1. Braced wall panel Types C and D shall have an aspect ratio ($H{:}L$) ≤ 2:1.
d. Subject to applicable seismic adjustment factors associated with Item 5 in Table R602.10.3(4).
e. Cob braced wall panel types indicated shall comply with Section AU106.11 and Table AU106.11(1).
f. Wall bracing lengths are based on a soil site class D. Interpolation of bracing lengths between S_{DS} values associated with the seismic design categories is allowable where a site-specific S_{DS} value is determined in accordance with Section 1613 of the *International Building Code*.
g. For total plaster thickness between 3 inches and 6 inches, the minimum total length of braced wall panels shall be multiplied by 1.2.

APPENDIX AU—COB CONSTRUCTION (MONOLITHIC ADOBE)

TABLE AU106.11(4)
BRACING REQUIREMENTS FOR COB BRACED WALL PANELS BASED ON SEISMIC DESIGN CATEGORY B

- SOIL CLASS D[f]
- TOTAL WALL HEIGHT = 10 FEET (INCLUDING STEM WALL AND BOND BEAM)
- COB WALL HEIGHT PER TABLE AU106.11(1)
- 15 PSF ROOF-CEILING DEAD LOAD[d]
- STORY LOCATION: ONE-STORY BUILDING
- SESIMIC DESIGN CATEGORY B
- 1.5″ PLASTER THICKNESS EACH SIDE[g]

MINIMUM TOTAL LENGTH (FEET) OF COB BRACED WALL PANELS REQUIRED ALONG EACH BRACED WALL LINE[a, b, c, d, e]

Braced Wall Line Spacing (feet)	Braced Wall Line Length (feet)	Min. Braced Wall Line % Openings	Min. Perpendicular Braced Wall Lines % Openings	Cob Braced Wall Panel[e] A, B	Cob Braced Wall Panel[e] C, D	Cob Braced Wall Panel[e] E
10	10	0	0	6.0	3.2	6.0
10	20	0	0	6.0	4.9	6.0
10	20	50	0	6.0	3.5	6.0
10	30	0	0	7.1	6.6	6.0
10	30	50	0	6.0	4.5	6.0
20	10	0	0	6.0[h]	4.9[h]	6.0
20	10	0	50	6.0	3.5	6.0
20	10	50	0	NP	4.2	NP
20	10	50	50	NP	3.0	NP
20	20	0	0	7.4	6.9	6.0
20	20	0	50	6.0	5.5	6.0
20	20	50	0	6.0	5.5	6.0
20	20	50	50	6.0	4.1	6.0
20	30	0	0	9.4	8.8	6.0
20	30	0	50	7.9	7.4	6.0
20	30	50	0	7.2	6.7	6.0
20	30	50	50	6.0	5.3	6.0
30	10	0	0	7.1	6.6	6.0
30	20	0	0	9.4	8.8	6.0
30	20	0	50	7.2	6.7	6.0
30	20	50	0	7.9	7.4	6.0
30	30	0	0	11.8	11.0	6.0
30	30	0	50	9.5	8.9	6.0
30	30	50	0	9.5	8.9	6.0
30	30	50	50	7.3	6.8	6.0

For SI: 1 inch = 25.4 mm, 1 foot = 304.8 mm, 1 pound per square foot = 0.0479 kPa.

NP = Not Permitted.

a. Interpolation is not permitted.

b. Braced wall panels shall be without openings.

c. Braced wall panel Types A, B and E shall have an aspect ratio $(H:L) \leq 1:1$. Braced wall panel Types C and D shall have an aspect ratio $(H:L) \leq 2:1$.

d. Subject to applicable seismic adjustment factors associated with Item 5 in Table R602.10.3(4).

e. Cob braced panel types indicated shall comply with Section AU106.11 and Table AU106.11(1).

f. Wall bracing lengths are based on a soil site class D. Interpolation of bracing lengths between S_{DS} values associated with the seismic design categories is allowable where a site-specific S_{DS} value is determined in accordance with Section 1613 of the *International Building Code*.

g. For total plaster thicknesses 3 inches to 6 inches, the minimum total length of braced wall panels shall be multiplied by 1.2.

h. Total plaster thicknesses shall be not greater than 3 inches. Substitute $^{15}/_{32}$″ roof sheathing and 10d at 6″ edge nailing for requirements in Table R602.3(1).

APPENDIX AU—COB CONSTRUCTION (MONOLITHIC ADOBE)

TABLE AU106.11(5)
BRACING REQUIREMENTS FOR COB BRACED WALL PANELS BASED ON SEISMIC DESIGN CATEGORY C

• SOIL CLASS D[f] • TOTAL WALL HEIGHT = 10 FEET (INCLUDING STEM WALL AND BOND BEAM) • COB WALL HEIGHT PER TABLE AU106.11(1) • 15 PSF ROOF-CEILING DEAD LOAD[d] • STORY LOCATION: ONE-STORY BUILDING • SEISMIC DESIGN CATEGORY C • 1.5″ PLASTER THICKNESS EACH SIDE[g]				MINIMUM TOTAL LENGTH (FEET) OF COB BRACED WALL PANELS REQUIRED ALONG EACH BRACED WALL LINE[a, b, c, d, e]		
Braced Wall Line Spacing (feet)	Braced Wall Line Length (feet)	Min. Braced Wall Line % Openings	Min. Perpendicular Braced Wall Lines % Openings	Cob Braced Wall Panel[e] A, B	Cob Braced Wall Panel[e] C, D	Cob Braced Wall Panel[e] E
10	10	0	0	8.3[h]	7.8[h]	6.0
10	10	0	50	6.5	6.1	6.0
10	10	25	0	7.4[h]	6.9[h]	6.0
10	10	50	50	NP	4.4	6.0
10	15	0	0	10.6	9.9	6.0
10	15	0	50	8.7	8.2	6.0
10	15	50	0	NP	7.3	6.0
10	15	50	50	6.0	5.6	6.0
10	20	0	0	12.8	11.9	6.0
10	20	0	50	11.0	10.2	6.0
10	20	50	0	9.1	8.5	6.0
10	20	50	50	7.3	6.8	6.0
15	10	25	0	NP	NP	6.0[h]
15	10	0	50	7.8	7.3	6.0
15	10	50	0	NP	NP	NP
15	10	50	50	NP	NP	NP
15	15	0	0	12.9	12.1	6.0
15	15	0	50	10.2	9.5	6.0
15	15	50	0	NP	NP	6.0
15	15	50	50	7.5	7.0	6.0
15	20	0	0	15.3	14.3	6.0
15	20	0	50	12.6	11.7	6.0
15	20	50	0	NP	NP	6.0
15	20	50	50	8.9	8.3	6.0
20	10	25	0	NP	NP	NP
20	10	0	50	9.1	8.5	6.0
20	10	50	0	NP	NP	NP
20	10	50	50	NP	NP	NP
20	15	0	0	NP	14.3[h]	6.0[h]
20	15	0	50	11.7[h]	10.9[h]	6.0
20	15	50	0	NP	NP	6.0[h]
20	15	50	50	NP	NP	6.0
20	20	0	0	17.8	16.7	6.9
20	20	0	50	14.2	13.3	6.0
20	20	50	0	NP	NP	6.0
20	20	50	50	NP	9.9	6.0

For SI: 1 inch = 25.4 mm, 1 foot = 304.8 mm, 1 pound per square foot = 0.0479 kPa.

NP = Not Permitted.

a. Interpolation is not permitted.

b. Braced wall panels shall be without openings.

c. Braced wall panel Types A, B and E shall have an aspect ratio ($H{:}L$) ≤ 1:1. Braced wall panel Types C and D shall have an aspect ratio ($H{:}L$) ≤ 2:1.

d. Subject to applicable seismic adjustment factors associated with Item 5 in Table R602.10.3(4).

e. Cob braced panel types indicated shall comply with Section AU106.11 and Table AU106.11(1).

f. Wall bracing lengths are based on a soil site class D. Interpolation of bracing lengths between S_{DS} values associated with the seismic design categories is allowable where a site-specific S_{DS} value is determined in accordance with Section 1613 of the *International Building Code*.

g. For total plaster thicknesses 3″ to 6″, multiply the minimum total length of braced wall panels by 1.2.

h. Total plaster thickness shall not be greater than 3 inches. Substitute $^{15}/_{32}$″ roof sheathing and 10d at 6″ edge nailing for requirements in Table R602.3(1).

APPENDIX AU—COB CONSTRUCTION (MONOLITHIC ADOBE)

AU106.11.1 Nonorthogonal braced wall panels. *Braced wall panels* at an angle to the orthogonal *braced wall lines* shall be considered to contribute to the minimum total braced wall lengths in Tables AU106.11(2), AU106.11(3), AU106.11(4) and AU106.11(5), as follows:

1. A *braced wall panel* not more than 45 degrees and greater than 30 degrees to an adjacent orthogonal braced wall line shall contribute 50 percent of its length to that line.

2. A *braced wall panel* not more than 30 degrees to an orthogonal braced wall line shall contribute 65 percent of its length to that line.

3. A *braced wall panel* greater than 45 degrees and not more than 60 degrees to an orthogonal braced wall line shall contribute 35 percent of its length to that line.

4. The angle of a curved *braced wall panel* to a braced wall line shall be determined with the chord of that section of wall, connecting the end points of the arc at the center of the wall.

❖ The braced wall panel Tables AU106.11(2), (3), (4) and (5) assume buildings with two sets of parallel braced wall lines orthogonal (perpendicular) to each other. See Section AU106.11. However, it is common for walls in cob buildings to be non-orthogonal.

The designer can count the full length of braced wall panels that are in line with an orthogonal direction toward the required length of braced wall panels along each braced wall line in the applicable table. Non-orthogonal braced wall panels are counted in both orthogonal directions, but their countable lengths are reduced because of their reduced ability to resist lateral loads relative to the chosen orthogonal lines. The percentage reduction depends on the wall's angle relative to each orthogonal line as required and described in Items 1–3.

Item 4 describes the method for determining the angle of a curved braced wall panel relative to a braced wall line by drawing a straight line connecting the ends of the center of the curved wall (the chord). See the dashed centerline in Figure AU106.11.3, where the shaded portion represents the extent of the curved wall utilized as a braced wall panel. The chord line is not shown.

AU106.11.2 Braced wall lines for buildings with curved walls. Buildings with curved *cob* walls shall contain two *braced wall lines* in two orthogonal directions. The spacing of the braced wall lines for wind design in Table AU106.11(2) and the spacing and length of the braced wall lines for seismic design in Tables AU106.11(3), AU106.11(4) and AU106.11(5) shall be the maximum widths of the building in the two orthogonal directions.

❖ It is common practice for North American cob buildings to contain curved walls.

Each building design must select two orthogonal directions with a set of braced wall lines to design the building's lateral load-resisting system (see Section AU106.11.1). This includes buildings with some or all curved walls and is intended to also include non-orthogonal walls (see Section AU106.11.1). This section provides a means of establishing the equivalent braced wall line spacing for buildings with curved and non-orthogonal walls. The spacing is the maximum exterior width of the building in each of two orthogonal directions chosen.

Curved braced wall panel lengths are determined per Section AU106.11.3 and are adjusted by their angle relative to the chosen orthogonal directions per Section AU106.11.1.

AU106.11.3 Radius, thickness and length of curved braced wall panels. *Cob* curved *braced wall panels* shall have an inside radius of not less than 5 feet (1524 mm), shall be of the thickness required in Table AU106.11(1) and of the length determined in accordance with Section AU106.11. The length of the curved wall shall be considered to be the length of the arc at the center of the wall, in accordance with Figure AU106.11.3 and determined by Equation AU-4.

$$ARC_C = 0.0175 \, RC \times A \qquad \text{(Equation AU-4)}$$

ARC_C = Length of arc at center of wall (in feet).

R_C = Radius at center of wall = $R_i + 0.5T$ (in feet).

R_i = Inside radius of wall (in feet).

T = Thickness of wall without *finish* (in feet).

A = Angle of extent of *braced wall panel* from the center of the arc (in degrees).

❖ To be counted as a braced wall panel, curved cob walls must have an inside radius of at least 5 feet. Walls with a smaller radius are less effective in resisting in-plane loads so are not considered in this appendix. The length of a curved braced wall panel is the length of its centerline arc and is determined by Equation AU-4. Section AU106.11.1, Item 4 describes how to find the equivalent straight line for curved walls to use in the chosen orthogonal braced wall panel scheme for the building.

FIGURE AU106.11.3. See page Appendix AU-29.

❖ See the commentary to Section AU106.11.3. The chord line mentioned in Section AU106.11.3 is not shown in this figure, but it would be a line connecting the ends of the dashed center line ARC_C.

AU106.12 Resistance to wind uplift forces. *Cob* walls that resist uplift forces from the *roof assembly*, as determined in accordance with Section R802.11, shall be in accordance with Table AU106.12.

❖ Cob walls subject to wind uplift forces according to Section R802.11 must have their bond beams anchored to the wall as specified in Table AU106.12. The required anchor embedment is shown in the table for the applicable combination of wall thickness and wind speed.

TABLE AU106.12. See page Appendix AU-29.

❖ See the commentary to Section AU106.12.

AU106.13 Post-and-beam with cob infill. Post-and-beam with *cob* infill wall systems shall be in accordance with an *approved* engineered design.

❖ Cob wall systems in which superimposed gravity loads are supported by posts and beams and not the cob wall are not covered in this appendix and thus require an approved engineered design.

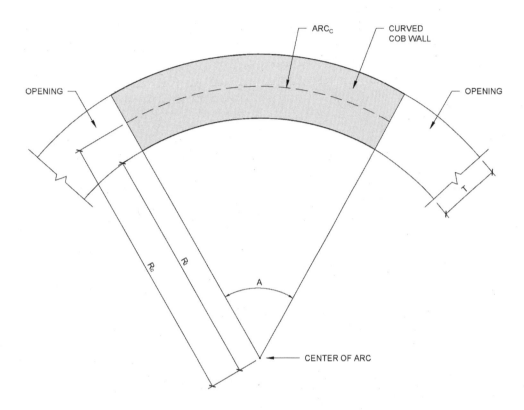

FIGURE AU106.11.3
CURVED BRACED WALL PANEL

TABLE AU106.12
ANCHORAGE OF BOND BEAMS FOR WIND UPLIFT

ANCHORS:
- $^5/_8$" ALL THREAD AT 12" O.C.[a, b]
- 2" × 2" × $^1/_4$" WASHERS AND NUT AT END IN COB
- 4" EMBEDMENT IN CONCRETE BOND BEAMS
- FULL PENETRATION THROUGH WOOD BOND BEAMS WITH 2" × 2" × $^1/_4$" WASHER AND NUT

WIND UPLIFT FORCE FROM TABLE R802.11 (PLF)	ANCHORAGE DEPTH IN INCHES, PER WALL WIDTH AND WIND UPLIFT FORCE		
	≤ 12" Wall Width[c]	≤ 16" Wall Width[c]	≤ 24" wall width[c]
< 75	16	12	12
< 100	24	16	12
< 150	48 o.c. continuous from foundation to bond beam[d]	24	16
< 200	48 o.c. continuous from foundation to bond beam[d]	48 o.c. continuous from foundation to bond beam[d]	24

For SI: 1 inch = 25.4 mm.

a. For wood bond beams a maximum of 6 inches from bond beam ends.

b. For minimum 6-inch by 8-inch concrete bond beams, at 18" o.c. for wind uplift forces less than 75 pounds per linear foot, and at 16" o.c. for wind uplift forces less than 100 pounds per linear foot.

c. Excluding finishes.

d. With 7-inch embedment in foundation, 4-inch embedment in concrete bond beam or full penetration through wood bond beam with 2-inch by 2-inch by $^1/_4$-inch washer and nut.

218 *Essential* COB CONSTRUCTION

APPENDIX AU—COB CONSTRUCTION (MONOLITHIC ADOBE)

AU106.14 Buttresses. *Cob buttresses* that are intended to provide out-of-plane wall bracing or additional capacity for *braced wall panels* shall be in accordance with an *approved* engineered design.

❖ Cob buttresses designed to aid out-of-plane or in-plane lateral wall strength and stability can be part of a cob building's lateral load-resisting system. However, cob buttresses require an approved engineered design.

SECTION AU107
COB FLOORS

AU107.1 Cob floors. *Cob* floors supported by *grade* shall be in accordance with an *approved* specification. Straw shall not be required in the material mix.

❖ Cob floors on grade, with or without straw, are permitted with specifications that have been approved by the building official.

The modern evolution and growing use of cob and other earthen floors in custom homes is a testament to their serviceability and aesthetic appeal. They also can reduce costs and environmental impact, especially where the floor is made from clay subsoil obtained from the site.

Cob floors are made of some or all of the materials described in Sections AU103.1, AU103.2, and AU103.3. They are generally composed of the following layers over undisturbed grade or well-compacted fill: a capillary break, a moisture barrier, an earthen base layer, a wear layer and a penetrating finish treatment.

Though cob floors have a history of successful raised floor applications, this section does not specifically include such installations. The primary differences include the lack of need for a capillary break and moisture barrier, the added weight to the raised floor structure and potential floor structure deflection under dynamic live loads that could induce cracking. The last two issues can be addressed with adequate structural design.

SECTION AU108
FIRE RESISTANCE

AU108.1 Fire-resistance rating. *Cob* walls are not fire-resistance rated.

❖ Although cob walls are not given a fire-resistance rating in this appendix, the materials described in the appendix for use in cob walls are the same as those used for centuries around the world to build wood-fired ovens, fireplaces and kilns. The absence of a fire rating for cob walls is due to the lack of specific fire tests required by the code for establishing a fire-resistance rating at the time this appendix was written.

Similarly, cob has not been classified as noncombustible because the specific testing in ASTM E136 required by the code had not been performed when this appendix was written. However, the long history of cob's use for high-heat, open-flame ovens and kilns clearly indicates its high fire resistance.

Guidance is available to fire engineers in proposing engineered fire-resistance equivalency ratings for cob construction. The most relevant comes from the Australian national regulatory agency following recent wildfire (bushfire) catastrophes there. The Australian Standard AS 3959—2009, *Construction of Buildings in Bushfire-Prone Areas*, was developed in response. Historic use of earthen wall systems, including cob and mud brick (adobe), where these fires occurred gave Australians direct experience with these buildings in intense firestorms. AS 3959 lists "earthwall, including mud brick" as one of only three wall materials not needing additional testing even in the most extreme and vulnerable bushfire zone, Bushfire Attack Level—Flame Zone. The standard stipulates that exposed components of external walls shall be of noncombustible material at least 90 mm (3.54 inches) thick. The only other materials listed as acceptable without additional testing for external walls are full masonry and concrete. The minimum 10-inch thickness for cob walls required in this appendix is almost three times the minimum thickness of the earth wall accepted by the Australian standard for its highest fire risk zones.

Additionally, *The Australian Earth Building Handbook*, HB 195—2002, Section 4.6 Fire Resistance Level, states, "In the absence of specific test data, the general fire resistance level (FRL) of earth walls satisfying the minimum thickness requirements outlined in Clause 4.3.4 may be taken as not greater than 120/120/120, or 90/90/90 where wall thickness is less than 200 mm." Clause 4.3.4 Structural Adequacy states, "Minimum recommended thicknesses for mud brick, stabilized pressed block and rammed earth are as follows: External walling—200 mm, Internal walling—125 mm. The minimum wall thickness for poured earth and cob wall construction is also recommended to be 200 mm, though in practice wall thickness will often exceed this value." The sets of numbers in the FRL represent minutes before failure for structural adequacy, integrity and insulation. Thus, Australia gives a 2-hour fire-resistance rating for a 200 mm (7.87 inches) earth wall, including cob.

AU108.2 Clearance to fireplaces and chimneys. *Cob* walls or other *cob* surfaces shall not require clearance to fireplaces and chimneys, except where clearance to noncombustibles is required by the manufacturer's instructions.

❖ Because cob does not burn or support combustion, this section treats cob walls as the code treats masonry and other noncombustible materials. Centuries of using cob to build ovens and kilns has influenced modern cob practitioners to commonly construct cob fireplaces in cob walls. Where specific clearances to noncombustible materials are specified in the manufacturer's instructions, those clearances are required.

SECTION AU109
THERMAL PERFORMANCE

AU109.1 Thermal characteristics. *Cob* walls shall be classified as mass walls in accordance with Section N1102.2.5 and shall meet the *R*-value requirements for mass walls in Table N1102.1.3.

❖ Cob walls are classified as mass walls in accordance with Section N1102.2.5 because the heat capacity of

2021 IRC® CODE and COMMENTARY

cob walls is greater than the 6 Btu/ft² × °F threshold in that section. The lowest heat capacity of a cob wall is 16 Btu/ft² × °F for the required minimum wall thickness of 10 inches and at the lowest practical density of 70 pcf. Thus, cob walls must meet the *R*-value requirements in the Mass Wall column of Table N1102.1.3. The first number in that column applies where there is no additional insulation or where additional insulation is applied to the exterior of the wall. The second number applies where additional insulation is applied to the interior of the wall and its *R*-value exceeds half of the insulating value of the cob wall itself.

In certain climates, buildings with cob or other mass walls that do not meet the *R*-value requirements in Table N1102.1.3 may exhibit a history of thermal performance equivalent to buildings with walls that meet the requirements. As allowed by Sections AU101.2 and AU101.3, such local conditions and history can be considered in evaluating equivalent compliance.

AU109.2 Thermal resistance. The unit *R*-value for *cob* walls with a density of 110 pounds per cubic foot (1762 kg/m³) shall be R-0.22 per inch of *cob* thickness. Walls that vary in thickness along their height or length shall use the average thickness of the wall to determine its *R*-value. The thermal resistance values of air films and finish materials or additional insulation shall be added to the *cob* wall's thermal resistance value to determine the *R*-value of the wall assembly.

❖ Cob's assigned unit *R*-value of 0.22 per inch with a density of 110 pcf was determined with an ASTM C1363 thermal resistance test at an independent laboratory in 2018. A density of 110 pcf is at the upper limit of the 70–110 pcf practical density range for cob. Therefore, the assigned *R*-value is conservative for all cob walls, even if the density is not known. Lower densities of cob, with more straw and/or less dense aggregate, will likely have a higher unit *R*-value. The unit *R*-value yielded by independent laboratory ASTM tests of lower density cob can be used in lieu of R-0.22 if reports are presented and a similar cob mix and density is utilized in a particular project.

The *R*-value of a cob wall assembly is determined by multiplying the unit *R*-value by the thickness of the cob wall in inches and adding the thermal resistance of air films and any finish or additional insulation. Use the average thickness of the wall when computing the *R*-value for cob walls that vary in thickness.

AU109.3 Additional insulation. Where insulating materials are added to the face of a *cob* wall, the combination of additional insulation and any associated connecting, weather-resisting or protective materials shall comply with Section AU104.1.2, Items 1–4.

❖ Adding insulation to the face of cob walls can allow them to more readily meet energy efficiency requirements in cold climates. This is allowed provided that the insulation assembly complies with the requirements in Section AU104.1.2, Items 1–4, for attachment or support, vapor permeance and weight limits.

APPENDIX AU—COB CONSTRUCTION (MONOLITHIC ADOBE)

SECTION AU110
REFERENCED STANDARDS

AU110.1 General. See Table AU110.1 for standards that are referenced in various sections of this appendix. Standards are listed by the standard identification with the effective date, the standard title and the section or sections of this appendix that reference the standard.

TABLE AU110.1
REFERENCED STANDARDS

STANDARD ACRONYM	STANDARD NAME	SECTIONS HEREIN REFERENCED
ASTM C5—10	*Standard Specification for Quicklime for Structural Purposes*	AU104.4.6
ASTM C141/ C141M—14	*Standard Specification for Hydrated Hydraulic Lime for Structural Purposes*	AU104.4.6
ASTM C206—14	*Standard Specification for Finishing Hydrated Lime*	AU104.4.6
ASTM C926	*Specification for Appliance of Portland Cement-based Plaster*	AU104.4.8
ASTM C1707—11	*Standard Specification for Pozzolanic Hydraulic Lime for Structural Purposes*	AU104.4.6
ASTM E2392/ E2392M—10	*Standard Guide for Design of Earthen Wall Building Systems*	AU104.4.3.2
ASTM BS1, ASTM BS EN 459—2015	Part 1: *Building Lime. Definitions, Specifications and Conformity Criteria*; Part 2: *Test Methods*	AU104.4.6

Bibliography

The following resource materials were used in the preparation of the commentary for this appendix:

AS 3959—2009, *Australian Standard Construction of Buildings in Bushfire Prone Areas* (incorporating Amendments 1, 2 and 3). Standards Australia, 2009.

ASTM E2392—10, *Standard Guide for Design of Earthen Wall Building Systems*. West Conshohocken, PA: ASTM International, 2010.

Brunello, Gabi, Jose Espinoza, Alex Golitz. *Cob Property Analysis*, Santa Clara University, Department of Civil Engineering, 2018.

Crimmel, Sukita Reay, James Thompson. *Earthen Floors: A Modern Approach to an Ancient Practice*. New Society Publishers, Gabriola Island, BC, 2014.

Devon Earth Building Association, *Cob Dwellings: Compliance with the Building Regulations 2000* (as amended), *The 2008 Devon Model*. Devon, England, 2008.

APPENDIX AU—COB CONSTRUCTION (MONOLITHIC ADOBE)

Eberhard, Daniel, Joseph Novara, Brandon Popovec. *Cob: A Sustainable Building Material,* Santa Clara University, Department of Civil Engineering, 2018.

Evans, Ianto, Michael G. Smith, Linda Smiley. *The Hand-Sculpted House*. Chelsea Green Publishing, White River Junction, VT, 2002.

NZS 4297—1998, *Engineering Design of Earth Buildings*. Standards New Zealand, 1998.

NZS 4298—1998, *Materials and Workmanship for Earth Buildings* (incorporates June 2000 Amendment). Standards New Zealand, 2000.

NZS 4299—1998, *Earth Buildings Not Requiring Specific Design* (incorporates December 1999 amendment). Standards New Zealand, 1998.

Volhard, Franz. *Light Earth Building: A Handbook for Building with Wood and Earth*. Birkhauser Verlag GmbH, Basel, Switzerland, 2016.

Walker, Peter. *Australian Earth Building Handbook:* HB 195—2002, Standards Australia, 2002.

Glossary

Because cob has an ages-old history in many parts of the world, and because it has recently been rediscovered and refined within small, isolated communities of interest, there is inconsistency in the terminology used to describe this building system. This glossary contains definitions for many of the technical words used throughout this book as well as our recommendations on which terms to use when there are multiple alternatives in common usage. Sometimes these preferences are intended to avoid ambiguity (for example, *earthen floor* rather than *cob floor*) and sometimes to reduce negative associations that might lead a building official or the general public to underestimate cob's strength and durability. Examples of the latter rationale include our preference for *straw-reinforced* over *unreinforced* and for *natural cob* over *unstabilized cob*. Language matters, especially when we are advocating for the greater adoption of a building system which has been greatly maligned in the past. Think, for example, of the connotations of *mud hut* or of *cob job,* which is defined as "to install or fabricate something with complete apathy to anyone's safety and industry standards of quality."

Adobe: A wall system constructed of sun-dried earthen blocks, stacked and mortared with earthen mortar. See *Monolithic adobe.*

Aggregate: Inert granular materials such as sand and gravel that are combined with a binder such as clay or Portland cement to make a concrete material.

Appendix AU: Model cob building code written by the Cob Research Institute and appended to the International Residential Code.

Aspect ratio: The ratio of a wall's length to its height.

ASTM Standards: ASTM International (formerly the American Society for Testing and Materials) is an organization made up of committees of relevant professionals who develop voluntary consensus standards that cover a wide range of materials, systems, and products to encourage consistency and safety among manufacturers. ASTM Standards are considered highly reliable and are often adopted into government regulations (such as building codes) in the US and around the world.

Bond beam: A structural beam connecting an entire building together at the top of the walls. Can also be referred to as a *top plate* or a *collector* when built out of wood.

Braced wall panel: See *Shear wall.*

Buckling: The sudden failure of a structural component such as a wall or column under high compressive stress.

Clay: A family of minerals known as *hydrous aluminum phyllosilicates* which share certain characteristics, including an extremely small particle size (less than 2 microns, or 0.002 mm), high dry strength, and the ability to bond other particles together in the presence of water. More loosely, *clay* refers to an inorganic soil containing clay particles and other minerals that can be used as the binder of other component materials in a mix of *cob* or of *clay plaster.*

Clay plaster: A mixture of clay or clay soil with sand and/or fibers such as chopped straw (and sometimes other ingredients, including pigments, stabilizers, etc.) and applied to a wall for protective and aesthetic purposes. Although the materials and mixes can be similar, we don't like the terms *cob plaster,* or *adobe plaster,* leaving *cob* and *adobe* to refer unambiguously to wall systems.

Clay slip: Clay or clay soil dissolved in water to be used as an ingredient in cob, clay plaster, clay paint, or for other purposes. It can be a thin consistency, like cream or paint, or a thicker consistency, like pudding.

Clay soil or subsoil: A soil containing enough clay to serve as the base ingredient for earthen building. The most appropriate building soils contain less-expansive clays (kaolinites) and little organic matter. They may also contain a large percentage of sand/aggregate.

Cob: A wall system consisting of clay soil mixed with water, straw, and often sand and/ or other aggregates. The wall is built while the mixture is still moist and plastic.

Cobber's thumb: A round-ended stick approximately 1" in diameter and 8" long used to integrate adjacent cob lumps and lifts as the wall is constructed.

Compressive strength: The capacity of a material to withstand compressive forces (typically gravity loads) without breaking.

Concrete: A composite material consisting of a binding medium (usually, but not necessarily, Portland cement) within which are embedded particles or fragments of aggregate. By this definition, cob is actually concrete!

CRI: The Cob Research Institute, a US nonprofit organization dedicated to researching cob and writing and promoting a cob building code.

Earthen building: Any building system using earth as a primary ingredient. Examples include cob, adobe, rammed earth, wattle-and-daub, earthen floors, and earthen plasters.

Earthen floor: A floor made of clay soil, aggregates, and (often) straw or other fibers, generally sealed and stabilized with a drying oil, such as linseed oil. We prefer *earthen floor* over terms such as *cob floor* or *adobe floor,* leaving *cob* and *adobe* to refer unambiguously to wall systems.

EPD: An Environmental Product Declaration is the summary of the data collected in the Life Cycle Assessment process.

Expansive clay: A clay that expands greatly in volume when it becomes hydrated and, therefore, shrinks greatly when it dries. Ilites and smectites (including bentonite and montmorillonite) are classes of expansive clay minerals. Less-expansive clays such as kaolinites are preferred for building.

Flexural strength: See *Modulus of rupture.*

Insulation: A material that resists the passage of heat by conduction. Most insulating materials are lightweight, containing many tiny pockets of trapped air.

IRC: The International Residential Code. A model code published by the International Code Council and used as a template for state and local building codes for one- and two-family residences throughout most of the US.

LCA: A Life Cycle Assessment is a standardized and systematic analysis of the environmental impacts of products or services during their entire life cycle. Building designers use them to help compare various options.

Lift: A layer of moist cob placed on a wall during construction. Lifts are typically between 6" and 24" high.

Lime: A building material derived from limestone or calcium carbonate. Limestone is heated in kilns to produce *quicklime* (calcium oxide) which is then slaked with water to make *hydrated lime,* or *builder's lime* (calcium hydroxide.) This material, available either as a dry powder or as a sticky putty, is commonly mixed with sand (and water, if dry) to make lime plaster or lime mortar. The calcium hydroxide then absorbs carbon dioxide from the atmosphere and reverts to calcium carbonate, which is hard and water-resistant but vapor permeable.

Load-bearing wall: A wall that supports more than its own weight, such as a roof or second floor. IRC Appendix AU, the model cob code, defines it as: "A cob wall that supports more than 100 pounds per linear foot (1459 N/m) of vertical load in addition to its own weight."

Modulus of rupture: Also known as *flexural strength* or *bending strength*, MOR is a measurement of a material's capacity to withstand bending forces without cracking. Technically, it is the maximum force that the tension side of a bending member can withstand.

Monolithic adobe: Synonymous with cob. This term emphasizes the close relationship between cob and adobe, which is better known in the US due to its long history in the American Southwest. The primary difference between the two techniques is that while adobe walls are made of blocks formed of unfired earth (usually with straw or other fiber added), cob walls are built from a similar earthen mix while still moist and plastic. Compared with adobe, cob walls have a more homogenous structure, with continuous straw reinforcing throughout, improving their ability to withstand lateral forces from wind and earthquakes. This is why cob is sometimes referred to as *monolithic* adobe.

Natural cob: A cob mix consisting of clay soil, water, straw, and sand or other aggregates only, without the addition of stabilizers such as Portland cement, lime, asphalt emulsion, etc. We prefer the term *natural* over *unstabilized* because the latter could be interpreted as meaning that the material is not stable.

Passive solar design: Design principles and strategies that allow a building to heat and cool itself through the proper placement of glazing, roof overhangs, thermal mass, and insulation.

Pcf: Pounds per cubic foot, a unit of density.

Psi: Pounds per square inch, a unit of pressure.

Reinforced: Containing fibers, bars, or mesh that distribute tensile forces through the material, reducing the likelihood of cracking or collapse. All cob walls are reinforced with straw. Some are further reinforced with metal bars (rebar) or meshes (plastic, steel, or natural fiber), especially in order to resist seismic forces.

Sand: Small particles of stone, generally between 0.002 and 0.25 inches (0.06–6 mm) in diameter. In a cob mix, sand adds compressive strength and reduces shrinkage. Sand is often found as a natural component of clay soils.

Shear wall: A wall designed and constructed to resist in-plane lateral seismic and wind forces. Synonymous with *braced wall panel*.

Silt: Very fine particles of stone, generally less than 0.002 inches (0.06 mm) in diameter. Think of it as natural rock dust. Due to its small size and usually rounded shape, silt adds little structural strength to a cob mix.

Stabilized: Containing one or more ingredients—such as Portland cement, lime, asphalt emulsion, and/or pozzolans—intended to improve an earthen mixture's water-resistance and strength. Cob, adobe, rammed earth, clay plasters, etc., are sometimes stabilized in this way. (See *Natural cob*.)

Steel-reinforced cob: Cob walls with steel bar or mesh reinforcing in addition to the straw microfiber reinforcing inherent in most cob mixes.

Straw: The dried stalks of cereal grains such as wheat, rice, rye, oats, and barley. These stalks are an agricultural byproduct of grain production and may be baled for easy transport and storage. Chemically, straw consists mainly of cellulose and lignin, which are low in nitrogen and therefore have little food value for microorganisms, making it slow to decay when kept dry. *Hay*, on the other hand, is harvested when the nitrogen content is higher to maximize nutritional value for livestock. Hay is therefore much quicker to decompose and should not be used as a building material.

Straw-reinforced cob: Cob wall system containing the straw microfiber reinforcing common to most cob mixes and no additional bar or mesh reinforcing. It is a common mistake to refer to these walls as *unreinforced.*

Thermal mass: A material with the capacity to absorb heat and store it, releasing it slowly over time. The most effective thermal mass materials are heavy solids (earth, stone, concrete, brick, etc.) or liquids (water, oil). When used in a building, thermal mass materials have the effect of reducing temperature fluctuations over time. Also called *thermal inertia.*

Waterproof: Describes a material that is impermeable to water in liquid or vapor form. For example, ceramic tile and Portland cement stucco are waterproof.

Water-resistant: Describes a material that can get wet without deteriorating. For example, lime plasters are water resistant, but still permeable to liquid water and vapor.

Vapor permeable: Describes a material's ability to allow water vapor to pass through it. Also imprecisely called *breathable.*

Recommended Resources

Cob Construction

Bee, Becky. *The Cob Builder's Handbook: You Can Hand-Sculpt Your Own Home.* Groundworks, 1997.

Devon Earth Building Association. Downloadable leaflets: www.devonearth building.com/leaflets.htm

Evans, Ianto, Michael G. Smith, and Linda Smiley. *The Hand-Sculpted House: A Practical and Philosophical Guide to Building a Cob Cottage.* Chelsea Green, 2002.

Smith, Michael G., ed. *The CobWeb Archive, Vol 1 1995–2002 and Vol 2 2003–2011.* Cob Cottage Company, 2012.

Rogue, Conrad. *House of Earth: A Complete Handbook for Earthen Construction.* House Alive, 2015.

Snell, Clarke and Tim Callahan. *Building Green: A Complete How-To Guide to Alternative Building Methods: Earth Plaster, Straw Bale, Cordwood, Cob, Living Roofs.* Lark Books, 2005.

Weismann, Adam and Katy Bryce. *Building with Cob: A Step-by-Step Guide.* Green Books, 2006.

Allied Natural Building Techniques

Baker-Laporte, Paula and Robert Laporte. *The EcoNest Home: Designing and Building a Light Straw Clay House.* New Society Publishers, 2015.

CASBA. *Straw Bale Building Details: An Illustrated Guide for Design and Construction.* New Society Publishers, 2019.

Crimmel, Sukita Reay and James Thomson. *Earthen Floors: A Modern Approach to an Ancient Practice.* New Society Publishers. 2014.

Denzer, Kiko. *Build Your Own Earth Oven.* 3rd ed. Hand Print Press, 2007.

Doleman, Lydia. *Essential Light Straw Clay Construction: The Complete Step-by-Step Guide.* New Society Publishers, 2017.

Doyle, Leslie. *Essential Green Roof Construction: The Complete Step-by-Step Guide.* New Society Publishers, 2021.

Elizabeth, Lynne and Cassandra Adams, eds. *Alternative Construction: Contemporary Natural Building Methods.* John Wiley & Sons, 2000.

Hart, Kelly. *Essential Earthbag Construction: The Complete Step-by-Step Guide.* New Society Publishers, 2015.

Holmes, Stafford and Bee Rowan. *Building with Lime Stabilized Soil.* Practical Action Publishing, 2021.

Hunter, Kaki and Donald Kiffmeyer. *Earthbag Building: The Tools, Tricks and Techniques.* New Society Publishers, 2004.

Kennedy, Joseph F., Michael G. Smith, and Catherine Wanek, eds. *The Art of Natural Building: Design, Construction, Resources,* 2nd ed. New Society Publishers, 2014.

Krahn, Tim. *Essential Rammed Earth Construction: The Complete Step-by-Step Guide.* New Society Publishers, 2019.

Magwood, Chris. *Essential Hempcrete Construction: The Complete Step-by-Step Guide.* New Society Publishers, 2016.

Magwood, Chris. *Making Better Buildings: A Comparative Guide to Sustainable Construction for Homeowners and Contractors.* New Society Publishers, 2014.

Volhard, Franz. *Light Earth Building: A Handbook for Building with Wood and Earth.* Birkhäuser, 2016.

Natural Plasters and Paints

Crews, Carole. *Clay Culture: Plasters, Paints and Preservation.* Gourmet Adobe Press, 2010.

Edwards, Lynn and Julia Lawless. *The Natural Paint Book: A Complete Guide to Natural Paints, Recipes, and Finishes.* Rodale, 2002.

Henderson, James. *Earth Render: The Art of Clay Plaster, Render and Paints.* Python Press, 2013.

Henry, Michael and Tina Therrien. *Essential Natural Plasters: A Guide to Materials, Recipes, and Use.* New Society Publishers, 2018.

Weismann, Adam and Katy Bryce. *Using Natural Finishes: Lime- and Earth-Based Plasters, Renders, and Paints: A Step-by-Step Guide.* Green Books, 2008.

Design, Engineering, and Building Science

King, Bruce. *The New Carbon Architecture: Building to Cool the Climate.* New Society Publishers, 2017.

Magwood, Chris. *Essential Sustainable Home Design: A Complete Guide to Goals, Options, and the Planning Process.* New Society Publishers, 2017.

Mazria, Edward. *The Passive Solar Energy Book: A Complete Guide to Passive Solar Home, Greenhouse and Building Design.* Rodale, 1979.

Minke, Gernot. *Building with Earth: Design and Technology of a Sustainable Architecture.* 4th ed. Birkhäuser, 2022.

Racusin, Jacob Deva. *Essential Building Science: Understanding Energy and Moisture in High Performance House Design.* New Society Publishers, 2017.

Racusin, Jacob Deva and Ace McArleton. *The Natural Building Companion: A Comprehensive Guide to Integrative Design and Construction.* Chelsea Green, 2012.

Wright, David. *Natural Solar Architecture: A Passive Primer.* Van Nostrand Reinhold, 1978.

Straube, John. Numerous articles published on The Building Science Corporation website: buildingscience.com/

Vernacular Building

Bourgeois, Jean-Louis and Carollee Pelos. *Spectacular Vernacular: The Adobe Tradition.* Aperture, 1989.

Dethier, Jean. *Down to Earth: Adobe Architecture: An Old Idea, A New Future.* Translated from the French by Ruth Eaton. Facts on File, 1982.

Morris, James and Suzanne Preston Blier. *Butabu: Adobe Architecture of West Africa.* Princeton Architectural Press, 2004.

Steen, Bill, Athena Steen, and Eiko Komatsu. *Built by Hand: Vernacular Buildings Around the World.* Gibbs Smith, 2003.

Organizations

Auroville Earth Institute, earth-auroville.com
Technical publications, research, and trainings (both online and in person at their campus in India) about many styles of earthen construction, especially compressed earth blocks.

Cob Cottage Company, cobcottage.com
First organization to teach cob workshops in North America.

Cob Research Institute (CRI), cobcode.org
Developers of cob building codes and educational materials. Many relevant documents on their website.

Cob Workshops, cobworkshops.org
International listing of hands-on workshops on cob and natural building.

CobBauge, cobbauge.eu
Research and development of energy-efficient earthen building systems.

CRAterre-EAG, craterre.org
International center for earthen architecture based in France.

Devon Earth Building Association, devonearthbuilding.com
Technical information related to cob conservation, maintenance, and construction.

Straw Clay Wood, strawclaywood.com
Michael G. Smith provides hands-on natural building workshops and consulting to owner-builders.

Verdant Structural Engineers, verdantstructural.com
Engineers specializing in natural building. Their website holds many useful reports and publications as well as material comparison information.

Endnotes

Chapter 1: The History of Cob

1. "Why Earthen Architecture May Be a Big Part of Our Future," *Getty Magazine*, 2022. www.getty.edu/news/why-earthen-architecture-may-be-a-big-part-of-our-future/

2. Morris, William. "The Cob Buildings of Devon 1: History, Building Methods and Conservation," Devon Historic Buildings Trust, 1993. devonearthbuilding.com/leaflets/cob_buildings_of_devon_1.pdf

3. Hamard, Erwan, et al. "Cob: A Vernacular Earth Construction Process in the Context of Modern Sustainable Building," in *Building and Environment* 106 (2016) 103–119.

4. Moquin, Michael. "Ancient Solutions for Future Sustainability: Building with Adobe, Rammed Earth, and Mud," in *The Adobe Journal*. www.irbnet.de/daten/iconda/CIB_DC24848.pdf

5. Ibid.

6. Moquin, Michael. "From Bis Sá Ani to Picuris: Early Pueblo Adobe Technology of New Mexico and the Southwest," *The Adobe Journal* 8:10–29 (Out of print quarterly periodical—worth trying to find back issues).

7. Niroumand, Hamed, M.F.M Zain, and Maslina Jamil. "Various Types of Earth Buildings," 2nd Cyprus International Conference on Educational Research, 2013.

8. Pieper, Richard. "Earthen Architecture in the Northern United States: European Traditions in Earthen Construction," in *Cultural Resource Management* Volume 22, No. 6, 1999, published by the National Park Service. npshistory.com/newsletters/crm/crm-v22n6.pdf

9. Morris, William. "The Cob Buildings of Devon 1: History, Building Methods and Conservation," Devon Historic Buildings Trust, 1993. devonearthbuilding.com/leaflets/cob_buildings_of_devon_1.pdf

10. Burke, Massey with Anthony Dente. "Permitting a Load-Bearing Cob Studio in Berkeley: A White Paper," Ecological Building Network, 2016. cobcode.s3.amazonaws.com/Cob+permit+white+paper.pdf

Chapter 2: Rationale and Appropriate Use

1. Inventory of Carbon and Energy (ICE). circularecology.com/embodied-carbon-footprint-database.html

2. King, Bruce. *The New Carbon Architecture.* New Society Publishers, 2017.

3. Assumes 1% (historic Devon and modern high-density mixes)–7% (testing figures from 70–75 pcf mixes) straw in cob mix by dry weight. Since the embodied energy and carbon of clay soil and sand are virtually the same, it is assumed that there is no sand in the mix, just clay soil and straw, since the numbers would be almost identical with any proportion of clay and sand. No embodied carbon is assigned to the mixing process. For a high-density mix the equation is: $(0.99 \times 0.083) + (0.01 \times 0.24) = 0.085$.

4. Dethier, Jean. *Down to Earth: Adobe Architecture, An Old Idea, A New Future.* translated by Ruth Eaton. Facts on File, 1983.

Chapter 3: Building Science

1. Goodvin, Christina, Gord Baird, and Ann Baird, "Cob Home Performance Report," 2011. buildwellsource.org/design/building-science/164-cob-home-performance-

report-c-goodvin-g-baird-a-baird-2011? highlight=WyJnb29kdmluIl0=

2. www.theearthbuildersguild.com/ performance-of-earthen-structures

3. Childs, K. W., G. E Courville, and E. L. Bales. "Thermal Mass Assessment: An Explanation of the Mechanisms by Which Building Mass Influences Heating and Cooling Energy Requirements," ORNL/ CON-97, 1983. www.osti.gov/servlets/ purl/5788833

4. Mantesi, Eirini, et al. "The Modelling Gap: Quantifying the Discrepancy in the Representation of Thermal Mass in Building Simulation," in *Building and Environment* 131, 2018. www.sciencedirect.com/ science/article/pii/S0360132317305851

5. www.ashrae.org/file%20library/technical %20resources/standards%20and%20 guidelines/standards%20addenda/169_ 2020_a_20211029.pdf

6. Hren, Stephen. "Experiments in Modernizing Cob Construction," in EarthUSA Proceedings 2022. Also see Stephen's website at www.kthonik.com

7. More on all figures and techniques described in this sidebar can be found in Steve Goodhew et al. "Improving the Thermal Performance of Earthen Walls to Satisfy Current Building Regulations," in *Energy & Buildings* 240, 2021. www.sciencedirect. com/science/article/abs/pii/S037877882 1001572

8. Straube, John. "BSD 138: Moisture and Materials," 2006. www.buildingscience. com/documents/digests/bsd-138- moisture-and-materials

9. Straube, John. "BSD 30: Rain Control Theory," 2010. www.buildingscience.com/ documents/digests/bsd030-rain-control- theory

10. Keefe, Larry. "The Cob Buildings of Devon 2, Repair and Maintenance," Devon Historic Buildings Trust, 1993. www.devonearth building.com/leaflets/the_cob_buildings_ of_devon_2.pdf

11. Goodvin, Christina, Gord Baird, and Ann Baird, "Cob Home Performance Report," 2011. buildwellsource.org/design/ building-science/164-cob-home- performance-report-c-goodvin-g-baird-a- baird-2011?highlight=WyJnb29kdkm luIl0=

12. NZS 4299 (1998): Earth Buildings Not Requiring Specific Design. (This Standard was revised in 2020, but the newer version is not available for free on the internet.) www.eastue.org/project/linea-adobe/ norme/NZD4299-1998-Earth_Buildings_ Not_Requiring_Specific_Design.pdf

13. See *The Natural Building Companion* by Jacob Deva Racusin and Ace McArleton for a thorough treatment of air fins and other air-control strategies.

Chapter 4: Materials

1. Avrami, Erica C., Hubert Guillaud, and Mary Hardy, eds. "Terra Literature Review: An Overview of Research in Earthen Architecture Conservation," Getty Conservation Institute, 2008. hdl.handle. net/10020/gci_pubs/terra_literature_ review

2. Ibid.

3. The Clay Minerals Society. www.clays.org/

Chapter 6: Mix Design and Testing

1. Hamard, Erwan, et al. "Cob: A Vernacular Earth Construction Process in the Context of Modern Sustainable Building," in *Building and Environment* 106 (2016).

2. Morris, William. "The Cob Buildings of Devon 1: History, Building Methods and Conservation," Devon Historic Buildings Trust, 1993. devonearthbuilding.com/leaflets/cob_buildings_of_devon_1.pdf

Chapter 7: Cob Building Design and Planning

1. International Wildland-Urban Interface Code (IWUIC). www.iccsafe.org/products-and-services/wildland-urban-interface-code/

2. An earthen oven is an excellent first project for a novice cob builder. We highly recommend Kiko Denzer's book *Build Your Own Earth Oven.* 3rd ed. Hand Print Press, 2007.

3. A *rocket mass heater* is a simple and very effective masonry heater that you can build yourself from cob, bricks, and scrap metal. There are two good guidebooks available: *Rocket Mass Heaters* by Ianto Evans and Leslie Jackson (3rd ed. Cob Cottage, 2014) and *The Rocket Mass Heater Builder's Guide* by Erica and Ernie Wisner (New Society Publishers, 2016).

4. When using passive solar design principles in the Southern Hemisphere, substitute "north" for "south" and vice versa in the following discussion.

5. A "Trombe wall" is a passive heating strategy most appropriate for cool, cloudy climates. It's a south-facing thermal mass wall covered on the outside with a glass skin, with a space between glass and mass through which air can be circulated into the building. This allows the mass wall to serve directly as a thermal collector while minimizing the heat loss associated with large windows. It is a potential application for exterior cob walls in climates where they would otherwise not be thermally appropriate.

6. Building Transparency: EC3 User Guide: The Embodied Carbon in Construction Calculator Tool, www.buildingtransparency.org/ec3-resources/ec3-user-guide/

7. Builders for Climate Action. The BEAM Estimator (Building Emissions Accounting for Materials) www.buildersforclimateaction.org/beam-estimator.html

8. Athena Sustainable Materials Institute. The Athena EcoCalculator, www.athenasmi.org/our-software-data/ecocalculator/

Chapter 8: Structural Engineering

1. Merriam-Webster.com Dictionary, Merriam-Webster, www.merriam-webster.com/dictionary/concrete. Accessed Oct 13, 2022.

2. King, Bruce. *The New Carbon Architecture.* New Society Publishers, 2017.

3. Harris, Cyril. *The Dictionary of Architecture and Construction.* 4th ed. McGraw-Hill Professional, 2005.

4. Gagg, Colin R. "Cement and Concrete as an Engineering Material: An Historic Appraisal and Case Study Analysis," *Engineering Failure Analysis,* 2014.

5. "Watershed Materials Reinvents the Concrete Block": www.metalocus.es/en/news/watershed-materials-reinvents-concrete-block

6. Morris, H.W., R. Walker, and T. Drupsteen. "Modern and Historic Earth Buildings: Observations of the 4th September 2010 Darfield Earthquake," in Proceedings of the Ninth Pacific Conference on Earthquake Engineering: Building an Earthquake-Resilient Society, 2011. cobcode.s3.amazonaws.com/Modern+and+historic+earth+buildings-+Observations+of+the+4th+September+2010+Darfield+earthquake.pdf

7. Dizhur, Dmytro, Marta Giaretton, and Jason Ingham. "Performance of Early Masonry, Cob and Concrete Buildings in the 14 November 2016 Kaikoura Earthquake," in *Bulletin of the New Zealand Society for Earthquake Engineering* 50(2), 2017. www.researchgate.net/publication/317551790_Performance_of_early_masonry_cob_and_concrete_buildings_in_the_14_November_2016_Kaikoura_earthquake

8. Dente, Anthony and Kevin Donahue. "Review of the Current State of Cob Structural Testing, Structural Design, the Drafting of Code Language, and Material Based Testing Challenges," Earth USA 2019. verdantstructural.com/EarthUSA2019_Dente_Donahue.pdf

9. Wright, David J. "Building from the Ground Up: Understanding and Predicting the Strength of Cob, an Earthen Construction Material," 2019. cobcode.s3.amazonaws.com/DissertationWright.pdf

10. "Cob Shake Table Test: University of British Columbia" cobcode.s3.amazonaws.com/Cob+Shake+Table+Test+-+University+of+British+Columbia.pdf

11. Saxton, R.H. "The Performance of Cob as a Building Material," *The Structural Engineer* 73, 1995. buildwellsource.org/materials3/materials-natural-rural/earth/cob/177-1995-the-performance-of-cob-as-a-building-material?highlight=WyJzYXh0b24iXQ==

12. NZS 4299 (1998): Earth Buildings Not Requiring Specific Design. (This Standard was revised in 2020, but the newer version is not available for free on the internet.) www.eastue.org/project/linea-adobe/norme/NZD4299-1998-Earth_Buildings_Not_Requiring_Specific_Design.pdf

13. NZS 4297 (1998): Engineering Design of Earth Buildings. (This Standard was revised in 2020, but the newer version is not available for free on the internet.) www.eastue.org/project/linea-adobe/norme/NZD4297-1998-Engineering_Design_of_Earth_Buildings.pdf

14. Dente, Anthony and Kevin Donahue. "Review of the Current State of Cob Structural Testing, Structural Design, the Drafting of Code Language, and Material Based Testing Challenges," Earth USA 2019. verdantstructural.com/EarthUSA2019_Dente_Donahue.pdf

15. verdantstructural.com/EarthUSAPoster.png

16. www.cobcode.org/cobcode-documents

Chapter 9: Building Codes and Permits

1. A detailed description of the process of an AMMR submittal is included in a paper Anthony and Ben Loescher, the president of The Earthbuilders Guild (TEG), presented at the EarthUSA 2022 Conference about an adobe project in Southern California. The paper, "Legal Adobe in California: A Pathway for Building Permits," was published in the proceedings of the EarthUSA 2022 Conference, and is also available on the VSE website: verdantstructural.com/EarthUSA2020_Loescher_Dente.pdf

2. Marin County Building Code Section 19.07:Carbon Concrete Requirements. library.municode.com/ca/marin_county/codes/municipal_code?nodeId=TIT 19 MACOBUCO_CH19.07CACORE_19.07.020DE

3. IRC Appendix AQ: Tiny Houses governs residential buildings of less than 400 ft² (37 m²). Some of its provisions (including

legal habitable lofts, lower minimum ceiling heights, etc.) could be relevant to small cob buildings.

4. TMS refers to The Masonry Society. This section on adobe and compressed earth block (CEB) masonry will be part of TMS 402: Building Code Requirements for Masonry Structures, which may be used alone and is also referenced by the IBC and other codes.

Chapter 12: Foundations, Floors, and Roofs

1. The relevant code section is Marin County Building Code Section 19.07: Carbon Concrete Requirements. library.municode.com/ca/marin_county/codes/municipal_code?nodeId=TIT19MACOBUCO_CH19.07CACORE_19.07.020DE

2. International Wildland-Urban Interface Code (IWUIC). www.iccsafe.org/products-and-services/wildland-urban-interface-code/

3. The Nubian Vault Association. www.lavout enubienne.org/?lang=en

Chapter 13: Mixing

1. Pollacek, Rob. "Thinking Big with Cob," *CobWeb* 22, Winter 2006. Collected in *The CobWeb Archive, Volume 2.*

2. Raduazo, Ed. "Tiller Cob," *CobWeb* 22, Winter 2006. Collected in *The CobWeb Archive, Volume 2.*

3. Jones, Chas T. "Of Mud and Machines," *CobWeb* 21, Spring/Summer 2006. Collected in *The CobWeb Archive, Volume 2.*

4. Gin, Yelda, Kamal Haddad, Wassim Jabi, et al. "Robotic 3D Printing with Earth: A Case Study for Optimization of 3D Printing Building Blocks," Proceedings of the IASS 2022 Symposium affiliated with APCS 2022 conference.

Chapter 14: Building Cob Walls

1. Morris, William. "The Cob Buildings of Devon 1: History, Building Methods and Conservation," Devon Historic Buildings Trust, 1993. devonearthbuilding.com/leaflets/cob_buildings_of_devon_1.pdf

Chapter 15: Finishes

1. Keefe, Larry. "The Cob Buildings of Devon 2: Repair and Maintenance," Devon Historic Buildings Trust, 1993.

2. Reynolds, Emily K. and Muramoto Makoto. "Circumstances Contributing to the Deterioration of Old Cob Structures in Japan," *Journal of Asian Architecture and Building Engineering* 21:5, 2021.

3. Devon Earth Building Association. "Appropriate Plasters, Renders and Finishes for Cob and Random Stone Walls in Devon," 2nd ed. 2002. www.devonearth building.com/leaflets/leaflet.pdf

4. Bourgeois, Jean-Louis and Carollee Pelos. *Spectacular Vernacular: The Adobe Tradition.* Aperture, 1989.

5. Builders for Climate Action. The BEAM Estimator (Building Emissions Accounting for Materials). www.buildersforclimate action.org/beam-estimator.html

Index

Page numbers in *italics* indicate tables and figures.

About the Authors

Anthony Dente, PE, LEED AP, is a licensed engineer and principal at Verdant Structural Engineers, a firm that has designed over 300 structures using natural-building wall systems such as straw bale, adobe, rammed earth, earthbag, and cob. As vice president of the Cob Research Institute, he was the lead engineer for the Cob Construction Appendix in the International Residential Code, as well as the Hemp-Lime (Hempcrete) Appendix, both the first of their kind in the US. Dente has advised, designed, and collaborated on numerous university research programs testing the structural behavior of natural materials, and he writes and lectures extensively about

appropriate use of environmentally sensitive building materials. He is also a principal at Verdant Building Products, and project lead for their prefabricated, carbon-storing, straw wall panels which were originally developed under the EPA SBIR grant program. Dente was recognized by the Constellation Prize for his Sustainable Engineering Practice and he continues as a visionary and leader in the natural and low-carbon building fields. He lives in the San Francisco Bay Area of California.

Michael G. Smith is a natural builder, trainer, designer, and consultant. In 1993, he co-founded the iconic Cob Cottage Company with Ianto Evans and Linda Smiley, reviving the ancient tradition of cob building. He is also a founding board member of the Cob Research Institute, where he helped write the first model building code for cob. Smith has led or been involved in over 100 natural building projects in North America and internationally. Over 30 years, he has taught hundreds of hands-on workshops on cob, straw bale, light straw-clay, natural plasters, earthen floors, and many other natural building techniques. His teaching and consulting work emphasize energy efficiency, empowerment of people through simple, accessible techniques, and the regenerative use of locally available materials. He is co-author of *The Hand-Sculpted House* and *The Art of Natural Building* and author of *The Cobber's Companion*. He lives on an organic farm near Sacramento, California. Find out more at strawclaywood.com.

Massey Burke is a natural materials design/build consultant and director of the California Straw Building Association. She has worked with cob since 2005 and has taught natural

building methods with nonprofits and educational organizations, including the University of San Francisco and UC Berkeley. She was project manager for the first permitted cob building in the San Francisco Bay Area and supported the writing of the recent cob building code as a board member of the Cob Research Institute. Burke's work in life-cycle assessment focuses on low-carbon building, with a specific interest in supply chains and bringing natural building materials into the urban fabric. She partners with organizations including Arup, StopWaste, and the Carbon Leadership Forum to generate technical information and help remove barriers to scaling-up natural, climate-positive building methods. Burke was a contributing author to *The New Carbon Architecture*. She lives in the San Francisco Bay Area of California.

Martin Hammer is a California architect who has practiced sustainable design throughout his 35-year career, with particular emphasis on

Martin Hammer

the design, testing, and construction of strawbale buildings. He has lectured and written widely on the subject, including as a contributing author of *Design of Straw Bale Buildings* and *Straw Bale Building Details*. Hammer is lead or co-author of five appendices to the International Residential Code: Strawbale Construction, Light Straw-Clay Construction, Cob Construction (Monolithic Adobe), Hemp-Lime (Hempcrete) Construction, and Tiny Houses. He is also co-director of Builders Without Borders, which facilitates the design and construction of sustainable shelter in places of need. Hammer has helped introduce straw bale construction and other sustainable building systems to earthquake-affected Pakistan, Haiti, and Nepal.

ABOUT NEW SOCIETY PUBLISHERS

New Society Publishers is an activist, solutions-oriented publisher focused on publishing books to build a more just and sustainable future. Our books offer tips, tools, and insights from leading experts in a wide range of areas.

We're proud to hold to the highest environmental and social standards of any publisher in North America. When you buy New Society books, you are part of the solution!

- This book is printed on **100% post-consumer recycled paper**, processed chlorine-free, with low-VOC vegetable-based inks (since 2002).
- Our corporate structure is an innovative employee shareholder agreement, so we're one-third employee-owned (since 2015)
- We've created a Statement of Ethics (2021). The intent of this Statement is to act as a framework to guide our actions and facilitate feedback for continuous improvement of our work
- We're carbon-neutral (since 2006)
- We're certified as a B Corporation (since 2016)
- We're Signatories to the UN's Sustainable Development Goals (SDG) Publishers Compact (2020–2030, the Decade of Action)

At New Society Publishers, we care deeply about *what* we publish—but also about *how* we do business.

To download our full catalog, sign up for our quarterly newsletter, and learn more about New Society Publishers, please visit newsociety.com

ENVIRONMENTAL BENEFITS STATEMENT

New Society Publishers saved the following resources by printing the pages of this book on chlorine free paper made with 100% post-consumer waste.

TREES	WATER	ENERGY	SOLID WASTE	GREENHOUSE GASES
74	**5,900**	**31**	**250**	**32,200**
FULLY GROWN	GALLONS	MILLION BTUs	POUNDS	POUNDS

Environmental impact estimates were made using the Environmental Paper Network Paper Calculator 4.0. For more information visit www.papercalculator.org